THE PLANTS OF BAXTER STATE PARK

THE PLANTS OF BAXTER STATE PARK

Glen Mittelhauser

by Glen H. Mittelhauser, Jensen Bissell, Don Cameron, Alison C. Dibble,

Arthur Haines, Jean Hoekwater, Marilee Lovit, and Aaron Megquier

University of Maine Press, Orono, Maine

in association with Baxter State Park, Friends of Baxter State Park,

and Maine Natural History Observatory

University of Maine Press
5729 Fogler Library
Orono, ME 04469-5729
http://umaine.edu/umpress/

Copyright © 2016 by the University of Maine Press. All rights reserved. No part of this publication may be reproduced or transmitted, in any form or by any means, without prior written permission from the University of Maine Press. Individual photographs are copyright by their photographers. Book design by Michael Alpert.

Front cover photograph: Center Ridge Trail at the Peak of the Ridges, one of the peaks of The Traveler, in the northern portion of Baxter State Park. Back cover photograph: Cushion-plant, *Diapensia lapponica* (see page 130). Both photographs by Glen Mittelhauser.

FIRST EDITION

20 19 18 17 16   1 2 3 4 5

Printed in China by Regent Publishing Services Limited. Paper used in this publication meets the minimum requirements of the American National Standard for Information Sciences—Permanence of Paper for Printed Library Materials, ansi z39.48–1984.

Library of Congress Cataloging-in-Publication Data

Names: Mittelhauser, Glen H. (Glen Howard), 1964– author.
Title: The plants of Baxter State Park / by Glen H. Mittelhauser, Jensen Bissell, Don Cameron, Alison C. Dibble, Arthur Haines, Jean Hoekwater, Marilee Lovit, and Aaron Megquier.
Description: Orono, Maine : University of Maine Press in association with Baxter State Park, Friends of Baxter State Park, and Maine Natural History Observatory, [2016] | Includes bibliographical references and index.
Identifiers: LCCN 2016015397 | ISBN 9780891011262 (pbk. : alk. paper)
Subjects: LCSH: Plants—Maine—Baxter State Park—Identification.
Classification: LCC QK164 .M58 2016 | DDC 581.9741—dc23
LC record available at https://lccn.loc.gov/2016015397

ISBN 978-0-89101-126-2

TABLE OF CONTENTS

Preface v

Sponsoring Organizations vii

Acknowledgments viii

Map of Baxter State Park x

Introduction 1

Literature Cited 8

Other Botanical Literature 9

Keys and Plant Family Photographs

    Terms and Definitions 10

    Master Key 11

    Plant Family Photographs 26

Species Accounts

    Tab-color Key for Species Accounts 48

    Wildflowers and Low Shrubs 49

    Trees and Tall Shrubs 305

    Ferns and Other Spore-producing Plants 349

    Sedges, Rushes, and Grasses 381

Glossary 449

Index 456

About the Authors 477

Six pages for Field Notes

Ruler (inside back cover)

PREFACE

Former Maine Governor Percival Proctor Baxter began his personal quest to preserve Katahdin for the people of Maine in 1931. More than 30 years later, he completed his dream and Baxter State Park became an iconic, and permanently protected, representation of the Maine landscape. Katahdin, Maine's highest mountain, dominates the Park. Many visitors use more than 215 miles of hiking trails to explore the magic of the Park's many lakes, ponds, and rivers, and 18 additional named peaks over 3,000 feet in elevation.

Percival Baxter's gift was not limited to the 201,018 acres he provided in trust to the people of Maine. It also included two additional key components to ensure successful management of the Park over time: independent funding and oversight. Endowment funds left by Governor Baxter allow Baxter State Park to operate independently of legislative or other outside funding. The Park receives no tax dollars from the State of Maine. The importance of this independent funding to the consistent and effective management of the Park cannot be overstated. Baxter also recognized the importance of independent oversight of the Park Trust. He established the Baxter State Park Authority in 1945, consisting of the Director of the Maine Forest Service, the Commissioner of Inland Fisheries and Wildlife, and the Maine Attorney General. For the past seventy years, this triumvirate has consistently managed the Park in accordance with the wishes of Governor Baxter as expressed in the Deeds of Trust and other communications.

The Park's mission statement is derived directly from the Deeds of Trust. Protecting the natural resources of this large landscape is clearly stated as the most important element in the Park's mission. It is difficult, if not impossible, to protect what you don't know you have. The work represented in this guide, *The Plants of Baxter State Park*, is critically important to the long-term management of the Park. This project exemplifies the great good that can come from the synergy of many minds. The persistence of Park Naturalist Jean Hoekwater, who advocated for this project for many years, combined with the support of Friends of Baxter State Park, created the necessary catalyst to launch the project. Many dedicated and capable volunteers contributed thousands of hours to the effort. Glen Mittelhauser of the Maine Natural History Observatory brought his extensive knowledge and strong work ethic as the lead botanist on the project. As a result of this monumental effort, Baxter State Park will be better protected in the years to come, and countless visitors will have an opportunity to enrich their experience and knowledge of Maine's largest wilderness.

– Jensen Bissell, Baxter State Park Director

SPONSORING ORGANIZATIONS

Baxter State Park

Baxter State Park is composed of more than 200,000 acres of wilderness and public forest. The Park was established by donations of land, in trust, by Percival P. Baxter beginning in 1931. Baxter State Park is governed by the Baxter State Park Authority, consisting of the Director of the Maine Forest Service, the Commissioner of Inland Fisheries and Wildlife, and the Maine Attorney General. The Park is funded independently through a combination of revenues from trusts, user fees, and the sale of forest products from the Park's Scientific Forest Management Area. There are more than 40 peaks and ridges in the Park. The trail system features more than 215 miles of trails popular with hikers, mountain climbers, and naturalists. Baxter State Park operates eight roadside campgrounds and two hike-in campgrounds as well as two dozen individual backcountry sites for backpackers. To learn more about the Park and to make a camping reservation, visit our web site at www.baxterstateparkauthority.com.

Friends of Baxter State Park

Friends of Baxter State Park is an independent citizen group with a mission to preserve, support, and enhance the wilderness character of Baxter State Park in the spirit of its founder, Governor Percival Baxter. We are thrilled to partner with Maine Natural History Observatory and Baxter State Park to publish *The Plants of Baxter State Park.* Since our founding in 2000, we have grown into a strong partner with Baxter State Park, with nearly 1,000 members, a full-time staff, and a wide range of programs in support of the Park's exemplary wilderness values. Our program areas span trail maintenance, infrastructure, search and rescue, youth programs, volunteerism, advocacy, publications, scientific research, and education. As a membership organization, our members participate in these programs and provide funding for our mission through annual membership dues. To learn more about Friends of Baxter State Park or to become a member, please contact us at PO Box 322, Belfast, ME 04915, or visit our website at www.friendsofbaxter.org.

Maine Natural History Observatory

Maine Natural History Observatory is a small nonprofit research organization located in Gouldsboro, Maine. Our mission is to advance the knowledge of Maine's natural history and resources. We coordinate efforts to inventory and monitor local and regional flora, fauna, and habitats of Maine, and compile and publish summaries of Maine's natural history. The Observatory also facilitates cooperation and exchange of information among organizations, agencies, and individuals conducting natural history research in Maine or caring for natural history collections. To learn more about our mission, current projects, and publications, visit the Observatory's web site at www.mainenaturalhistory.org.

ACKNOWLEDGMENTS

This field guide was produced with the help of many people. We thank the dedicated staff of Baxter State Park for support of the project over many years, particularly Rick Morrill, former Resource Manager at Baxter State Park, who helped us with creating GIS databases of our plant observations and assisted with field work. We thank our exceptional interns, Abbe Urban, Matt Dickinson, and Jordan Chalfant, who assisted with field work for portions of this project. We thank all of the volunteers who contributed their time and effort to join us on one or more field inventory trips in the Park: Barbara Brown, Collin Cunning-Wolfe, Harry Cunningham, Bart DeWolf, Diane Freelove, Connie Gatz, Scott Holste, Anne Huntington, Donna Kausen, Megan Leach, Adrienne Leppold, Jerry Longcore, Caitlin McDonough MacKenzie, Aleta McKeage, Candace McKellar, Barry Millman, Sue Millman, Celeste Mittelhauser, Pepin Mittelhauser, Dawn Morgan, Sarah O'Malley, Kit Pfeiffer, Dakota Smith, Rick Speer, Lois Stack, Keith Williams, and Lynne Zimmerman. They helped document and photograph plants and habitats throughout the Park. Without them we could not have completed this guide. Thanks to Maggie Barr, Janet Christrup, and Matt Dickinson for their research on species of selected plant families. We thank Janet Christrup, Eric Doucette, Diane Freelove, Barbara Grunden, Don Hudson, Jerry Longcore, Tom Vining, and Jill Weber for their careful reviews of this guide. Eben Sypitkowski, Resource Manager at Baxter State Park, created the map.

We are grateful to the numerous individuals and organizations who gave us financial support. Without their support, this guide would not have been possible. Special thanks to Richard and Ellen Klain for their exceptionally generous donation in 2013 that helped put the project on firm financial ground. We thank the following organizations and foundations who helped to fund portions of the production of this guide: Baxter State Park, Conservation and Research Foundation, Friends of Baxter State Park, Garden Maine, Hummingbird Farm, John Sage Foundation, Maine Community Foundation's Fund for Maine Land Conservation, Maine Natural Areas Program, Maine Natural History Observatory, Maine Outdoor Heritage Fund, Margaret Burnham Charitable Trust, The Natural History Center, Norcross Wildlife Foundation, The Home Depot Foundation, Stewards LLC, University of Maine, and The Waterman Fund. We also thank the following individuals who made significant donations to the project: Anonymous, Matt Arsenault, Margaret Rigg Atwood, Scott and Amey Bailey, Jeff Beckley and Sarah Brandon, Chris J. and Dorothy Beeuwkes, Kenneth Brattlie, Thomas Byther, Sally Campbell and Mark Rogers, Vivian and Jim Chaplin, Cloe Chunn and David Thanhauser, The Daicey Goddesses, Ronald and Shirley Davis, Gary A. and Teresa K. Dean, Alison C. Dibble, Douglas Dolan, Ted Elliman, Richard Fournier, Carol and Ira Garber, Jim Garland and Carol Andreae, John and Katie Greenman, Barbara and Charlie Grunden, Susan Hayward, James and Patricia Hinds, Jean Hoekwater, Marc and Laurie Howlett, Don Hudson and Phine Ewing, Charlie Jacobi, Helen Koch, Bruce J. Leavitt MD, Patricia Ledlie, Joan Sarles Lee, David Little and Mikki Jones-Little, Jerry and Joyce Longcore, Nancy McReel, Joshua Nagine, Betsy Newcomer, Moira O'Neill, Edward Barrett and Nancy Orr, Eliot Paine, Beedy Parker, Marsten and Lori Renn Parker, Deb Piot, Nishanta Rajakaruna, Doug and Laurie Rich, Vicki Richardson, Janet Saetta, Kate Proctor, Grant Seba, Ann and David

Simmons, Ann Slayton, Rick Speer and Judith Frost, Lois Stack, Kent Tableman, Steuart and Linda Thomsen, Cindy Tibbetts, Sheila Unvala and Joseph Huber, Jill Weber, Bill and Molly Webster, Heidi Welch and Patricia Tierney, Linda Welch, and Bill Zoellick and Pauline Angione. In addition, we thank the many additional individuals who helped this project with smaller donations.

We thank Christopher Campbell for opening the University of Maine Herbarium for our research. Special thanks also to the staff of Harvard University Herbaria, New England Botanical Club Herbarium, Maine State Museum, and College of the Atlantic Herbarium for allowing us to consult their collections.

Obtaining photographs of all species included in this guide was a big challenge. We particularly thank Keith Williams, Barre Hellquist, Jill Weber, Sally Rooney, Matt Arsenault, Eric Doucette, Garth Holman, Tony Reznicek, and Ted Elliman who helped with plant identifications and helped us track down photos of elusive species. We thank Susan Aiken, Matt Arsenault, Sean Blaney, Michael Burzynski, Gerald Carr, Jordan Chalfant, René Charest, Collin Cunning-Wolfe, Harry Cunningham, Bart DeWolf, Matt Dickinson, Carol Gracie, Craig Greene (posthumously), Don Hudson, Donna Kausen, Megan Leach, Adrienne Leppold, Jerry Longcore, Maine Forest Service, John Msaunder, Caitlin McDonough MacKenzie, Larry Mellichamp, Barry Millman, Celeste Mittelhauser, Pepin Mittelhauser, Robbin Moran, Dawn Morgan, Sarah O'Malley, Harald Pauli, Charles Peirce, Diane Peirce, Sally Rooney, Paul Rothrock, Jason Sachs, Russ Schipper, Brad Slaughter, Dakota Smith, David Smith, Robert Smith, Paul Sokoloff, Rick Speer, Phil Sturman, Abbe Urban, Jill Weber, Lynne Zimmerman, and Alexey Zinovjev for generously allowing us to use their photos in this guide.

Above all, we are grateful to our families for their encouragement and support throughout this project.

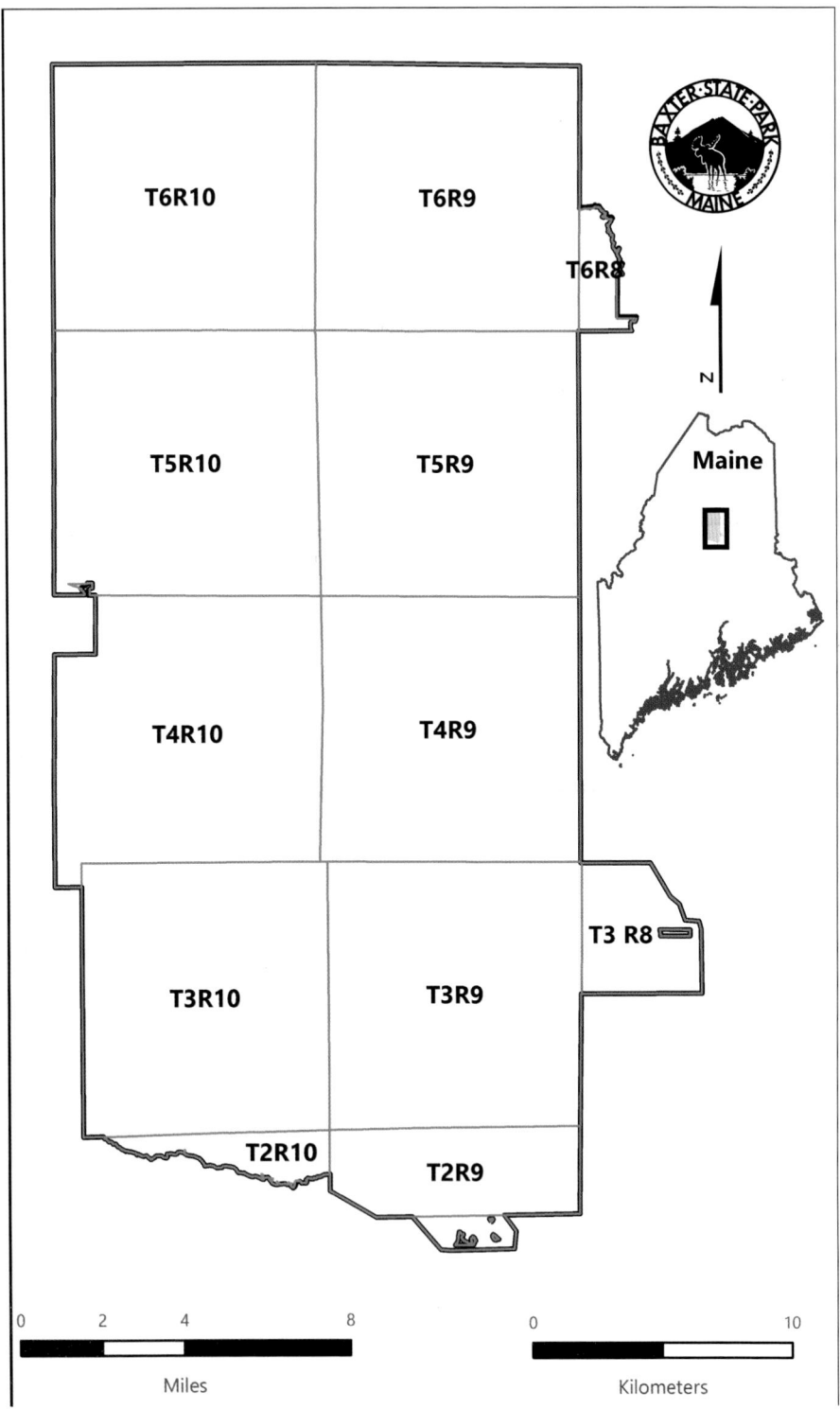

x THE PLANTS OF BAXTER STATE PARK

INTRODUCTION

Baxter State Park contains more than 200,000 acres of wilderness and public forest in Piscataquis County, a majority of this donated by former Maine Governor Percival Baxter. The Park has more than 40 peaks and ridges, numerous streams, more than 75 lakes and ponds, many waterfalls, wetlands, and bogs, and more than 215 miles of trails. The most notable mountains are the Katahdin massif, the highest mountain in Maine, in the southern portion of the Park, and the Traveler range in the northern part. Katahdin granite is typical on the rugged mountains to the south, while rhyolite and sedimentary rock are typically found to the north. Glacial features, in the form of kettle ponds, eskers, cirques, moraines, and erratics, are found throughout the Park. The Park has an extensive and complex fire history. The most significant fire occurred in 1903 and burned multiple townships. The climate is typically cool and moist, with an average annual precipitation of about 1 meter, including approximately 2.5 meters of snow.

Habitats in the Park are diverse and range from large expanses of unfragmented lowland forest communities and broad areas of complex wetlands and ponds to more than 4,500 acres of alpine habitat above treeline. The soils are typically acidic, with much gravel and sand rich in quartz and feldspar. With its extensive alpine zone, Katahdin hosts populations of 20 rare vascular plants found nowhere else in Maine, including one species, *Micranthes foliolosa*, not found elsewhere in the eastern United States. Despite the large number of rare plants and the great diversity of habitats, a complete inventory of the vascular plants growing in Baxter State Park has never been conducted until now. Prior to this project, vegetation was reasonably well documented in the alpine and subalpine zones, but there was no overall plant list for the Park.

The goal of this project was to inventory the vascular plants of Baxter State Park and make that information available to Park staff and visitors. This guide represents one of the ways we are sharing this information. We hope our book will help visitors to Baxter State Park–as well as all those who live, work, and recreate anywhere in the Maine woods–develop a deeper appreciation for plants. This inventory will provide an important baseline in a changing world, especially for the Park's lower-elevation forests and wetlands that have never been carefully surveyed. There is no other intact natural area of this size in the northern forest region of New England that approaches the wild character of Baxter and that has in place such a comprehensive system for resource management. Above all, we hope this book will help visitors enjoy their time in the Park and support the continued protection of Baxter State Park's rare and exemplary plant communities and wilderness values.

History of botanical exploration in Baxter State Park

The Katahdin region has a long and rich history of botanical exploration, which began in the 1800s and continued into the early 1900s. These early botanical explorations, focusing on Katahdin, included the work of Bailey (1837), Churchill (1901), Fernald (1901), Williams (1901), Harvey (1903, 1909), Norton (1924), Blake (1926a, 1926b), Thurber (1926), Laski (1927), Stebbins (1927, 1929), Ewer (1930), and Allard and Leonard (1945). Norton (1924) noted that with the possible exception of Mount Desert Island, no section of Maine "boasts of a better botanical literature" than Katahdin. Specimens collected during these early trips can still be found at the University of Maine Herbarium, Harvard University Herbaria, and in other collections.

In the later 1900s, researchers continued the investigation of alpine vegetation on Katahdin. Projects or field checks at alpine areas within the Park were conducted by May and Davis (1978), Hudson (1985, 1987), Hudson et al. (1985), Dibble et al. (1990), Cogbill and Hudson (1990), and Clark (1999). In addition, botany volunteers have continued to work with the Park to check known rare plant populations, especially on Katahdin. Don Hudson (1985) conducted the first Park-wide inventory of plants in alpine and old-growth areas and compiled his results to produce the first checklist of plants in the Park that included areas away from Katahdin. There have also been numerous botanists who studied mosses, lichens, and liverworts on Katahdin, summarized by Dibble et al. (2009), Hinds et al. (2009), and Miller (2009).

Methods and results

To compile a list of vascular plants growing in Baxter State Park, we relied on three primary research methods: literature review, herbarium research, and field surveys. We began by completing an exhaustive search of the published literature and unpublished reports for data on plants found in the Park. We compiled this information into a plant list by township within the Park. We also carried out research at several herbaria within New England to find specimens from past explorations in Baxter State Park. These included the University of Maine Herbarium, Harvard University Herbaria, College of the Atlantic Herbarium, and the Maine State Museum. When combined with the results of the literature review, this brought the existing number of documented species in the Park to nearly 550.

From 2011 to 2014, we conducted field inventories in the Park with the goal of adding to this baseline list of species. The field surveys used a citizen science approach developed by Mittelhauser and Dibble in collaboration with Baxter State Park staff. Research teams spent more than 4,200 hours in the field documenting the flora of the Park. Each team consisted of a professional botanist working with one to five volunteers. Teams developed plant lists for each area explored during the field surveys. They took more than 17,000 digital photos as vouchers of plant observations, as well as GPS points to document where uncommon and rare species were found. Teams also gathered data on the abundance of each species, and the habitats in which each occurs. As research progressed, we used a gap analysis approach to focus our efforts on unexplored areas and habitat types.

Over the course of four years, we documented approximately 300 species that had not previously been reported from Baxter State Park, bringing the total flora of Baxter State Park to 857 species. This represents about a third of the 2,490 plant species known to grow wild in Maine, and over half of Maine's 1,527 native plant species. These are remarkable totals for a Park that covers less than one percent of Maine's land area. This level of diversity, coupled with the Park's forever wild management, suggests that Baxter State Park will have a lead role to play in the long-term conservation of Maine's flora.

The flora of Baxter State Park is notable for its large percentage of native species. Of the 857 total species found in the Park, 745, or 87%, are considered native to Maine. This is a significantly higher percentage of native species than for Maine's overall flora, which is about 61% native (Haines 2011). Only seven species found in Baxter State Park—less than 1% of the Park's flora—appear on the list of invasive or potentially invasive species maintained by the Maine Natural Areas Program. These include purple loosestrife (*Lythrum salicaria*), common reed (*Phragmites australis*), wood blue grass (*Poa nemoralis*), creeping thistle (*Cirsium arvense*), common thistle (*Cirsium vulgare*), bishop's goutweed (*Aegopodium podagraria*), and coltsfoot (*Tussilago farfara*). This is an exceptionally small number of invasive species for an area with 65,000 visitors annually. Unfortunately, it is very likely to increase over time.

This guide includes 86 species of trees and tall shrubs; 508 species of wildflowers and low shrubs; 60 ferns and other spore-producing plants; and 203 graminoids, including sedges, rushes, and grasses. Many of the plants in this guide are common and widely distributed in Maine, such as eastern white pine (*Pinus strobus*) and Canada mayflower (*Maianthemum canadense*). Other plants found in the Park have extremely limited distribution, including moss-plant (*Harrimanella hypnoides*) and naked-bulbil small-flowered-saxifrage (*Micranthes foliolosa*), which are found only on Katahdin.

Sixty-nine species of plants in Baxter are listed as rare, threatened, or endangered by the Maine Natural Areas Program. These plants typically occur in very specific habitats, such as exposed alpine ridges, enriched northern hardwood forest, or circumneutral bedrock outcrops.

The most speciose genus of plants found in the park is *Carex*, in the sedge family, with 88 species, more than 10% of the total species in the Park. The Park includes extensive populations of common species such as greater bladder sedge (*Carex intumescens*), as well as rarities in Maine like Bigelow's sedge (*Carex bigelowii*) and hop sedge (*Carex lupulina*). For a more detailed reference on sedge identification, see *Sedges of Maine: A Field Guide to Cyperaceae* by Arsenault et al. (2013).

Part of the reason for Baxter State Park's plant diversity is the wide range of elevations found in the Park, from approximately 570 feet near the Abol Deadwater to 5,267 feet at the summit of Katahdin. Exposed alpine ridges and cirque headwalls have a climate similar to the Arctic, vastly different from protected, south-facing slopes at lower elevations. Because of this, Baxter State Park contains several species at the northern or southern limit of their ranges.

This field guide presents our current understanding of the plants growing in Baxter State Park. Although it represents thousands of hours of careful research, we hope you view this guide as a beginning rather than an ending. Plant communities change over time as new species arrive and other species can no longer find suitable habitat. Despite the best efforts of Baxter State Park staff, new invasive species will likely arrive and establish a foothold. Ecosystems in Baxter State Park will change in response to global climate change, although we do not yet know exactly how. The data generated by this project will provide a baseline for measuring that change, and a tool for understanding it.

Botanizing in the Park

Our field surveys for this guide covered a small portion of the land area of Baxter State Park. Much of the area of the Park that we did not survey is covered by large patches of common natural communities, such as northern hardwood forest, that tend to include the same plant species, acre after acre. In the portions of the Park that we were not able to survey, there are likely specialized habitats that may harbor interesting or rare plants, including plants that have not been documented in the Park. With this guide in hand, and with the utmost attention to minimizing your impact, we hope that you find some of these plants, leave them undisturbed, and submit your observations to us for the next edition of this guide.

While it may be appealing to botanize in remote terrain, you can also learn about plants, and contribute to the Park's flora, by simply identifying the plants next to your car in the parking lot. When new species migrate into Baxter State Park, disturbed habitats may be the first places they appear. These include roadsides, trailsides, ditches, parking lots, and gravel pits. Many of the species in this guide, particularly the nonnative species, were first documented in these areas.

While intensive botanical work on the Katahdin massif began well over a century ago, it is the Park's good fortune that most of the field work to create this guide was carried out during the past three years. Modern botanists understand the need to limit the environmental impacts that were once considered to be a normal part of research. Today's botanists take fewer plant specimens and leave fewer traces of their passing than the botanical expeditions of past eras. While we collected a small number of voucher specimens to create this guide, we identified as many plants as possible *in situ*, and collected only when absolutely necessary for identification.

While on research expeditions in the Park, our field teams worked under a Special Use Permit. This is required for anyone seeking exemption from a Baxter State Park regulation that prohibits the collection of any natural or cultural item in the Park. Identifying plants *in situ* meant carrying field guides, well-armored digital tablets, and sophisticated digital cameras mile after mile through some of the wildest weather and terrain in New England. It was well worth it to reduce the impact of the project on Park ecosystems. We hope everyone using this guide will botanize with the same care and respect for Baxter State Park and its exemplary wilderness values.

When botanizing in Baxter State Park, please follow these guidelines:

- Do not pick or collect plants in the Park. Take a photograph instead to serve as a voucher of your observation.
- Familiarize yourself with Baxter State Park rules and regulations before your visit.
- Use trails whenever possible to minimize trampling vegetation. Try to hike in the middle of the trail corridor, even if it means getting your boots wet. Hiking on the sides of trails can kill plants. This is particularly important above treeline, where hiking trails traverse several natural communities with critical habitat for rare animals and plants.
- If you are leaving the trails, especially in wetlands or above treeline, please restrict your party to no more than three people. When possible, rock hop to avoid stepping on plants or habitat where plants may be able to colonize.
- Try to visit when soils are not saturated or experiencing the freeze-thaw cycles of spring and fall.

How This Guide is Organized

This guide includes details on how to identify all of the wildflowers, trees, shrubs, ferns and other spore-producing vascular plants, and grasses, sedges, and rushes that grow within the boundaries of Baxter State Park. Take this book into the field to identify plants as you find them. If you don't have the guide handy or the time to identify a plant in the field, take a few sketches or photographs rather than collecting. It is illegal to collect plants within the boundary of Baxter State Park.

We include identification keys to and thumbnail images of all known vascular plant families and genera in Baxter State Park. These tools should help you identify an unknown plant. We separated plants into four groups: wildflowers and low shrubs (pages 49–304); trees and tall shrubs (pages 305–348); ferns and other spore-producing vascular plants (pages 349–380); and sedges, grasses, and rushes (pages 381–448). Low shrubs are defined as woody plants less than or equal to 2 m tall, often with several main stems. Tall shrubs are woody plants greater than 2 m tall, often with several main stems. Here, Shadbush (*Amelanchier* spp.) and Willows (*Salix* spp.), except for dwarf alpine species, are included in the tall shrub category even if they are less than 2 m tall. Each of the four groups of plants in this guide is marked by a band of color located on the outside edge of each page. Within these groups, plant families are arranged alphabetically with genera and species alphabetically listed within each family. Scientific plant names used in this field guide follow *Flora Novae Angliae* (Haines 2011) and subsequent changes (Haines 2015). Varieties and subspecies are not considered in this guide.

- **Nonnative species:** An asterisk (*) is placed before the scientific name of any plant not native to Maine.
- **Species description:** Plant descriptions are largely nontechnical, with occasional use of some technical terms where necessary. A glossary of the technical terms begins on page 446. Italicized text in the species description indicates key information used to distinguish the species from other similar looking species. Measurements are in meters (m), centimeters (cm), or millimeters (mm). A ruler in both metric and English units is printed on the last page of the book.
- **Occurrence:** We estimated plant abundance and distribution within Baxter State Park based on our field work between 2011 and 2014. Abundance estimates are Park-wide, and distribution within the Park is based on the number of townships where a species was seen. The abundance categories are: **common** – species with a wide distribution, occurring in large numbers; **occasional** – species with a scattered distribution, sometimes found in large numbers; **uncommon** – species not widely distributed, not often encountered; **rare** – species usually restricted to small areas, specialized habitats, or consisting of one or a few populations.
- **Notes:** This section includes reference to similar looking species and additional information that readers may find interesting. Any reference to a plant's edibility or medicinal use is listed for interest only; please do not pick wild plants in Baxter State Park.
- **Other names:** Other common names and older scientific names are listed at the end of the section for each species.
- **Plant photographs:** For each species, we included multiple photographs of the characters that are useful for identification as well as photos of plants taken in Baxter State Park. All photos are marked with the date taken and initials of the photographer. Photos taken within Baxter State Park also include details on the location in the Park where the photo was taken. Uncommon or rare plant photographs taken from within the Park are noted with only township details. Photographers and their initials are as follows: SA – Susan Aiken © Canadian Museum of Nature; MA – Matt Arsenault; SB – Sean Blaney; MB – Michael Burzynski; DC – Don Cameron; GC – Gerald Carr; JC – Jordan Chalfant; RC – René Charest; CC – Collin Cunning-Wolfe; HC – Harry Cunningham; BD – Bart

DeWolf; AD – Alison C. Dibble; MD – Matt Dickinson; CG – Carol Gracie; CWG – Craig Greene; AH – Arthur Haines; JH – Jean Hoekwater; DH - Don Hudson; DK – Donna Kausen; MEL – Megan Leach; AL – Adrienne Leppold; JL – Jerry Longcore; ML – Marilee Lovit; MFS – Maine Forest Service; JM – John Maunder; CMM – Caitlin McDonough MacKenzie; AM – Aaron Megquier; LM – Larry Mellichamp; BM – Barry Millman; CM – Celeste Mittelhauser; GM – Glen Mittelhauser; PM – Pepin Mittelhauser; RM – Robbin Moran; DM – Dawn Morgan; SO – Sarah O'Malley; HP – Harald Pauli; CP – Charles Peirce; DP – Diane Peirce; SR – Sally Rooney; PR – Paul Rothrock; JS – Jason Sachs; RS – Russ Schipper; BS – Brad Slaughter; DS – Dakota Smith; DGS – David Smith; RWS – Robert Smith; PCS – Paul C. Sokoloff © Canadian Museum of Nature; RAS – Rick Speer; PS – Phil Sturman; AU – Abbe Urban; JW – Jill Weber; LZ – Lynne Zimmerman; and AZ – Alexey Zinovjev.

New Plant Records
If you think that you have found a species not included in this guide that is growing within Baxter State Park, please send details to Glen Mittelhauser via the contact link at www.mainenaturalhistory.org or mail the information to Maine Natural History Observatory, 317 Guzzle Road, Gouldsboro, ME 04607. Please include the following details: your name; digital photographs including images of plant parts needed to identify the plant; the names of the people who checked the identification; the scientific name of the plant; a detailed description of where the plant was found, including a sketch map or coordinates recorded with a GPS unit; the day, month, and year when this plant was observed; additional information on the habitat, abundance of the species at the site, and other species present. Thank you! If you encounter a species listed as having a Maine Rarity Ranking, please report it to the Maine Natural Areas Program at www.maine.gov/doc/nrimc/mnap/ or 93 State House Station, Augusta, ME 04333-0093.

Identifying an Unknown Plant
Plants in this guide are arranged by family. Identifying an unknown plant will become easier as you learn the diagnostic features of each family. Several methods can be used to identify a plant ranging from scanning the pages of thumbnail photographs to using a series of tablular keys that rely on simple vegetative and flower characters. Identification methods are described below.

- **Use the index:** If you know the common or scientific name of the plant, look up the name in the index (beginning on page 456), go to the referenced page number, and confirm your identification.
- **Browse the plant family photos:** Turn to page 26 and scan the thumbnail photgraphs of all plant families and genera to find those that match your specimen. Then, look up the family (arranged alphabetically within each of four groups) and skim through the species pages.
- **Examine photos relevant to your plant:** Thumbing through the entire book is an inefficient way to identify species, but it is a great way to become familiar with other plants to be watching for during your hikes.
- **Use the keys:** We have included a **Master Key** and 16 **Tabular Keys** that rely on

simple vegetative and flower characters to identify the family and genus of an unknown plant. We highly recommend using these keys for identifying unknown plants. Take some time to become familiar with how these keys work and give them a try on a plant that you already know. To use the keys, take the following steps:

1. First, use the **Master Key** on pages 11–12. Pick the row that best describes plant category and leaf characteristics. Refer to the right-most column for the appropriate key and page to go to next. Consult the terms and definitions table on page 10 or the glossary beginning on page 449 to become familiar with the terms used in the keys.
2. Go to the appropriate **Key** and again find the row that best describes the flower or other characters of your plant. In the right-most column, family names are in bold letters and genera are in italics.
3. Confirm or eliminate families (in bold) by looking up each family in the **Plant Family Photographs** section, which starts on page 26. Read the family descriptions and look at the photos that represent the genera within the families. Choose the family for which the description and photos best match your specimen. Do the same to narrow down the choices to one or a few genera.
4. Look up the family and the genera on the species pages (the first group starting on page 49). Consult the photographs and species description. Pay particular attention to the notes section, which will refer you to other similar looking species.

Some groups of plants do not have flowers (e.g., ferns and other spore-producing plants, conifers) or may have flowers without showy petals (e.g., sedges, rushes, and grasses). Identification of these plants can be a little intimidating at first, but with practice you will sharpen your observation skills, and may find it fun and rewarding to identify some of the features that characterize these groups.

LITERATURE CITED

Allard, H.A., and E.C. Leonard. 1945. Plants collected on and around Mount Katahdin in Maine. Castanea 10:46–53.

Arsenault, M., G.H. Mittelhauser, D. Cameron, A.C. Dibble, A. Haines, S.C. Rooney, and J.E. Weber. 2013. *Sedges of Maine: a field guide to Cyperaceae*. University of Maine Press. ISBN: 9780891011231.

Bailey, J.W. 1837. Account of an excursion to Mount Katahdin, in Maine. American Journal of Science and Arts 32:20–34.

Blake, J. 1926a. An excursion to Mount Katahdin. Maine Naturalist 6:71–73.

Blake, J. 1926b. A second excursion to Mount Katahdin. Maine Naturalist 6:74–83.

Churchill, J.R. 1901. A botanical excursion to Mount Katahdin. Rhodora 3:147–160.

Clark, D.L. 1999. Alpine flora of two study sites on Mount Katahdin, Maine. M.S. thesis, University of Maine, Orono.

Cogbill, C., and W.D. Hudson, Jr. 1990. Alpine Vegetation Study. Unpublished report to Baxter State Park.

Dibble, A.C., S.C. Rooney, H. Hinds, W.D. Hudson, Jr., and B.A. Sorrie. 1990. Rediscovery of some rare plants on Mt. Katahdin. Rhodora 92:38–41.

Dibble, A.C., N.G. Miller, J.W. Hinds, and A.M. Fryday. 2009. Lichens and bryophytes of the alpine and subalpine zones of Katahdin, Maine, I: Overview, ecology, climate and conservation aspects. The Bryologist 112:651–672.

Ewer, J. 1930. Botanical explorations at Katahdin. Maine Naturalist 10:87–98.

Fernald, M.L. 1901. The vascular plants of Mount Katahdin. Rhodora 3:166–177.

Haines, A. 2011. *Flora Novae Angliae*. Yale University Press. ISBN: 9780300171549.

Harvey, L.H. 1903. An ecological excursion to Mt. Ktaadn. Rhodora 5:41–52.

Harvey, L.H. 1909. The floristic composition of the vascular flora of Mount Ktaadn, Maine. Michigan Academy of Sciences 11:37–47.

Hinds, J.W., A.M. Fryday, and A.C. Dibble. 2009. Lichens and bryophytes of the alpine and subalpine zones on Katahdin, Maine, II. Lichens. The Bryologist 112:673–703.

Hudson, W.D., Jr. 1985. The preliminary vascular flora of Baxter State Park. Maine State Planning Office, Critical Areas Program Misc. Report #37.

Hudson, W.D., Jr. 1987. The reproductive biology of *Saxifraga stellaris* var. *comosa* on Mt. Katahdin, Maine. Rhodora 89:1–12.

Hudson, W.D., Jr., R. Cannarella, L. Garnett, and K. Huntington. 1985. Old-growth forests, subalpine forest, and alpine areas in Baxter State Park. Maine State Planning Office, Critical Areas Program Misc. Report #31.

Laski, J.K. 1927. Dr. Young's botanical expedition to Mount Katahdin. Maine Naturalist 7:38–62.

May, D.E., and R.B. Davis. 1978. Alpine tundra vegetation on Maine mountains. Maine State Planning Office Misc. Report #36.

Miller, N.G. 2009. Lichens and bryophytes of the alpine and subalpine zones on Katahdin, Maine, III. Bryophytes. The Bryologist 112:704–748.

Norton, A.H. 1924. Some of the more conspicuous plants of Mount Ktaadn. Maine Naturalist 4:77–82.

Stebbins, G.L. 1927. Two plants from Mt. Katahdin. Rhodora 29:15.

Stebbins, G.L. 1929. Some interesting plants from Mt. Katahdin. Rhodora 31:142–143.

Thurber, G. 1926. Notes of an excursion to Mount Katahdin. *Maine Naturalist* 6:134–151.
Williams, E.F. 1901. Floras of Mt. Washington and Mt. Katahdin. *Rhodora* 3:160–165.

OTHER BOTANICAL LITERATURE

Technical floras, guides, and reference books

Crow, G.E., and C.B. Hellquist. 2000. *Aquatic and wetland plants of northeastern North America*. 2 volumes. ISBN: 029916330x, 029916280x.

Fernald, M.L. 1950. *Gray's manual of botany, 8th edition*. ISBN: 0931146097.

Gawler, S., and A. Cutko. 2010. *Natural landscapes of Maine: a guide to natural communities and ecosystems*. ISBN: 9780615347394.

Gleason, H.A., and A. Cronquist. 1991. *Manual of vascular plants of northeastern United States and adjacent Canada*. ISBN: 0893273651.

Haines, A. 2001. *The genus Viola of Maine: a taxonomic and ecological reference*. ISBN: 0966487427.

Haines, A. 2003. *The families Huperziaceae and Lycopodiaceae of New England: a taxonomic and ecological reference*. ISBN: 096648746.x

Haines, A. 2015. Synonymized Checklist of New England Tracheophytes. Accessed November 13, 2015. www.arthurhaines.com/tracheophyte-checklist.

Haines, A., and T.F. Vining. 1998. *Flora of Maine: a manual for identification of native and naturalized vascular plants of Maine*. ISBN: 0966487400.

Harris, J.G., and M.W. Harris. 1999. *Plant identification terminology: an illustrated glossary*. ISBN: 096402215x.

Holmgren, N.H. 1998. *The illustrated companion to Gleason and Cronquist's manual – illustrations of the vascular plants of northeastern United States and adjacent Canada*. ISBN: 0893273996.

Voss, E.G., and A.A. Reznicek. 2012. *Field manual of Michigan flora*. ISBN: 9780472118113.

Photographic or illustrated field guides

Clemants, S., and C. Gracie. 2006. *Wildflowers in the field and forest: a field guide to the northeastern United States*. ISBN: 0195150058.

Cobb, B., E. Farnsworth, and C. Lowe. 2005. *Ferns of northeastern and central North America*. ISBN: 0618394060.

Maine Forest Service. 2008. *Forest trees of Maine*. ISBN: 9781882190614.

Mittelhauser, G.H., L.L. Gregory, S.C. Rooney, and J.E. Weber. 2010. *The plants of Acadia National Park*. ISBN: 9780891011200.

Newcomb, L. 1977. *Newcomb's wildflower guide*. ISBN: 0316604429.

Online resources

A digital flora of Newfoundland and Labrador vascular plants. www.digitalnaturalhistory.com/flora.htm

Arctic flora of Canada and Alaska. www.arcticplants.myspecies.info/

Go Botany. https://gobotany.newenglandwild.org/

Maine Natural Areas Program rare plant fact sheets. http://www.maine.gov/dacf/mnap/index.html

TERMS AND DEFINITIONS USED IN KEYS

Category	Wildflowers	Herbaceous plants producing flowers and fruit.
	Low shrubs	Woody plants less than or equal to 2 m tall, often with several main stems arising at or near the ground.
	Tall shrubs	Woody plants greater than 2 m tall, often with several, self-supporting main stems arising at or near the ground. All Shadbush (*Amelanchier* spp.) and Willows (*Salix* spp.), except for dwarf alpine species, are included in this category even if they are less than 2 m tall.
	Trees	Woody plants greater than 2 m tall, often with a single, self-supporting main stem arising at or near the ground.
	Conifers	Trees or shrubs bearing needle-like or scale-like leaves and bearing cones.
	Ferns and other spore-producing plants	Spore-producing vascular plants such as ferns, horsetails, firmosses, quillworts, clubmosses, and moonworts.
Leaf Arrangement	Opposite or whorled	Plants with 2 (opposite) or more (whorled) leaves at each node along the stem.
	Alternate	Plants with only 1 leaf at each node along the stem.
	Basal only	Leaves growing only from the base of the plant; without leaves along the stem.
	Absent at flowering	No leaves present during flowering.
Leaf Type	Simple	Leaves not divided into distinct leaflets, though the margins may be toothed or lobed.
	Divided	Leaves deeply cut into distinct leaflets; divisions reaching the main vein.
Leaf Margin	Entire	Margin of the leaves without teeth or lobes; occasionally species with only basal lobes included here.
	Toothed or lobed	Margin of the leaves with more or less regular serrations, notches, points, or lobes.
Flowers	Bilaterally symmetric	Type of flower that can be divided through the center into two equal parts by only one plane.
	Radially symmetric	Type of flower with petal-like parts radiating from a common point, like the spokes of a wheel.
	Petal-like part	Whorl of showy floral blades or rays surrounding the flower or flower cluster.

MASTER KEY

Category	Leaf arrangement	Leaf type	Leaf margin	Go to
Wildflowers & low shrubs	Absent at flowering			Key 1 Page 13
	Basal only	Simple	Entire	Key 2 Page 14
		Simple	Toothed or lobed	Key 3 Page 15
		Divided	Various	Key 4 Page 15
	Opposite or whorled	Simple	Entire	Key 5 Page 16
		Simple	Toothed or lobed	Key 6 Page 17
		Divided	Various	Key 7 Page 17
	Alternate	Simple	Entire	Key 8 Page 18
		Simple	Toothed or lobed	Key 9 Page 20
		Divided	Various	Key 10 Page 21

Master Key continued on next page (trees, tall shrubs, ferns & other spore-producing plants, and graminoids).

MASTER KEY (CONTINUED)

Category	Leaf arrangement	Leaf type	Go to
Trees & tall shrubs	Opposite or whorled	Various	Key 11, Page 22
	Alternate	Various	Key 12, Page 22
Ferns & other spore-producing plants	Various	Various	Key 13, Page 23
Graminoids - grasses, sedges, and rushes	Various	Various	Key 14, Page 24

KEY 1. WILDFLOWERS & LOW SHRUBS (PP. 49–304), LEAVES ABSENT AT FLOWERING

Flower type		Primary flower color
Symmetry	Number of petal-like parts	
Bilateral	Various	Pink to red: **Orchidaceae** (pp. 34–35) – *Arethusa, Corallorhiza*; **Orobanchaceae** (p. 35) – *Epifagus* Yellow to orange: **Lentibulariaceae** (pp. 32–33) – *Corallorhiza, Utricularia* Blue to purple: **Lentibulariaceae** (pp. 32–33) – *Utricularia*; **Orobanchaceae** (p. 35) – *Epifagus*
Radial	5	Mostly white: **Ericaceae** (pp. 30–31) – *Monotropa, Hypopitys*; **Orobanchaceae** (p. 35) – *Orobanche* Yellow to orange: **Ericaceae** (pp. 30–31) – *Hypopitys*
	7 or more	Mostly white: **Asteraceae** (pp. 27–28) – *Petasites* Yellow to orange: **Asteraceae** (pp. 27–28) – *Tussilago*

KEY 2. WILDFLOWERS & LOW SHRUBS (PP. 49–304), LEAVES BASAL ONLY, SIMPLE, MARGINS ENTIRE

Flower type		Primary flower color
Symmetry	Number of petal-like parts	
Bilateral	Various	Mostly white: Campanulaceae (p. 29) – *Lobelia*; Ericaceae (pp. 30–31) – *Pyrola*; Hydrocharitaceae (p. 32) – *Vallisneria*; Orchidaceae (pp. 34–35) – *Cypripedium, Goodyera, Platanthera, Spiranthes* Pink to red: Ericaceae (pp. 30–31) – *Pyrola*; Orchidaceae (pp. 34–35) – *Arethusa, Calopogon, Calypso, Cypripedium* Blue to purple: Campanulaceae (p. 29) – *Lobelia*; Pontederiaceae (p. 36) – *Pontederia* Green to brown: Orchidaceae (pp. 34–35) – *Platanthera*
Radial	3	Mostly white: Alismataceae (p. 26) – *Alisma, Sagittaria* Yellow to orange: Xyridaceae (p. 39) – *Xyris*
Radial	4	Mostly white: Brasicaceae (pp. 28–29) – *Cardamine*
Radial	5	Mostly white: Diapensiaceae (p. 30) – *Diapensia*; Droseraceae (p. 30) – *Drosera*; Ericaceae (pp. 30–31) – *Pyrola*; Menyanthaceae (p. 33) – *Nymphoides* Pink to red: Ericaceae (pp. 30–31) – *Pyrola*; Sarraceniaceae (p. 38) – *Sarracenia* Yellow to orange: Nymphaeaceae (p. 34) – *Nuphar*; Ranunculaceae (pp. 36–37) – *Ranunculus* Green to brown: Ericaceae (pp. 30–31) – *Pyrola*
Radial	6	Mostly white: Acoraceae (p. 26) – *Acorus* Pink to red: Alliaceae (p. 26) – *Allium* Yellow to orange: Hemerocallidaceae (p. 32) – *Hemerocallis*; Liliaceae (p. 33) – *Clintonia, Erythronium* Blue to purple: Iridaceae (p. 32) – *Sisyrinchium* Green to brown: Acoraceae (p. 26) – *Acorus*
Radial	7 or more	Mostly white: Nymphaeaceae (p. 34) – *Nymphaea* Pink to red: Asteraceae (pp. 27–28) – *Pilosella* Yellow to orange: Asteraceae (pp. 27–28) – *Pilosella* Green to brown: Asteraceae (pp. 27–28) – *Omalotheca*
Indistinguishable		Acoraceae (p. 26) – *Acorus*; Araceae (p. 27) – *Calla, Lemna*; Eriocaulaceae (p. 31) – *Eriocaulon*; Plantaginaceae (pp. 35–36) – *Plantago*

KEY 3. WILDFLOWERS & LOW SHRUBS (PP. 49–304), LEAVES BASAL ONLY, SIMPLE, MARGINS TOOTHED OR LOBED

Flower type		Primary flower color
Symmetry	Number of petal-like parts	
Bilateral	Various	Mostly white: Ericaceae (pp. 30–31) – *Pyrola*; **Hydrocharitaceae** (p. 32) – *Vallisneria*; **Violaceae** (p. 39) – *Viola* Pink to red: Ericaceae (pp. 30–31) – *Pyrola* Yellow to orange: Violaceae (p. 39) – *Viola* Blue to purple: Violaceae (p. 39) – *Viola*
Radial	3	Mostly white: **Alismataceae** (p. 26) – *Sagittaria*
Radial	5	Mostly white: Ericaceae (pp. 30–31) – *Moneses, Orthilia, Pyrola*; **Ranunculaceae** (pp. 36–37) – *Anemone*; **Rosaceae** (pp. 37–38) – *Rubus*; **Saxifragaceae** (p. 38) – *Micranthes, Tiarella* Pink to red: Ericaceae (pp. 30–31) – *Pyrola*; **Ranunculaceae** (pp. 36–37) – *Anemone*; **Sarraceniaceae** (p. 38) – *Sarracenia* Yellow to orange: Ericaceae (pp. 30–31) – *Orthilia*; **Ranunculaceae** (pp. 36–37) – *Ranunculus*; **Saxifragaceae** (p. 38) – *Mitella* Blue to purple: Ranunculaceae (pp. 36–37) – *Anemone* Green to brown: Saxifragaceae (p. 38) – *Mitella*
Radial	6	Mostly white: Ranunculaceae (pp. 36–37) – *Anemone* Pink to red: Ranunculaceae (pp. 36–37) – *Anemone* Blue to purple: Ranunculaceae (pp. 36–37) – *Anemone*
Radial	7 or more	Mostly white: Asteraceae (pp. 27–28) – *Petasites*; **Papaveraceae** (p. 35) – *Sanguinaria* Pink to red: Asteraceae (pp. 27–28) – *Pilosella* Yellow to orange: Asteraceae (pp. 27–28) – *Hieracium, Pilosella, Scorzoneroides, Taraxacum, Tussilago*
Indistinguishable		Araceae (p. 27) – *Calla, Lemna*; **Plataginaceae** (pp. 35–36) – *Plantago*

KEY 4. WILDFLOWERS & LOW SHRUBS (PP. 49–304), LEAVES BASAL ONLY, DIVIDED, MARGINS VARIOUS

Flower type		Primary flower color
Symmetry	Number of petal-like parts	
Bilateral	Various	Mostly white: **Papaveraceae** (p. 35) – *Dicentra*; **Fabaceae** (p. 31) – *Trifolium*
Radial	5	Mostly white: **Apiaceae** (p. 26) – *Aralia*; **Menyanthaceae** (p. 33) – *Menyanthes*; **Oxalidaceae** (p. 35) – *Oxalis*; **Ranunculaceae** (pp. 36–37) – *Coptis*; **Rosaceae** (pp. 37–38) – *Fragaria* Pink to red: Oxalidaceae (p. 35) – *Oxalis*
Radial	6	Mostly white: Ranunculaceae (pp. 36–37) – *Coptis*
Radial	7 or more	Mostly white: Ranunculaceae (pp. 36–37) – *Coptis*
Indistinguishable		Araceae (p. 27) – *Arisaema*

KEY 5. WILDFLOWERS & LOW SHRUBS (PP. 49–304), LEAVES OPPOSITE OR WHORLED, SIMPLE, MARGINS ENTIRE

Flower type		
Symmetry	Number of petal-like parts	Primary flower color
Bilateral	Various	Mostly white: Orobanchaceae (p. 35) – *Melampyrum*; Plantaginaceae (pp. 35–36) – *Gratiola, Veronica* Pink to red: Caprifoliaceae (p. 29) – *Symphoricarpos*; Orchidaceae (pp. 34–35) – *Neottia*; Orobanchaceae (p. 35) – *Agalinis* Yellow to orange: Caprifoliaceae (p. 29) – *Lonicera*; Plantaginaceae (pp. 35–36) – *Gratiola, Linaria* Blue to purple: Lamiaceae (p. 32) – *Prunella*; Plantaginaceae (pp. 35–36) – *Veronica* Green to brown: Orchidaceae (pp. 34–35) – *Neottia*
Radial	2	Green to brown: Elatinaceae (p. 30) – *Elatine*
Radial	3	Mostly white: Hydrocharitaceae (p. 32) – *Elodea*; Melanthiaceae (p. 33) – *Trillium*; Rubiaceae (p. 38) – *Galium* Pink to red: Melanthiaceae (p. 33) – *Trillium* Green to brown: Elatinaceae (p. 30) – *Elatine*
Radial	4	Mostly white: Caryophyllaceae (p. 29) – *Sagina*; Cornaceae (p. 30) – *Chamaepericlymenum*; Onagraceae (p. 34) – *Epilobium*; Plantaginaceae (pp. 35–36) – *Veronica*; Rubiaceae (p. 38) – *Galium, Mitchella* Pink to red: Caprifoliaceae (p. 29) – *Symphoricarpos*; Onagraceae (p. 34) – *Epilobium* Yellow to orange: Saxifragaceae (p. 38) – *Chrysosplenium* Blue to purple: Gentianaceae (p. 31) – *Halenia*; Onagraceae (p. 34) – *Epilobium*; Plantaginaceae (pp. 35–36) – *Veronica*; Rubiaceae (p. 38) – *Houstonia* Green to brown: Gentianaceae (p. 31) – *Halenia*; Rubiaceae (p. 38) – *Galium*; Saxifragaceae (p. 38) – *Chrysosplenium*
Radial	5	Mostly white: Apocynaceae (pp. 26–27) – *Apocynum*; Caryophyllaceae (p. 29) – *Cerastium, Moehringia, Mononeuria, Saponaria, Silene, Stellaria*; Diapensiaceae (p. 30) – *Diapensia* Pink to red: Apocynaceae (pp. 26–27) – *Apocynum, Asclepias*; Caprifoliaceae (p. 29) – *Symphoricarpos*; Caryophyllaceae (p. 29) – *Gypsophila, Saponaria, Silene, Spergularia*; Ericaceae (pp. 30–31) – *Kalmia*; Hypericaceae (p. 32) – *Hypericum*; Lythraceae (p. 33) – *Lythrum*; Orobanchaceae (p. 35) – *Agalinis*; Portulacaceae (p. 36) – *Claytonia* Yellow to orange: Caprifoliaceae (p. 29) – *Lonicera*; Hypericaceae (p. 32) – *Hypericum*; Myrsinaceae (pp. 33–34) – *Lysimachia*
Radial	6	Mostly white: Myrsinaceae (pp. 33–34) – *Lysimachia* Pink to red: Lythraceae (p. 33) – *Lythrum* Yellow to orange: Asteraceae (pp. 27–28) – *Bidens*; Liliaceae (p. 33) – *Lilium, Medeola*; Myrsinaceae (pp. 33–34) – *Lysimachia* Blue to purple: Asteraceae (pp. 27–28) – *Eutrochium* Green to brown: Liliaceae (p. 33) – *Medeola*
Radial	7 or more	Mostly white: Myrsinaceae (pp. 33–34) – *Lysimachia* Yellow to orange: Asteraceae (pp. 27–28) – *Arnica, Bidens*
Indistinguishable		Plantaginaceae (pp. 35–36) – *Callitriche, Hippuris*; Viscaceae (p. 39) – *Arceuthobium*

KEY 6. WILDFLOWERS & LOW SHRUBS (PP. 49–304), LEAVES OPPOSITE OR WHORLED, SIMPLE, MARGINS TOOTHED OR LOBED

Flower type		Primary flower color
Symmetry	Number of petal-like parts	
Bilateral	Various	Mostly white: **Lamiaceae** (p. 32) – *Galeopsis, Lycopus, Mentha*; **Onagraceae** (p. 34) – *Circaea*; **Orobanchaceae** (p. 35) – *Euphrasia, Melampyrum*; **Plantaginaceae** (pp. 35–36) – *Chelone, Gratiola, Veronica* Pink to red: **Caprifoliaceae** (p. 29) – *Symphoricarpos*; **Lamiaceae** (p. 32) – *Galeopsis, Stachys, Teucrium* Yellow to orange: **Caprifoliaceae** (p. 29) – *Diervilla*; **Orobanchaceae** (p. 35) – *Rhinanthus*; **Plantaginaceae** (pp. 35–36) – *Gratiola* Blue to purple: **Lamiaceae** (p. 32) – *Glechoma, Mentha, Prunella, Scutellaria, Stachys, Teucrium*; **Orobanchaceae** (p. 35) – *Euphrasia*; **Phrymaceae** (p. 35) – *Mimulus*; **Plantaginaceae** (pp. 35–36) – *Veronica*; **Verbenaceae** (p. 39) – *Verbena*
Radial	2	Mostly white: **Onagraceae** (p. 34) – *Circaea*
Radial	4	Mostly white: **Lamiaceae** (p. 32) – *Lycopus*; **Onagraceae** (p. 34) – *Epilobium*; **Plantaginaceae** (pp. 35–36) – *Veronica* Pink to red: **Melastomataceae** (p. 33) – *Rhexia*; **Onagraceae** (p. 34) – *Epilobium* Yellow to orange: **Saxifragaceae** (p. 38) – *Chrysosplenium* Blue to purple: **Plantaginaceae** (pp. 35–36) – *Veronica* Green to brown: **Saxifragaceae** (p. 38) – *Chrysosplenium*
Radial	5	Mostly white: **Asteraceae** (pp. 27–28) – *Ageratina*; **Ericaceae** (pp. 30–31) – *Chimaphila*; **Lamiaceae** (p. 32) – *Lycopus* Pink to red: **Caprifoliaceae** (p. 29) – *Linnaea, Symphoricarpos* Yellow to orange: **Caprifoliaceae** (p. 29) – *Diervilla* Blue to purple: **Verbenaceae** (p. 39) – *Verbena*
Radial	6	Yellow to orange: **Asteraceae** (pp. 27–28) – *Bidens*
Radial	7 or more	Yellow to orange: **Asteraceae** (pp. 27–28) – *Bidens* Blue to purple: **Asteraceae** (pp. 27–28) – *Eutrochium*
Indistinguishable		**Asteraceae** (pp. 27–28) – *Ageratina, Bidens, Eupatorium, Eutrochium*; **Hydrocharitaceae** (p. 32) – *Najas*; **Urticaceae** (p. 39) – *Boehmeria, Pilea, Urtica*

KEY 7. WILDFLOWERS & LOW SHRUBS (PP. 49–304), LEAVES OPPOSITE OR WHORLED, DIVIDED, MARGINS VARIOUS

Flower type		Primary flower color
Symmetry	Number of petal-like parts	
Radial	4	Mostly white: **Brassicaceae** (pp. 28–29) – *Cardamine*; **Ranunculaceae** (pp. 36–37) – *Clematis* Blue to purple: **Ranunculaceae** (pp. 36–37) – *Clematis*
Radial	5	Mostly white: **Apiaceae** (p. 26) – *Panax*
Indistinguishable		**Asteraceae** (pp. 27–28) – *Bidens*; **Haloragaceae** (p. 32) – *Myriophyllum*

KEY 8. WILDFLOWERS & LOW SHRUBS (PP. 49–304), LEAVES ALTERNATE, SIMPLE, MARGINS ENTIRE

Flower type		Primary flower color
Symmetry	Number of petal-like parts	
Bilateral	Various	Mostly white: Orchidaceae (pp. 34–35) – *Platanthera, Spiranthes* Pink to red: Orchidaceae (pp. 34–35) – *Epipactis, Platanthera, Pogonia* Blue to purple: Campanulaceae (p. 29) – *Lobelia*; Ericaceae (pp. 30–31) – *Rhododendron*; Orchidaceae (pp. 34–35) – *Epipactis, Platanthera*; Orobanchaceae (p. 35) – *Castilleja, Epifagus*; Plataginaceae (pp. 35–36) – *Veronica* Yellow to orange: Orchidaceae (pp. 34–35) – *Cypripedium*; Orobanchaceae (p. 35) – *Castilleja*; Plantaginaceae (pp. 35–36) – *Linaria*; Scrophulariaceae (p. 39) – *Verbascum* Green to brown: Orchidaceae (pp. 34–35) – *Coeloglossum, Epipactis, Malaxis, Platanthera*
Radial	3	Mostly white: Ericaceae (pp. 30–31) – *Empetrum* Pink to red: Cistaceae (p. 30) – *Lechea*; Nymphaeaceae (p. 34) – *Brasenia* Yellow to orange: Asteraceae (pp. 27–28) – *Solidago*; Scheuchzeriaceae (p. 38) – *Scheuchzeria*; Xyridaceae (p. 39) – *Xyris* Blue to purple: Iridaceae (p. 32) – *Iris* Green to brown: Scheuchzeriaceae (p. 38) – *Scheuchzeria*
Radial	4	Mostly white: Brassicaceae (pp. 28–29) – *Boechera, Borodinia, Cardamine, Draba, Lepidium*; Ericaceae (pp. 30–31) – *Gaultheria, Monotropa*; Onagraceae (p. 34) – *Epilobium*; Polygonaceae (p. 36) – *Persicaria* Ruscaceae (p. 38) – *Maianthemum* Pink to red: Ericaceae (pp. 30–31) – *Hypopitys, Vaccinium*; Onagraceae (p. 34) – *Chamaenerion, Epilobium*; Polygonaceae (p. 36) – *Persicaria* Yellow to orange: Asteraceae (pp. 27–28) – *Nabalus, Solidago*; Ericaceae (pp. 30–31) – *Hypopitys*; Onagraceae (p. 34) – *Oenothera*; Thymelaeaceae (p. 39) – *Dirca* Blue to purple: Brassicaceae (pp. 28–29) – *Boechera*; Comandraceae (p. 30) – *Geocaulon* Green to brown: Comandraceae (p. 30) – *Geocaulon*
Radial	5	Mostly white: Boraginaceae (p. 28) – *Myosotis*; Ericaceae (pp. 30–31) – *Arctostaphylos, Chamaedaphne, Epigaea, Gaultheria, Harrimanella, Monotropa, Rhododendron, Vaccinium*; Polygonaceae (p. 36) – *Bistorta, Fallopia, Persicaria* Pink to red: Convolvulaceae (p. 30) – *Calystegia*; Ericaceae (pp. 30–31) – *Andromeda, Arctostaphylos, Epigaea, Gaylussacia, Hypopitys, Vaccinium*; Polygonaceae (p. 36) – *Bistorta, Persicaria* Yellow to orange: Asteraceae (pp. 27–28) – *Nabalus, Solidago*; Ericaceae (pp. 30–31) – *Hypopitys*; Nymphaeaceae (p. 34) – *Nuphar*; Ranunculaceae (pp. 36–37) – *Ranunculus*; Scrophulariaceae (p. 39) – *Verbascum* Blue to purple: Boraginaceae (p. 28) – *Myosotis*; Campanulaceae (p. 29) – *Campanula*; Comandraceae (p. 30) – *Geocaulon*; Ericaceae (pp. 30–31) – *Rhododendron* Green to brown: Amaranthaceae (p. 26) – *Chenopodium*; Comandraceae (p. 30) – *Geocaulon*

Continued next page

KEY 8. WILDFLOWERS & LOW SHRUBS (PP. 49–304), LEAVES ALTERNATE, SIMPLE, MARGINS ENTIRE (CONTINUED)

Flower type		Primary flower color
Symmetry	Number of petal-like parts	
Radial	6	Mostly white: **Asteraceae** (pp. 27–28) – *Doellingeria*; **Ruscaceae** (p. 38) – *Maianthemum*; **Scheuchzeriaceae** (p. 38) – *Scheuchzeria* Pink to red: **Liliaceae** (p. 33) – *Streptopus*; **Polygonaceae** (p. 36) – *Rumex* Yellow to orange: **Asteraceae** (pp. 27–28) – *Nabalus, Solidago*; **Colchicaceae** (p. 30) – *Uvularia*; **Liliaceae** (p. 33) – *Streptopus*; **Melanthiaceae** (p. 33) – *Veratrum*; **Ranunculaceae** (pp. 36–37) – *Ranunculus*; **Ruscaceae** (p. 38) – *Polygonatum* Blue to purple: **Iridaceae** (p. 32) – *Iris* Green to brown: **Melanthiaceae** (p. 33) – *Veratrum*; **Ruscaceae** (p. 38) – *Polygonatum*; **Scheuchzeriaceae** (p. 38) – *Scheuchzeria*
	7 or more	Mostly white: **Asteraceae** (pp. 27–28) – *Anaphalis, Doellingeria, Erigeron, Nabalus, Solidago, Symphyotrichum*; **Nymphaeaceae** (p. 34) – *Nymphaea* Pink to red: **Asteraceae** (pp. 27–28) – *Erigeron, Euthamia, Pilosella, Oclemena* Yellow to orange: **Asteraceae** (pp. 27–28) – *Hieracium, Nabalus, Pilosella, Rudbeckia, Solidago, Tragopogon* Blue to purple: **Asteraceae** (pp. 27–28) – *Oclemena, Symphyotrichum*
Indistinguishable		**Asteraceae** (pp. 27–28) – *Antennaria, Arctium, Gnaphalium, Omalotheca*; **Cistaceae** – *Lechea*; **Ericaceae** (pp. 30–31) – *Empetrum*; **Haloragaceae** (p. 32) – *Myriophyllum*; **Myricaceae** (p. 33) – *Morella*; **Orobanchaceae** (p. 35) – *Castilleja*; **Polygonaceae** (p. 36) – *Rumex*; **Potamogetonaceae** (p. 36) – *Potamogeton*; **Salicaceae** (p. 38) – *Salix*; **Typhaceae** (p. 39) – *Sparganium, Typha*

KEY 9. WILDFLOWERS & LOW SHRUBS (PP. 49–304), LEAVES ALTERNATE, SIMPLE, MARGINS TOOTHED OR LOBED

Flower type		Primary flower color
Symmetry	Number of petal-like parts	
Bilateral	Various	Pink to red: Campanulaceae (p. 29) – *Lobelia* Yellow to orange: Balsaminaceae (p. 28) – *Impatiens*; Scrophulariaceae (p. 39) – *Verbascum*; Violaceae (p. 39) – *Viola* Blue to purple: Campanulaceae (p. 29) – *Lobelia*; Pontederiaceae (p. 36) – *Pontederia*; Violaceae (p. 39) – *Viola*
Radial	3	Yellow to orange: Asteraceae (pp. 27–28) – *Solidago*
Radial	4	Mostly white: Brassicaceae (pp. 28–29) – *Boechera, Borodinia, Capsella, Draba, Lepidium*; Onagraceae (p. 34) – *Epilobium* Pink to red: Onagraceae (p. 34) – *Chamaenerion, Epilobium* Yellow to orange: Asteraceae (pp. 27–28) – *Nabalus, Solidago*; Brassicaceae (pp. 28–29) – *Barbarea, Rorippa, Sisymbrium*; Onagraceae (p. 34) – *Oenothera* Blue to purple: Brassicaceae (pp. 28–29) – *Boechera*
Radial	5	Mostly white: Apiaceae (p. 26) – *Hydrocotyle, Sanicula*; Ericaceae (pp. 30–31) – *Arctous, Gaultheria, Harrimanella, Moneses, Vaccinium*; Grossulariaceae (p. 31) – *Ribes*; Rosaceae (pp. 37–38) – *Aronia, Spiraea*; Saxifragaceae (p. 38) – *Saxifraga* Pink to red: Ericaceae (pp. 30–31) – *Vaccinium*; Grossulariaceae (p. 31) – *Ribes*; Malvaceae (p. 33) – *Malva*; Rosaceae (pp. 37–38) – *Spiraea* Yellow to orange: Asteraceae (pp. 27–28) – *Nabalus, Solidago*; Grossulariaceae (p. 31) – *Ribes*; Ranunculaceae (pp. 36–37) – *Ranunculus*; Rhamnaceae (p. 37) – *Rhamnus* Blue to purple: Campanulaceae (p. 29) – *Campanula*; Ericaceae (pp. 30–31) – *Phyllodoce*; Grossulariaceae (p. 31) – *Ribes* Green to brown: Amaranthaceae (p. 26) – *Chenopodium*; Apiaceae (p. 26) – *Sanicula*
Radial	6	Mostly white: Asteraceae (pp. 27–28) – *Eurybia* Pink to red: Polygonaceae (p. 36) – *Rumex* Yellow to orange: Asteraceae (pp. 27–28) – *Nabalus, Solidago*
Radial	7 or more	Mostly white: Asteraceae (pp. 27–28) – *Erigeron, Eurybia, Leucanthemum, Nabalus, Oclemena, Solidago, Symphyotrichum* Pink to red: Asteraceae (pp. 27–28) – *Erigeron, Nabalus, Oclemena* Yellow to orange: Asteraceae (pp. 27–28) – *Hieracium, Lactuca, Packera, Rudbeckia, Solidago, Sonchus* Blue to purple: Asteraceae (pp. 27–28) – *Cirsium, Eurybia, Lactuca, Oclemena, Symphyotrichum*
Indistinguishable		Asteraceae (pp. 27–28) – *Cirsium, Erechtites*; Myricaceae (p. 33) – *Comptonia, Morella, Myrica*; Rhamnaceae (p. 37) – *Rhamnus*; Salicaceae (p. 38) – *Salix*; Urticaceae (p. 39) – *Laportea*

KEY 10. WILDFLOWERS & LOW SHRUBS (PP. 49–304), LEAVES ALTERNATE, DIVIDED, MARGINS VARIOUS

Flower type		Primary flower color
Symmetry	Number of petal-like parts	
Bilateral	Various	Mostly white: **Apiaceae** (p. 26) – *Aegopodium, Daucus, Heracleum*; **Fabaceae** (p. 31) – *Trifolium* Pink to red: **Fabaceae** (p. 31) – *Trifolium*; **Papaveraceae** (p. 35) – *Capnoides* Yellow to orange: **Fabaceae** (p. 31) – *Lotus, Trifolium*; **Lentibulariaceae** (pp. 32–33) – *Utricularia*; **Papaveraceae** (p. 35) – *Capnoides* Blue to purple: **Fabaceae** (p. 31) – *Trifolium, Vicia*
Radial	4	Mostly white: **Brassicaceae** (pp. 28–29) – *Capsella, Cardamine*; **Ranunculaceae** (pp. 36–37) – *Actaea* Yellow to orange: **Brassicaceae** (pp. 28–29) – *Barbarea, Rorippa, Sisymbrium*; **Papaveraceae** (p. 35) – *Chelidonium*
Radial	5	Mostly white: **Anacardiaceae** (p. 26) – *Toxicodendron*; **Apiaceae** (p. 26) – *Aegopodium, Aralia, Carum, Cicuta, Conioselinum, Daucus, Heracleum, Osmorhiza, Panax, Sanicula, Sium*; **Asteraceae** (pp. 27–28) – *Achillea*; **Ranunculaceae** (pp. 36–37) – *Actaea, Ranunculus*; **Rosaceae** (pp. 37–38) – *Drymocallis, Geum, Rubus, Sibbaldia* Pink to red: **Malvaceae** (p. 33) – *Malva*; **Ranunculaceae** (pp. 36–37) – *Aquilegia*; **Rosaceae** (pp. 37–38) – *Comarum, Rosa* Yellow to orange: **Apiaceae** (p. 26) – *Pastinaca, Zizia*; **Oxalidaceae** (p. 35) – *Oxalis*; **Ranunculaceae** (pp. 36–37) – *Aquilegia, Ranunculus*; **Rosaceae** (pp. 37–38) – *Agrimonia, Dasiphora, Geum, Potentilla* Blue to purple: **Rosaceae** (pp. 37–38) – *Comarum* Green to brown: **Apiaceae** (p. 26) – *Sanicula*; **Vitaceae** (p. 39) – *Parthenocissus*
Radial	6	Mostly white: **Ranunculaceae** (pp. 36–37) – *Actaea* Yellow to orange: **Berberidaceae** (p. 28) – *Caulophyllum*; **Rosaceae** (pp. 327–38) – *Potentilla* Green to brown: **Berberidaceae** (p. 28) – *Caulophyllum*
Radial	7 or more	Mostly white: **Ranunculaceae** (pp. 36–37) – *Actaea*
Indistinguishable		**Asteraceae** (pp. 27–28) – *Matricaria*; **Fabaceae** (p. 31) – *Trifolium*; **Ranunculaceae** (pp. 36–37) – *Thalictrum*

KEY 11. TREES & TALL SHRUBS (PP. 305–348), LEAVES NEEDLE-LIKE OR BROAD AND OPPOSITE, OR WHORLED

Leaf type	Leaf(let) margin	Fruit type or cones
Needle-like		Berry-like: Cupressaceae (p. 41) – *Juniperus*; Taxaceae (p. 42) – *Taxus* Seeds borne in woody cones: Cupressaceae (p. 41) – *Thuja*; Pinaceae (p. 41) – *Abies, Larix, Picea, Pinus, Tsuga*
Simple	Entire	Berry-like: Adoxaceae (p. 40) – *Viburnum*; Cornaceae (p. 40) – *Swida* Capsule containing multiple seeds: Oleaceae (p. 41) – *Syringa*
Simple	Toothed	Berry-like: Adoxaceae (p. 40) – *Viburnum*
Simple	Lobed	Berry-like: Adoxaceae (p. 40) – *Viburnum* Samara-like schizocarp: Sapindaceae (p. 42) – *Acer*
Divided	Entire	Winged Samara: Oleaceae (p. 41) – *Fraxinus*
Divided	Toothed	Berry-like: Adoxaceae (p. 40) – *Sambucus* Winged Samara: Oleaceae (p. 41) – *Fraxinus*

KEY 12. TREES & TALL SHRUBS (PP. 305–348), LEAVES ALTERNATE

Leaf type	Leaf(let) margin	Fruit type
Simple	Entire	Berry-like: Aquifoliaceae (p. 40) – *Ilex*; Cornaceae (p. 40) – *Swida*; Ericaceae (p. 41) – *Vaccinium* Capsule containing multiple seeds: Salicaceae (p. 42) – *Salix*
Simple	Toothed	Berry-like: Aquifoliaceae (p. 40) – *Ilex*; Ericaceae (p. 41) – *Vaccinium*; Rosaceae (pp. 41–42) – *Amelanchier, Crataegus, Malus, Prunus* Nut inside a husk-like covering: Betulaceae (p. 40) – *Corylus, Ostrya*; Fagaceae (p. 41) – *Fagus* Capsule containing multiple seeds: Hamamelidaceae (p. 41) – *Hamamelis*; Salicaceae (p. 42) – *Populus, Salix* Dry nutlet: Malvaceae (p. 41) – *Tilia* Winged samara: Betulaceae (p. 40) – *Alnus, Betula*; Ulmaceae (p. 42) – *Ulmus*
Simple	Lobed	Nut in a cup-like base: Fagaceae (p. 41) – *Quercus*
Divided	Entire	Bristly pod: Fabaceae (p. 41) – *Robinia*
Divided	Toothed	Berry-like: Rosaceae (pp. 41–42) – *Sorbus* Hairy, dry drupe: Anacardiaceae (p. 40) – *Rhus*

KEY 13. FERNS AND OTHER SPORE-PRODUCING PLANTS (PP. 349–380)

Aquatic	Blade length	Leaf division	Spore-bearing structures
Fully submerged	Varies	Entire, without lobes	Sporangia borne at base of leaf-like sporophylls: Huperziaceae (p. 44) – *Huperzia*; Isoëtaceae (p. 44) – *Isoëtes*
Not fully submerged	Less than 1 cm		Sporangia borne in terminal clusters or spore cones: Equisetaceae (p. 43) – *Equisetum*; Lycopodiaceae (p. 44) – *Dendrolycopodium, Diphasiastrum, Lycopodiella, Lycopodium, Spinulum*
	Varies	Lobed	Spore-bearing leaves very different from sterile leaves: Dryopteridaceae (p. 43) – *Dryopteris*; Equisetaceae (p. 43) – *Equisetum*; Onocleaceae (p. 44) – *Onoclea*; Woodsiaceae (p. 45) – *Woodsia* Spore-bearing leaves similar to sterile leaves: Dryopteridaceae (p. 43) – *Dryopteris*; Equisetaceae (p. 43) – *Equisetum*; Polypodiaceae (p. 45) – *Polypodium*; Woodsiaceae (p. 45) – *Woodsia* Leaf divided into a sterile portion (trophophore) and a stalked reproductive portion (sporophore): Ophioglossaceae (p. 44) – *Botrychium*
		Once (or 1.5 times) divided (pinnate)	Spore-bearing leaves very different from sterile leaves: Dryopteridaceae (p. 43) – *Dryopteris, Polystichum*; Onocleaceae (p. 44) – *Matteuccia, Onoclea*; Osmundaceae (p. 45) – *Osmundastrum*; Woodsiaceae (p. 45) – *Woodsia* Spore-bearing leaves similar to sterile leaves: Aspleniaceae (p. 43) – *Asplenium*; Blechnaceae (p. 43) – *Woodwardia*; Dryopteridaceae (p. 43) – *Dryopteris, Polystichum*; Osmundaceae (p. 45) – *Osmunda*; Polypodiaceae (p. 45) – *Polypodium*; Thelypteridaceae (p. 45) – *Parathelypteris, Thelypteris*; Woodsiaceae (p. 45) – *Cystopteris, Deparia, Woodsia* Leaf divided into a sterile portion (trophophore) and a stalked reproductive portion (sporophore): Ophioglossaceae (p. 44) – *Botrychium*
		Twice (or 2.5 times) divided (bipinnate)	Spore-bearing leaves very different from sterile leaves: Dryopteridaceae (p. 43) – *Dryopteris*; Onocleaceae (p. 44) – *Matteuccia*; Osmundaceae (p. 45) – *Osmundastrum*; Woodsiaceae (p. 45) – *Woodsia* Spore-bearing leaves similar to sterile leaves: Blechnaceae (p. 43) – *Woodwardia*; Dennstaedtiaceae (p. 43) – *Dennstaedtia*; Dryopteridaceae (p. 43) – *Dryopteris, Polystichum*; Osmundaceae (p. 45) – *Osmunda*; Thelypteridaceae (p. 45) – *Parathelypteris, Phegopteris, Thelypteris*; Woodsiaceae (p. 45) – *Athyrium, Cystopteris, Deparia, Gymnocarpium, Woodsia* Leaf divided into a sterile portion (trophophore) and a stalked reproductive portion (sporophore): Ophioglossaceae (p. 44) – *Botrychium*
		Thrice divided (tripinnate) or more	Spore-bearing leaves very different from sterile leaves: Dryopteridaceae (p. 43) – *Dryopteris*; Woodsiaceae (p. 45) – *Gymnocarpium, Woodsia* Spore-bearing leaves similar to sterile leaves: Dennstaedtiaceae (p. 43) – *Dennstaedtia, Pteridium*; Dryopteridaceae (p. 43) – *Dryopteris*; Osmundaceae (p. 45) – *Osmunda*; Pteridaceae (p. 45) – *Adiantum*; Woodsiaceae (p. 45) – *Gymnocarpium, Woodsia* Leaf divided into a sterile portion (trophophore) and a stalked reproductive portion (sporophore): Ophioglossaceae (p. 44) – *Botrychium*

KEY 14. SEDGES, RUSHES, AND GRASSES (PP. 381–448)

Flowers	Leaves	Stems	Fruit	Go to
with 1 scale	3-ranked, alternate; sheaths closed	solid, not jointed, usually 3-sided	1-seeded achene	Sedges (Cyperaceae): Key 15
with 6 tepals	2-ranked, basal only; sheaths closed	solid, not jointed, round	many-seeded capsule	Rushes (Juncaceae): Key 15
with 2 bracts	2-ranked, alternate; sheaths usually open	hollow at internodes, jointed, round	1-seeded grain	Grasses (Poacaea): Key 16

KEY 15. SEDGES (CYPERACEAE; PP. 381–419) AND RUSHES (JUNCACEAE; PP. 420–425)

Fruit	Scales of spikelets	# of spikelets per stem	Perianth bristles
Enclosed in a sac-like structure (perigynium)	Spirally arranged or in 3 or more ranks	1 or more	Perianth bristles absent: Cyperaceae (p. 46) – *Carex*
Not enclosed in a sac-like structure	2 ranked, given spikelets a flattened appearance	1 or more	Perianth bristles absent: Cyperaceae (p. 46) – *Cyperus* Perianth bristles present: Cyperaceae (p. 46) – *Dulichium*
	Spirally arranged or in 3 or more ranks, giving spikelets a rounded appearance	1	Perianth bristles greatly exceeding floral scales: Cyperaceae (p. 46) – *Eriophorum, Trichophorum* Perianth bristles not greatly exceeding floral scales: Cyperaceae (p. 46) – *Eleocharis, Schoenoplectus, Trichophorum*
		2 or more	Perianth bristles greatly exceeding floral scales: Cyperaceae (p. 46) – *Scirpus, Eriophorum* Perianth bristles not greatly exceeding floral scales: Cyperaceae (p. 46) – *Rhynchospora, Bulbostylis, Cladium, Scirpus*
A many seeded capsule	Spirally arranged or in 3 or more ranks	1 or more	Perianth bristles absent: Juncaceae (p. 46) – *Juncus, Luzula, Oreojuncus*

KEY 16. GRASSES (POACEAE; PP. 426–448)

Glumes	# of florets per spikelet	Glumes or Lemmas or both	Leaf ligule
Both glumes as long as or longer than florets	1	without an awn	Ligule is a membrane without hairs: Poaceae (pp. 46–47) – *Agrostis, Cinna, Digitaria, Milium, Phalaris* Ligule is a membrane with fine hairs: Poaceae (pp. 46–47) – *Agrostis* Ligule is a membrane with fine hairs: Poaceae (pp. 46–47) – *Dichanthelium, Panicum*
		with an awn	Ligule is a membrane without hairs: Poaceae (pp. 46–47) – *Agrostis, Alopecurus, Calamagrostis, Cinna, Phleum* Ligule is a membrane with fine hairs: Poaceae (pp. 46–47) – *Agrostis, Muhlenbergia, Oryzopsis, Panicum*
	2 or more	without an awn	Ligule is a membrane without hairs: Poaceae (pp. 46–47) – *Anthoxanthum, Digitaria, Elymus* Ligule is a membrane with fine hairs: Poaceae (pp. 46–47) – *Phragmites* Ligule is a membrane with fine hairs: Poaceae (pp. 46–47) – *Dichanthelium, Panicum, Phragmites*
		with an awn	Ligule is a membrane without hairs: Poaceae (pp. 46–47) – *Anthoxanthum, Deschampsia, Elymus, Trisetum, Vahlodea* Ligule is a membrane with fine hairs: Poaceae (pp. 46–47) – *Anthoxanthum, Arrhenatherum, Phragmites* Ligule is a membrane with fine hairs: Poaceae (pp. 46–47) – *Danthonia, Panicum, Phragmites*
Neither glume as long as or longer than florets	1	without an awn	Ligule is a membrane without hairs: Poaceae (pp. 46–47) – *Digitaria* Ligule is a membrane with fine hairs: Poaceae (pp. 46–47) – *Muhlenbergia* Ligule is a membrane with fine hairs: Poaceae (pp. 46–47) – *Dichanthelium*
		with an awn	Ligule is a membrane without hairs: Poaceae (pp. 46–47) – *Brachyelytrum, Dichanthelium, Muhlenbergia* Ligule is a membrane with fine hairs: Poaceae (pp. 46–47) – *Muhlenbergia, Oryzopsis, Panicum*
	2 or more	without an awn	Ligule is a membrane without hairs: Poaceae (pp. 46–47) – *Bromus, Dactylis, Digitaria, Elymus, Glyceria, Poa, Torreyochloa* Ligule is a membrane with fine hairs: Poaceae (pp. 46–47) – *Dichanthelium, Festuca, Phragmites, Poa* Ligule is a membrane with fine hairs: Poaceae (pp. 46–47) – *Dichanthelium, Panicum, Phragmites*
		with an awn	Ligule is a membrane without hairs: Poaceae (pp. 46–47) – *Bromus, Dactylis, Deschampsia, Elymus, Festuca, Schizachne, Trisetum* Ligule is a membrane with fine hairs: Poaceae (pp. 46–47) – *Festuca, Phragmites* Ligule is a membrane with fine hairs: Poaceae (pp. 46–47) – *Danthonia, Panicum, Phragmites, Poa*
Absent	1	without an awn	Ligule is a membrane without hairs: Poaceae (pp. 46–47) – *Leersia*

PLANT FAMILY PHOTOGRAPHS • WILDFLOWERS & LOW SHRUBS

Acoraceae ▶ (p. 49) – Flowers minute, 3-parted, aggregated in spikes; leaves alternate, clustered at base.

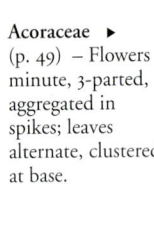
Acorus

Alismataceae ▶ (pp. 49–51) – Aquatic; flowers 3-parted; leaves entire or with basal lobes, erect or floating; emergent leaves with long petioles; sap milky.

Alisma   *Sagittaria*

Alliaceae ▶ (p. 51) – Flowers with 6 tepals, 6 stamens; leaves basal or alternate, the veins parallel; leaves and stem with onion odor.

Allium

Amaranthaceae ▶ (p. 52) – Flowers with 4 or 5 sepals, 4 or 5 stamens.

Chenopodium

Anacardiaceae ▶ (pp. 52–53) – Shrubs; flowers with 5 petals, 5 sepals, 5 or 10 stamens; leaves alternate, pinnately compound.

Toxicodendron

Apiaceae ▶ (pp. 53–62) – Flowers in compound umbels, with 5 petals, 5 stamens, 2 stigmas; flower stalks usually hollow; petioles generally with inflated bases.

Aegopodium   *Aralia*

Carum   *Cicuta*   *Conioselinum*   *Daucus*   *Heracleum*

Hydrocotyle   *Osmorhiza*   *Panax*   *Pastinaca*   *Sanicula*

Apocynaceae ▶ (pp. 62–63) – Tubular flowers with 5 connate sepals, 5 connate petals, 5 stamens; leaves opposite; latex milky.

Sium   *Zizia*   *Apocynum*   *Asclepias*

26 THE PLANTS OF BAXTER STATE PARK

PLANT FAMILY PHOTOGRAPHS • WILDFLOWERS & LOW SHRUBS

Apocynaceae (continued; pp. 62–63)

Asclepias

Araceae ▶ (pp. 64–65) – Flowers tiny, in a dense spike enclosed in a large sheathing bract; leaves alternate; sap milky or watery; aromatic.

Arisaema

Calla

Lemna

Asteraceae ▶ (pp. 65–106) – Inflorescence a dense head of small florets; disc florets tubular and tipped with 5 small teeth; ray florets with a strap-like petal.

Achillea

Ageratina

Anaphalis

Antennaria

Arctium

Arnica

Bidens

Bidens

Cirsium

Doellingeria

Erechtites

Erigeron

Erigeron

Eupatorium

Eurybia

Euthamia

Eutrochium

Gnaphalium

Hieracium

Lactuca

Lactuca

Leucanthemum

Matricaria

PLANT FAMILY PHOTOGRAPHS • WILDFLOWERS & LOW SHRUBS

Asteraceae (continued; pp. 65–106)

Nabalus   *Nabalus*   *Oclemena*   *Omalotheca*

Omalotheca   *Packera*   *Petasites*   *Pilosella*   *Pilosella*

Rudbeckia   *Scorzoneroides*   *Solidago*   *Solidago*   *Sonchus*

Symphyotrichum   *Symphyotrichum*   *Taraxacum*   *Tragopogon*   *Tussilago*

Balsaminaceae ▶ (p. 107) – Flowers bilaterally symmetric, with a spur; seeds explosively dispersed from a linear capsule.

Berberidaceae ▶ (p. 107) – Flowers yellow, with 6 sepals and 6 petals; inner bark bright yellow.

Impatiens

Caulophyllum

Boraginaceae ▶ (p. 108) – Flowers in curved or forked cymes, 5 united petals with 5 adnate stamens, 5 sepals; leaves alternate, simple.

Brassicaceae ▶ (pp. 108–114) – Flowers in a raceme; 4 sepals, 4 petals (narrowed at base and wider toward apex), 4 tall and 2 short stamens; leaves alternate or basal.

Myosotis   *Barbarea*   *Boechera*

PLANT FAMILY PHOTOGRAPHS • WILDFLOWERS & LOW SHRUBS

Brassicaceae (continued; pp. 108–114)

Borodinia   *Capsella*   *Cardamine*   *Draba*

Lepidium   *Rorippa*   *Sisymbrium*

Campanulaceae ▶ (pp. 115–117) – Flowers bell-shaped or bilaterally symmetric in Lobelia, 5-parted, with 5 stamens; leaves alternate, simple; sap milky.

Campanula

Caprifoliaceae ▶ (pp. 118–120) – Flowers with 5 united petals, 5 small sepals, 5 stamens, an inferior ovary, and long styles; flowers and fruits often in pairs; leaves opposite.

Lobelia   *Lobelia*   *Diervilla*   *Linnaea*

Caryophyllaceae ▶ (pp. 121–127) – Flowers in forked cymes, with 5 sepals and 5 petals; styles 2–5; leaves opposite, entire.

Lonicera   *Lonicera*   *Symphoricarpos*   *Cerastium*

Gypsophila   *Moehringia*   *Mononeuria*   *Sagina*   *Saponaria*

Silene   *Silene*   *Spergularia*   *Stellaria*

PLANT FAMILY PHOTOGRAPHS • WILDFLOWERS & LOW SHRUBS

Cistaceae ▶ (p. 127) – Flowers with 3 or 5 sepals and petals; leaves often with stellate and glandular hairs.

Lechea

Colchicaceae ▶ (p. 128) – Flowers with 6 tepals and 6 stamens; leaves alternate, with parallel veins.

Uvularia

Comandraceae ▶ (p. 128) – Flowers with 4–6 petals, no sepals; leaves alternate, entire.

Geocaulon

Convolvulaceae ▶ (p. 129) – Vines; flowers tubular, with 5 distinct sepals and 5 connate petals, with fold lines on petals.

Calystegia

Cornaceae ▶ (p. 129) – Shrubs; with petal-like bracts; flowers small, with 4 or 5 petals; leaves opposite or whorled.

Chamaepericlymenum

Diapensiaceae ▶ (p. 130) – Herbs or shrublets; flowers with 5 petals; leaves alternate, evergreen, entire.

Diapensia

Droseraceae ▶ (pp. 130–131) – Flowers with 5 sepals, 5 petals, 5 stamens; insects trapped by sticky hairs.

Drosera

Elatinaceae ▶ (p. 131) – Aquatic; petals and sepals usually 4 (2–5); leaves opposite or whorled, simple.

Elatine

Ericaceae ▶ (pp. 132–150) – Flowers urn-shaped, with 4 or 5 basally connate sepals; leaves without stipules.

Andromeda   *Arctostaphylos*   *Arctous*   *Chamaedaphne*

Chimaphila   *Empetrum*   *Epigaea*   *Gaultheria*   *Gaylussacia*

30   THE PLANTS OF BAXTER STATE PARK

PLANT FAMILY PHOTOGRAPHS • WILDFLOWERS & LOW SHRUBS

Ericaceae (continued; pp. 132–150)

Harrimanella   *Hypopitys*   *Kalmia*   *Kalmia*

Moneses   *Monotropa*   *Orthilia*   *Phyllodoce*   *Pyrola*

Rhododendron   *Rhododendron*   *Rhododendron*   *Vaccinium*   *Vaccinium*

Eriocaulaceae ▶ (p. 150) – Flowers with 2 or 3 sepals, 2 or 3 petals, 2–6 stamens; leaves in basal rosette.

Eriocaulon

Fabaceae ▶ (pp. 151– 154) – Flowers bilaterally symmetric consisting of upper banner petals and lower wings and a keel; leaves often pinnately compound.

Lotus   *Trifolium*

Gentianaceae ▶ (p. 155) – Flowers with 4 or 5 sepals, 4 or 5 connate petals, 4 or 5 stamens; leaves opposite.

Trifolium   *Trifolium*   *Vicia*   *Halenia*

Grossulariaceae ▶ (pp. 155–158) – Flowers with 5 small petals, 5 stamens, 2 styles; leaves palmately lobed; berries with sepals attached.

Ribes   *Ribes*   *Ribes*   *Ribes*

PLANT FAMILY PHOTOGRAPHS • WILDFLOWERS & LOW SHRUBS

Haloragaceae ▶ (pp. 158–160) – Aquatic; flowers greenish; submersed leaves finely dissected.

Myriophyllum   *Myriophyllum*

Hemerocallidaceae ▶ (p. 161) – Flowers showy, red-orange or yellow; leaves basal only, hairless.

Hemerocallis

Hydrocharitaceae ▶ (pp. 161–162) – Aquatic; flowers with 3 sepals, 3 white petals (or absent), 2 or more stamens.

Elodea   *Najas*   *Vallisneria*

Hypericaceae ▶ (pp. 163–168) – Flowers with 2–5 sepals, 4 or 5 petals, numerous stamens; leaves opposite or whorled.

Hypericum   *Hypericum*   *Hypericum*   *Hypericum*

Iridaceae ▶ (pp. 168–169) – Flowers with 3 petals, 3 showy sepals, 3 stamens; capsules 3-chambered.

Iris   *Sisyrinchium*

Lamiaceae ▶ (pp. 170–176) – Flowers bilaterally symmetric, with 2 upper and 3 lower lobes; leaves opposite or whorled, often aromatic; stems square.

Galeopsis

Glechoma   *Lycopus*   *Mentha*   *Prunella*

Scutellaria

Stachys   *Teucrium*

Lentibulariaceae ▶ (pp. 176–180) – Aquatic; flowers bilaterally symmetric; trapped prey visible in bladders.

Utricularia

Utricularia

32   THE PLANTS OF BAXTER STATE PARK

PLANT FAMILY PHOTOGRAPHS • WILDFLOWERS & LOW SHRUBS

Lentibulariaceae (continued; pp. 176–180)

Utricularia

Liliaceae ▸ (pp. 181–183) – Flowers with 6 tepals, 6 stamens; carpel with a 3-parted stigma; leaves with parallel veins; nectaries at base of tepals.

Clintonia

Erythronium

Lilium

Medeola

Streptopus

Lythraceae ▸ (p. 184) – Petals often rugose; leaves opposite or whorled; stem square.

Lythrum

Malvaceae ▸ (p. 184) – Flowers tubular, with 3–5 partially connate sepals, and 5 petals; stamens numerous, forming a tube surrounding the pistil.

Malva

Melanthiaceae ▸ (pp. 185–186) – Flowers with 3 or 4 sepals, 3 or 4 petals, 6–8 stamens; leaves whorled.

Trillium

Trillium

Veratrum

Melastomataceae ▸ (p. 186) – Flowers with 3–5 sepals, 3–5 petals, 6–10 stamens; leaves opposite.

Rhexia

Menyanthaceae ▸ (p. 187) – Aquatic; leaves alternate, divided into 3 leaflets; petioles sheathing.

Menyanthes

Nymphoides

Myricaceae ▸ (pp. 188–189) – Shrubs; leaves alternate; wood and leaves aromatic.

Comptonia

Morella

Myrica

Myrsinaceae ▸ (pp. 189–191) – Flowers with 5 basally connate sepals, 5 petals, 5 stamens.

Lysimachia

Lysimachia

PLANT FAMILY PHOTOGRAPHS • WILDFLOWERS & LOW SHRUBS

Myrsinaceae (continued; pp. 189–191)

Lysimachia   *Lysimachia*

Nymphaeaceae ▶ (pp. 192–193) – Aquatic; showy flowers emerge above water surface; leaves large and floating.

Brasenia

Nuphar   *Nymphaea*

Onagraceae ▶ (pp. 194–200) – Flowers with 2–4 sepals, 2–4 petals, 4 or 8 stamens, 4-lobed stigma.

Chamaenerion   *Circaea*

Epilobium   *Epilobium*   *Oenothera*

Orchidaceae ▶ (pp. 201–214) – Flowers bilaterally symmetric; leaves with parallel veins.

Arethusa

Calopogon   *Calypso*   *Coeloglossum*   *Corallorhiza*   *Corallorhiza*

Cypripedium   *Cypripedium*   *Epipactis*   *Goodyera*   *Malaxis*

Neottia   *Neottia*   *Platanthera*   *Platanthera*   *Platanthera*

PLANT FAMILY PHOTOGRAPHS • WILDFLOWERS & LOW SHRUBS

Orchidaceae (continued; pp. 201–214)

Platanthera   *Platanthera*   *Platanthera*   *Pogonia*

Orobanchaceae ▶ (pp. 215–219) – Parasitic; flowers with 5 sepals, 5 petals, 4 stamens; leaves fleshy.

Spiranthes   *Agalinis*   *Castilleja*   *Epifagus*

Euphrasia   *Euphrasia*   *Melampyrum*   *Orobanche*   *Rhinanthus*

Oxalidaceae ▶ (pp. 219–220) – Flowers with 5 sepals, 5 petals, 10 stamens (outer whorl shorter than inner).

Papaveraceae ▶ (pp. 220–222) – Flowers and petals showy; stamens numerous; stems with milky or yellow sap.

Oxalis   *Oxalis*   *Capnoides*

Phrymaceae ▶ (p. 222) – Flowers bilaterally symmetric, with 4 sepals and petals; leaves simple, opposite, toothed.

Chelidonium   *Dicentra*   *Sanguinaria*   *Mimulus*

Plantaginaceae ▶ (pp. 223–229) – Petals and sepals usually 4 or 5; stigma usually 2-lobed; leaves various; stipules absent.

Callitriche   *Chelone*   *Gratiola*   *Gratiola*

PLANT FAMILY PHOTOGRAPHS   35

PLANT FAMILY PHOTOGRAPHS • WILDFLOWERS & LOW SHRUBS

Plantaginaceae (continued; pp. 223–229)

Hippuris   *Linaria*   *Plantago*

Polygonaceae ▶ (pp. 230–236) – Flowers with 5 or 6 petal-like sepals, 3–9 stamens, 3 styles; nodes swollen; leaves alternate, simple, entire.

Veronica   *Veronica*   *Veronica*   *Bistorta*

Fallopia   *Persicaria*   *Persicaria*   *Persicaria*   *Persicaria*

Pontederiaceae ▶ (p. 237) – Aquatic; flowers with 6 tepals, 6 stamens; leaves alternate, with sheathing bases.

Rumex   *Rumex*   *Rumex*   *Pontederia*

Portulacaceae ▶ (p. 237) – Flowers with 2 sepal-like bracts and 5 petals; stems usually red to purple.

Potamogetonaceae ▶ (pp. 238–245) – Aquatic; flowers often 4-parted; inflorescence a spike, without subtending bracts.

Claytonia   *Potamogeton*

Ranunculaceae ▶ (pp. 246–255) – Flowers with multiple carpels at center; leaves alternate or basal.

Potamogeton   *Potamogeton*   *Actaea*   *Anemone*

PLANT FAMILY PHOTOGRAPHS • WILDFLOWERS & LOW SHRUBS

Ranunculaceae (continued; pp. 246–255)

Aquilegia — *Aquilegia* — *Clematis* — *Clematis*

Coptis — *Ranunculus* — *Ranunculus* — *Ranunculus*

Rhamnaceae ▶ (p. 255) – Shrubs; flowers with 4 or 5 sepals, 4 or 5 petals, 4 or 5 stamens; capsule 3-parted.

Ranunculus — *Thalictrum* — *Thalictrum* — *Rhamnus*

Rosaceae ▶ (pp. 256–273) – Flowers with 5 sepals, 5 petals, numerous stamens, numerous styles; leaves or leaflets toothed.

Agrimonia — *Aronia* — *Comarum* — *Dasiphora*

Drymocallis — *Fragaria* — *Geum* — *Geum* — *Geum*

Geum — *Potentilla* — *Potentilla* — *Rosa* — *Rosa*

PLANT FAMILY PHOTOGRAPHS 37

PLANT FAMILY PHOTOGRAPHS • WILDFLOWERS & LOW SHRUBS

Rosaceae (continued; pp. 256–273)

Rubus — *Sibbaldia* — *Spiraea* — *Spiraea*

Rubiaceae ▶ (pp. 273–279) – Flowers with 3–5 connate petals, 4 or 5 stamens; leaves opposite or whorled.

Galium — *Galium* — *Houstonia* — *Mitchella*

Ruscaceae ▶ (pp. 280–282) – Flowers with 6 tepals, 6 stamens; leaves alternate or basal.

Salicaceae ▶ (pp. 283–284) – Shrubs; staminate and carpellate flowers on separate plants; petals absent; leaves alternate (see also pp. 337–344 for larger shrubs).

Maianthemum — *Polygonatum* — *Salix*

Sarraceniaceae ▶ (p. 285) – Insectivorous; found in bogs and marshes; flowers with 4 or 5 sepals, 5 petals; leaves tubular.

Salix — *Salix* — *Sarracenia*

Saxifragaceae ▶ (pp. 286–288) – Flowers with 5 distinct sepals, 5 distinct petals, 5 or 10 stamens, 2 styles; leaves without stipules.

Chrysosplenium — *Micranthes* — *Micranthes* — *Mitella*

Scheuchzeriaceae ▶ (p. 289) – Found in bogs; flowers in racemes, with 6 petals; leaves in 2 vertical rows; sheaths with ligules.

Saxifraga — *Tiarella* — *Scheuchzeria* — *Scheuchzeria*

PLANT FAMILY PHOTOGRAPHS • WILDFLOWERS & LOW SHRUBS

Scrophulariaceae ▶ (p. 289) – Flowers bilaterally symmetric (occasionally nearly radially symmetric), with 2 upper lobes and 3 lower lobes.

Verbascum

Thymelaeaceae ▶ (p. 290) – Shrubs; flowers with sepals and petal-like appendages on rim of floral tube.

Dirca

Typhaceae ▶ (pp. 290–293) – Aquatic; inflorescence a dense spike; leaves narrow, alternate.

Sparganium   *Sparganium*   *Sparganium*

Typha

Urticaceae ▶ (pp. 294–295) – Staminate and carpellate flowers separate, in long clusters originating from the leaf axils.

Boehmeria   *Laportea*   *Pilea*

Verbenaceae ▶ (p. 296) – Flowers with 5 basally connate petals; leaves opposite or whorled; stem square in cross section.

Urtica   *Verbena*

Violaceae ▶ (pp. 296–302) – Flowers bilaterally symmetric, nodding, with 5 distinct petals; capsules 3-valved.

Viola

Viola   *Viola*   *Viola*

Viscaceae ▶ (p. 302) – Parasitic; flowers unisexual, with 3 sepals, no petals; leaves usually opposite.

Arceuthobium

Vitaceae ▶ (p. 303) – Flowers with 4 or 5 small sepals, petals often connate at tip, 4 or 5 stamens.

Parthenocissus

Xyridaceae ▶ (pp. 303–304) – Flowers with 3 sepals, 3 petals, 3 stamens.

Xyris

PLANT FAMILY PHOTOGRAPHS • TREES & TALL SHRUBS

Adoxaceae ▶ (pp. 305–309) – Flowers with 3–5 sepals, 5 petals, 4 or 5 stamens; leaves opposite.

Anacardiaceae ▶ (pp. 309–310) – Flowers with 5 petals, 5 sepals, 5 or 10 stamens; leaves alternate, pinnately compound.

Aquifoliaceae ▶ (pp. 310–311) – Flowers with 4 or 5 sepals, 4–8 petals, 4–6 stamens; leaves alternate.

Betulaceae ▶ (pp. 312–316) – Flowers of separate staminate and carpellate catkins without petals; leaves alternate, simple, toothed.

Cornaceae ▶ (pp. 317–318) – Inflorescence with petaloid bracts; flowers small, with 4 or 5 petals; leaves opposite or whorled.

40 THE PLANTS OF BAXTER STATE PARK

PLANT FAMILY PHOTOGRAPHS • TREES & TALL SHRUBS

Cupressaceae ▶ (pp. 318–319) – Leaves small, scale-like, aromatic.

Juniperus *Thuja*

Ericaceae ▶ (pp. 319–320) – Flowers urn-shaped, with 4 or 5 connate sepals; leaves without stipules.

Vaccinium

Fabaceae ▶ (p. 320) – Flowers bilaterally symmetric, with upper banner petals, side wings, and keel; leaves often pinnately compound.

Robinia

Fagaceae ▶ (p. 321) – Leaves alternate, with straight, pinnate veins; fruit a nut partially enclosed in a woody cupule.

Fagus *Quercus*

Hamamelidaceae ▶ (p. 322) – Leaves alternate, with deciduous stipules; hairs stellate; fruit a woody capsule.

Hamamelis

Malvaceae ▶ (p. 322) – Flowers tubular, with 3–5 partially connate sepals, and 5 petals; stamens numerous, forming a tube surrounding the pistil.

Tilia *Tilia*

Oleaceae ▶ (pp. 323–324) – Flowers with 4 connate sepals, 4 connate petals, 2 distinct short stamens; leaves opposite.

Fraxinus *Fraxinus* *Syringa*

Pinaceae ▶ (pp. 325–329) – Leaves needle-like, solitary or in fascicles of 2–5.

Abies *Larix* *Picea* *Pinus*

Rosaceae ▶ (pp. 330–336) – Flowers with 5 sepals, 5 petals, numerous stamens, numerous styles; leaves or leaflets toothed.

Tsuga *Amelanchier* *Amelanchier*

PLANT FAMILY PHOTOGRAPHS • TREES & TALL SHRUBS

Rosaceae (continued; pp. 330–336)

Crataegus *Malus* *Malus* *Prunus*

Prunus *Prunus* *Sorbus* *Sorbus*

Salicaceae ▶ (pp. 337–344) – Staminate and carpellate flowers on separate plants; petals absent; leaves alternate (see also pp. 283–284 for drarf alpine species).

Populus *Populus* *Populus* *Salix*

Salix *Salix*

Sapindaceae ▶ (pp. 345–347) – Leaves palmately lobed; seeds paired, with wings.

Acer

Acer *Acer* *Acer*

Taxaceae ▶ (p. 348) – Leaves needle-shaped, evergreen; aril berry-like, red or green.

Taxus

Ulmaceae ▶ (p. 348) – Leaves asymmetric at base, alternate.

Taxus *Taxus* *Ulmus*

42 THE PLANTS OF BAXTER STATE PARK

PLANT FAMILY PHOTOGRAPHS • FERNS & OTHER SPORE-PRODUCING PLANTS

Aspleniaceae ▶ (p. 349) – Indusia narrow; veins do not reach the margins; petioles at base with 2 vascular bundles.

Blechnaceae ▶ (p. 350) – Leaves glossy green, with veins in an interconnected net-like pattern; sori in single, chain-like row.

Dennstaedtiaceae ▶ (pp. 350–351) – Forming large, tenacious colonies; sori very small, at margin of leafules.

Dryopteridaceae ▶ (pp. 351–356) – Indusia round to kidney-shaped; teeth on leaf segments often bristle-tipped; petioles with 3–7 vascular bundles at base.

Equisetaceae ▶ (pp. 357–358) – Fertile stalks with cone-like spore-producing structures; infertile stalks usually with whorls of branches at nodes.

Asplenium   *Asplenium*
Woodwardia   *Dennstaedtia*
Pteridium   *Pteridium*   *Pteridium*   *Dryopteris*
Dryopteris   *Dryopteris*   *Dryopteris*   *Dryopteris*
Dryopteris   *Polystichum*   *Polystichum*   *Polystichum*
Equisetum   *Equisetum*   *Equisetum*   *Equisetum*

PLANT FAMILY PHOTOGRAPHS • FERNS & OTHER SPORE-PRODUCING PLANTS

Huperziaceae ▶ (p. 359) – Upright, evergreen; sporophylls (modified leaves) green, unstalked.

Isoëtaceae ▶ (pp. 360–361) – Aquatic; spore sack located at base of each leaf; leaves hollow and quill-like.

Lycopodiaceae ▶ (pp. 362–367) – Spores in cone-like structures; leaves tiny, with a solitary vein.

Onocleaceae ▶ (p. 368) – Leaves with interconnected, net-like veins; spores green.

Ophioglossaceae ▶ (pp. 369–371) – Spores produced on a stalk overtopping the leaf; sporangia thick-walled.

Huperzia *Huperzia* *Isoëtes*
Isoëtes *Isoëtes* *Dendrolycopodium*
Diphasiastrum *Diphasiastrum* *Diphasiastrum* *Diphasiastrum* *Lycopodiella*
Lycopodium *Lycopodium* *Lycopodium* *Spinulum* *Spinulum*
Matteuccia *Matteuccia* *Onoclea* *Onoclea*
Botrychium *Botrychium* *Botrychium* *Botrychium*

44   THE PLANTS OF BAXTER STATE PARK

PLANT FAMILY PHOTOGRAPHS • FERNS & OTHER SPORE-PRODUCING PLANTS

Osmundaceae ▶ (pp. 371–372) – Fertile leaves or leaflets brown at maturity; leaves compound.

Osmunda   *Osmunda*   *Osmundastrum*   *Osmundastrum*

Polypodiaceae ▶ (p. 373) – Leaves evergreen, growing in dense colonies on rocks; sori round.

Pteridaceae ▶ (p. 374) – Leaves clustered; leafules fan-shaped, the margins forming false indusia.

Polypodium   *Adiantum*   *Adiantum*

Thelypteridaceae ▶ (pp. 374–376) – Leaves pubescent with needle-like, transparent hairs; petioles with 2–7 vascular bundles in cross section at base.

Parathelypteris   *Parathelypteris*   *Phegopteris*

Woodsiaceae ▶ (pp. 377–380) – Leaves with veins not reaching margins of leaf segments; petioles with 2 vascular bundles in cross section at base.

Thelypteris   *Thelypteris*   *Athyrium*

Athyrium   *Athyrium*   *Cystopteris*   *Deparia*   *Deparia*

Gymnocarpium   *Woodsia*   *Woodsia*

PLANT FAMILY PHOTOGRAPHS • SEDGES, RUSHES, & GRASSES

Cyperaceae ▶ (pp. 381–419) – Stem bases usually triangular in cross section, solid, never jointed; leaves arranged in 3 vertical rows; fruit a single-seeded nut.

Bulbostylis — *Carex* — *Carex* — *Carex*

Cladium — *Cyperus* — *Dulichium* — *Eleocharis* — *Eriophorum*

Rhynchospora — *Rhynchospora* — *Schoenoplectus* — *Scirpus* — *Trichophorum*

Juncaceae ▶ (pp. 420–425) – Stem bases round in cross section, solid, never jointed; leaves typically basal only and spirally arranged; fruit a many-seeded capsule.

Trichophorum — *Juncus* — *Juncus* — *Juncus*

Poaceae ▶ (pp. 426–448) – Stem bases usually round in cross section, hollow, jointed; leaves arranged in 2 vertical rows; fruit a single-seeded grain.

Luzula — *Luzula* — *Oreojuncus* — *Agrostis*

Alopecurus — *Anthoxanthum* — *Anthoxanthum* — *Arrhenatherum* — *Brachyelytrum*

46   THE PLANTS OF BAXTER STATE PARK

PLANT FAMILY PHOTOGRAPHS • SEDGES, RUSHES, & GRASSES

Poaceae
(continued;
pp. 426–448)

Bromus	*Calamagrostis*	*Cinna*	*Dactylis*
Danthonia	*Deschampsia*	*Dichanthelium*	*Digitaria*	*Elymus*
Elymus	*Festuca*	*Festuca*	*Glyceria*	*Glyceria*
Leersia	*Milium*	*Muhlenbergia*	*Muhlenbergia*	*Oryzopsis*
Panicum	*Phalaris*	*Phleum*	*Phragmites*	*Poa*
Poa	*Schizachne*	*Torreyochloa*	*Trisetum*	*Vahlodea*

PLANT FAMILY PHOTOGRAPHS   47

TAB-COLOR KEY FOR SPECIES ACCOUNTS

Wildflowers and low shrubs .... Page 49

Trees and tall shrubs .... Page 305

Ferns and other spore-producing plants .... Page 349

Sedges, rushes, and grasses .... Page 381

* BEFORE SCIENTIFIC NAME = NONNATIVE

ACORACEAE • SWEETFLAG FAMILY ▼

Several-veined Sweetflag • *Acorus americanus*

Perennial, aquatic herb of shallow water, up to 1.5 m tall. **Flowers** yellowish brown, numerous, with 6 tepals, in a club-shaped cluster (spadix) 3.3–7.4 cm long and 4.7–10 mm wide. **Leaves** *crowded at base, erect, with a sweet smell when crushed, 3–12 mm wide*, with prominent raised midvein, other major veins also raised.
OCCURRENCE: Historical record, last documented in 1982, from northern portion of Park.
OTHER NAMES: Beewort

ALISMATACEAE • WATER-PLANTAIN FAMILY ▼

Northern Water-plantain • *Alisma triviale*

Emergent, perennial herb of shallow and slow-moving water, 0.1–1 m tall. **Flowers** *white, 3-parted*, 7–13 mm wide; petals much longer than sepals. **Leaves** basal only, on a long petiole.
OCCURRENCE: Rare, with scattered distribution.
OTHER NAMES: Large Water-plantain, *Alisma brevipes*

WILDFLOWERS AND LOW SHRUBS 49

ALISMATACEAE • WATER-PLANTAIN FAMILY

Northern Arrowhead • *Sagittaria cuneata*

Submerged to emergent, perennial herb of ponds, streams, and pools, up to 1 m tall. **Flowers** white, up to 2.5 cm wide; *petals 7–10 mm long; filaments glabrous.* **Leaves** variable depending on water depth, *broadly to narrowly sagittate, with basal lobes; floating leaves produced.* **Fruit** an achene, 1.8–2.6 mm long; *beak 0.1–0.4 mm long.*
OCCURRENCE: Uncommon, with scattered distribution.
NOTES: See *S. graminea* and *S. latifolia*.

Grass-leaved Arrowhead • *Sagittaria graminea*

Emergent, perennial herb of ponds, streams, and pools, up to 1 m tall. **Flowers** white, up to 2.3 cm wide; *filaments with hairs.* **Leaves** simple, *unlobed*, narrowly lance-shaped; *floating leaves never produced.* **Fruit** an achene, 1.5–2.8 mm long; *beak 0.1–0.2 mm long.*
OCCURRENCE: Rare, with scattered distribution.
NOTES: See *S. cuneata* and *S. latifolia*.

ALISMATACEAE • WATER-PLANTAIN FAMILY

Common Arrowhead • *Sagittaria latifolia*

Emergent, perennial herb of ponds, streams, and pools, up to 45 cm tall. **Flowers** white, up to 4 cm wide; *petals 10–20 mm long; filaments glabrous.* **Leaves** variable, *broadly to narrowly sagittate, usually with basal lobes; floating leaves never produced.* **Fruit** an achene, 2.5–3.5 mm long; *beak 1–2 mm long.*
OCCURRENCE: Occasional, widely distributed.
NOTES: See *S. cuneata* and *S. graminea*.
OTHER NAMES: Broad-leaved Arrowleaf, Wapato, Duck-potato

ALLIACEAE • ONION FAMILY ▼

Wild Chives • *Allium schoenoprasum*

Perennial herb of wet meadows and shorelines, 20–60 cm tall. **Flowers** *pink to light purple, united at base into umbels atop long stems.* **Leaves** basal only, usually 2, erect, round in cross section, *hollow,* 2–7 mm wide.
OCCURRENCE: Very rare, documented from only one township in Park.
NOTES: Some populations may have been planted.
OTHER NAMES: *Allium sibiricum*

WILDFLOWERS AND LOW SHRUBS

AMARANTHACEAE • AMARANTH FAMILY ▼

White Goosefoot • *Chenopodium album*

Annual herb of waste places, 0.2–2 m tall. **Flowers** with 5 green tepals, keeled along midrib, giving flower and fruit a star-shaped appearance. **Leaves** alternate, *up to 2-times as long as wide*, usually with basal lobes and often with teeth on margin.
OCCURRENCE: Historical record, last documented in 1982, from northern portion of Park.
OTHER NAMES: Lamb's Quarters, *Chenopodium lanceolatum*

ANACARDIACEAE • CASHEW FAMILY ▼

Poison-ivy • *Toxicodendron radicans*

Deciduous shrub, *typically over 2 m tall*. **Flowers** whitish, 5-parted; *inflorescence branched, usually with more than 25 flowers*. **Leaves** divided with 3 leaflets, usually shiny. **Stems** woody, much-branched, climbing or trailing, *with clinging aerial roots*. **Fruit** a berry-like drupe *with sparse hairs*.
OCCURRENCE: Uncommon, with scattered distribution.
NOTES: See *T. rydbergii*. Contact with any part of this plant may cause a skin rash.
OTHER NAMES: Common Poison-ivy, *Rhus radicans*

ANACARDIACEAE • CASHEW FAMILY

Western Poison-ivy • *Toxicodendron rydbergii*

Deciduous shrub, *typically up to 1 m tall*. **Flowers** whitish, 5-parted; *inflorescence typically unbranched, usually with fewer than 25 flowers*. **Leaves** divided, with 3 leaflets, usually shiny. **Stems** woody, with few if any branches for lowest 5–60 cm, spreading to erect, *without aerial roots*. **Fruit** a berry-like drupe *without hairs*.
OCCURRENCE: Uncommon, with scattered distribution.
NOTES: See *T. radicans*. Contact with any part of this plant may cause a skin rash.
OTHER NAMES: *Rhus radicans* var. *rydbergii*, *Toxicodendron radicans* var. *rydbergii*

APIACEAE • CARROT FAMILY ▼

Bishop's Goutweed • **Aegopodium podagraria*

Perennial carpet-forming herb of roadsides and riparian forests, up to 0.9 m tall. **Flowers** white, 5-parted, 3–4 mm wide; umbels 6–12 cm wide, compound. **Leaves** *alternate, two or more times pinnately divided; uppermost leaflets distinct, 1–4 cm wide*, sharply toothed, acuminate.
OCCURRENCE: Very rare, documented from only one township in Park.
NOTES: A variegated form with white and green leaves occurs in the Park. Species has potential to spread aggressively in some natural habitats.
OTHER NAMES: Herb Gerard

WILDFLOWERS AND LOW SHRUBS 53

APIACEAE • CARROT FAMILY

Bristly Sarsaparilla • *Aralia hispida*

Perennial herb of dry, open woods, 0.2–1.2 m tall. **Flowers** white, 5-parted, up to 2.5 mm wide, borne in 5–25 umbels; *inflorescence terminal on leafy stem.* **Leaves** alternate, long-stalked, two or more times pinnately divided; uppermost leaflets distinct, 0.8–4 cm wide. **Stems** woody, *armed with sharp bristles at base.* **Fruit** a dark purple berry, 5–8 mm wide.
OCCURRENCE: Occasional, widely distributed.
NOTES: This species commonly colonizes burned or otherwise denuded sites.

APIACEAE • CARROT FAMILY

Wild Sarsaparilla • *Aralia nudicaulis*

Perennial herb of forests, 20–40 cm tall. **Flowers** greenish white, 5-parted, up to 4 mm wide, borne in 2–3 umbels on a long stalk; *inflorescence arising from rhizome*. **Leaves** basal only, *solitary*, long-stalked, two or more times pinnately divided; uppermost leaflets distinct, 5–15 cm wide. **Stems** woody at ground level, *without bristles at base*. **Fruit** a purple-black berry, 4–10 mm wide.
OCCURRENCE: Common, widely distributed.

American Spikenard • *Aralia racemosa*

Perennial herb of rich soils, up to 2 m tall. **Flowers** greenish white, 5-parted, up to 2 mm wide, borne in numerous umbels; *inflorescence terminal on leafy stem*. **Leaves** alternate, *enormous*, two or more times pinnately divided; uppermost leaflets distinct, up to 15 cm wide. **Stems** *herbaceous, without sharp bristles at base*. **Fruit** a red berry, turning darker with age, 4–6 mm wide.
OCCURRENCE: Rare, with scattered distribution.
NOTES: Plant of high-pH soils, usually associated with hardwoods.
OTHER NAMES: Spikenard

WILDFLOWERS AND LOW SHRUBS

APIACEAE • CARROT FAMILY

Caraway • *Carum carvi*

Biennial herb of fields, roadsides, and disturbed sites, 0.2–0.8 m tall. **Flowers** white, 5-parted, *in umbels with 7–14 primary branches, with narrow bracts beneath umbel*; petals less than 2 mm long. **Leaves** alternate, very finely divided, the largest with uppermost segments 5–15 mm long and less than 1 mm wide. **Stems** glabrous.
OCCURRENCE: Historical record, last documented in 1938, from one township in Park.

Bulblet-bearing Water-hemlock • *Cicuta bulbifera*

Perennial herb of wetlands, 0.2–1.8 m tall. **Flowers** white, 5-parted, in umbels 2–5 cm wide. **Leaves** alternate, two or more times pinnately divided, *with veins ending at sinuses of margin*; uppermost leaflets distinct, usually lobed, *up to 5 mm wide; bulblets often present in axils of upper leaves*. **Stems** purplish.
OCCURRENCE: Occasional, widely distributed.
NOTES: All parts of plant extremely poisonous.
OTHER NAMES: Bulbiferous Water-hemlock

APIACEAE • CARROT FAMILY

Spotted Water-hemlock • *Cicuta maculata*

Perennial herb of wetlands, 0.5–1.8 m tall. **Flowers** white, 5-parted, in umbels 5–12 cm wide. **Leaves** alternate, two or more times pinnately divided, *with veins ending at sinuses of margin*; uppermost leaflets distinct, *wider than 6 mm; bulblets in axils of leaves absent.* **Stems** with a whitish bloom, *usually mottled or striped with purple*, especially at nodes.
OCCURRENCE: Historical record, last documented in 1984, from northern portion of Park.
NOTES: Roots and stems extremely poisonous.
OTHER NAMES: Spotted Cowbane, Musquash-root

Chinese Hemlock-parsley • *Conioselinum chinense*

Perennial herb of *cool, shaded wetlands*, up to 1.5 m tall. **Flowers** white, 5-parted, in terminal umbels up to 15 cm wide. **Leaves** alternate, *two or more times divided, with numerous, finely divided leaf segments; leaflets 1–2 cm wide.* **Stems** *glabrous, unspotted.*
OCCURRENCE: Rare, with scattered distribution.

WILDFLOWERS AND LOW SHRUBS 57

APIACEAE • CARROT FAMILY

Wild Carrot • *Daucus carota*

Biennial herb of dry fields and disturbed sites, 0.4–1 m tall. **Flowers** white, 5-parted, 2–3 mm wide; umbels 3–10 cm wide, *with a red or purple flower usually found at center; bracts under umbel deeply lobed.* **Leaves** alternate, *two or more times divided, with numerous, finely divided leaf segments.* **Stems** *covered with bristles.*
OCCURRENCE: Historical record, last documented in 1982, from northern portion of Park.
OTHER NAMES: Queen Anne's Lace, Bird's Nest

American Cow-parsnip • *Heracleum maximum*

Perennial herb of river banks, fields, roadside ditches, and subalpine meadows, up to 3 m tall. **Flowers** white, 5-parted; terminal umbel often 10–20 cm wide with 15–30 rays. **Leaves** alternate, *once pinnately divided,* the largest 20–50 cm wide; *leaflets stalked,* lobed, coarsely toothed, *very pubescent beneath.*
OCCURRENCE: Occasional, widely distributed.
NOTES: Sap may cause rash on skin after exposure to sun.
OTHER NAMES: *Heracleum lanatum*

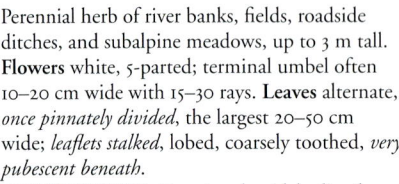

APIACEAE • CARROT FAMILY

American Marsh-pennywort • *Hydrocotyle americana*

Perennial herb of moist and marshy areas of low fields and along streams, up to 30 cm tall. **Flowers** white, 5-parted; *umbels simple, sessile or nearly so*, with 2–7 flowers. **Leaves** alternate, *light green, round, shiny, with scalloped edges*; petiole attached at blade sinus. **Stems** *creeping, rooting at nodes*.
OCCURRENCE: Occasional, widely distributed.
OTHER NAMES: American Water-pennywort

Bland Sweet-cicely • *Osmorhiza claytonii*

Perennial herb of hardwood forests at middle to higher elevations, 30–90 cm tall. **Flowers** white, 5-parted, 3–4 mm wide. **Leaves** alternate, *two or more times pinnately divided, with veins ending at tip of teeth on margin; uppermost leaflets distinct, 10–40 mm wide*.
OCCURRENCE: Occasional, widely distributed.

WILDFLOWERS AND LOW SHRUBS

APIACEAE • CARROT FAMILY

Dwarf Ginseng • *Panax trifolius*

Perennial herb of hardwood forests and floodplains, 30–90 cm tall. **Flowers** white, 5-parted, 3–4 mm wide. **Leaves** *in a whorl of 3 divided leaves, each with 3 leaflets*; leaflets 2.5–8 cm long and 0.6–2 cm wide, sessile.
OCCURRENCE: Very rare, documented from only one township in Park.
OTHER NAMES: Fairy Spuds

Wild Parsnip • **Pastinaca sativa*

Biennial herb of fields and roadsides, up to 2 m tall. **Flowers** yellow, 5-parted, *1.5 mm wide, in umbels 5–10 cm wide*. **Leaves** alternate, *once pinnately divided; largest 3–25 cm wide*.
OCCURRENCE: Historical record, last documented in 1982, from northern portion of Park.
NOTES: Sap may be irritating to the skin and cause blistering.

APIACEAE • CARROT FAMILY

Maryland Sanicle • *Sanicula marilandica*

Perennial herb of moist forests, often associated with rich soils and rocky slopes, up to 1.2 m tall. **Flowers** greenish white, 5-parted. **Leaves** alternate, *once palmately divided, with 3–5 leaflets*; leaflets sharply toothed.
OCCURRENCE: Rare, with scattered distribution.
OTHER NAMES: *Sanicula canadensis*

Water-parsnip • *Sium suave*

Emergent, perennial herb of shallow water and wet banks, 0.3–2 m tall. **Flowers** white, 5-parted; umbels 2–12 cm wide, with 10–25 rays. **Leaves** alternate, *once pinnately divided or further divided when submerged*; leaflets 2–15 cm long and 0.5–4 cm wide.
OCCURRENCE: Occasional, widely distributed.

WILDFLOWERS AND LOW SHRUBS   61

APIACEAE • CARROT FAMILY

Common Golden Alexanders • *Zizia aurea*

Perennial herb of riparian forests, river shorelines, and wet meadows, up to 1 m tall. **Flowers** *yellow*, 5-parted. **Leaves** alternate, *once pinnately divided*; leaflets irregularly toothed.
OCCURRENCE: Very rare, documented from only one township in Park.

APOCYNACEAE • MILKWEED FAMILY ▼

Spreading Dogbane • *Apocynum androsaemifolium*

Perennial herb of dry fields and roadsides, 20–90 cm tall. **Flowers** *pink or white, with pink stripes, bell-shaped, nodding*, 6–10 mm long, *with flaring petals*. **Leaves** opposite, entire, spreading or sometimes drooping. **Stems** reddish, with milky sap. **Seeds** *2.5–3 mm long.*
OCCURRENCE: Occasional, widely distributed.
NOTES: May be poisonous to livestock.
OTHER NAMES: Indian Hemp

APOCYNACEAE • MILKWEED FAMILY

Swamp Milkweed • *Asclepias incarnata*

Perennial herb of marshes, wet fields, borders of swamps and shorelines, up to 1.5 m tall. **Flowers** pink to red or purple; pedicels 0.8–2.4 cm long. **Leaves** opposite, up to 7 cm wide. **Stems** with milky sap. **Fruit** *a smooth follicle, on erect pedicels.*
OCCURRENCE: Rare, with scattered distribution.

Common Milkweed • *Asclepias syriaca*

Perennial herb of fields, roadsides, and disturbed habitats, up to 2 m tall. **Flowers** pink to red or purple; pedicels 3–10 cm long. **Leaves** opposite, up to 10 mm wide. **Stems** stout, with milky sap. **Fruit** *a rough follicle, with sharp projections, on deflexed pedicels.*
OCCURRENCE: Uncommon, with scattered distribution.
NOTES: Silkweed, Silky Milkweed

Jack-in-the-pulpit • *Arisaema triphyllum*

Perennial herb of wet, rich woods, swamps, and bogs, up to 1 m tall. **Flowers** in a club-shaped cluster (spadix) *surrounded by a green- and purple-striped bract (spathe)*. **Leaves** *divided*; leaflets 3, green on both sides, with parallel veins. **Fruit** a red berry.
OCCURRENCE: Occasional, widely distributed.
OTHER NAMES: Indian Turnip

26 June – DK

29 May Kidney Pond area – LZ

23 September – GM

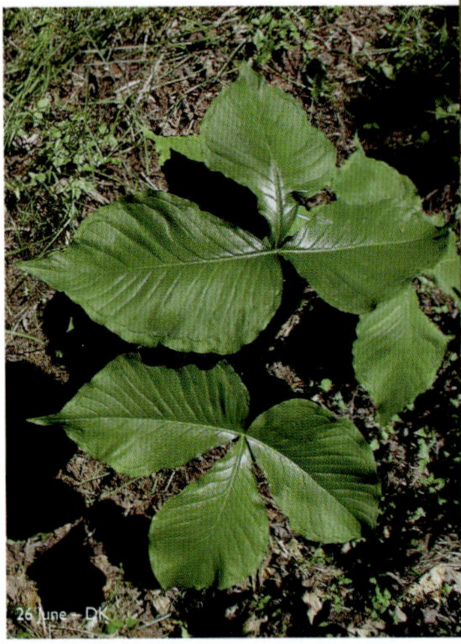

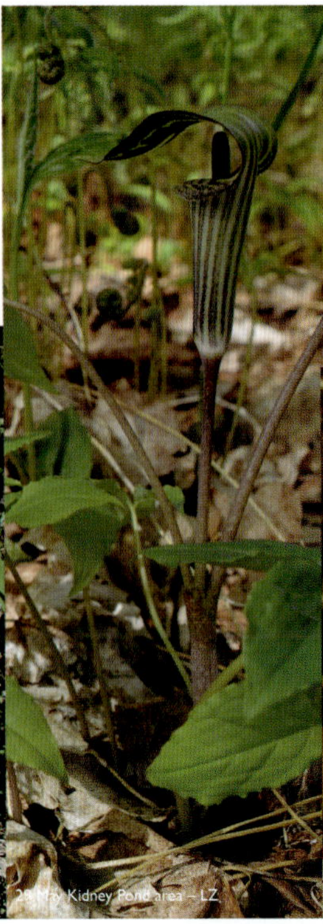

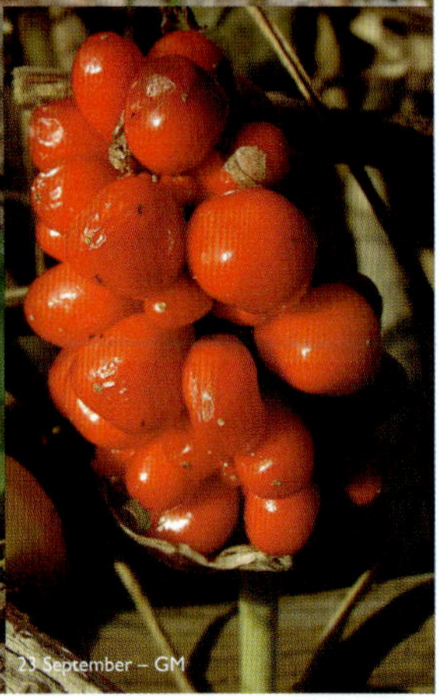

Wild Calla • *Calla palustris*

Perennial herb of wet areas, up to 40 cm tall. **Flowers** in a fleshy spike (spadix), 6-parted, greenish, the lowest bisexual, the uppermost often staminate only; *spadix with an adjacent showy, white, persistent bract (spathe)*. **Leaves** *simple*, round to heart-shaped. **Fruit** a red berry.
OCCURRENCE: Occasional, widely distributed.
OTHER NAMES: Water Arum

29 August – GM

28 July Daicey Pond area – GM

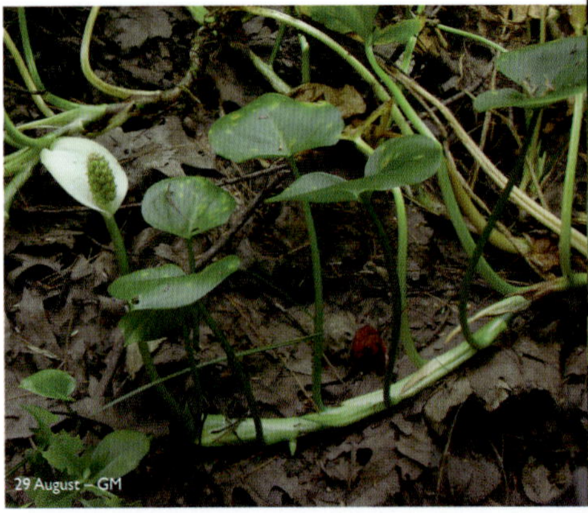

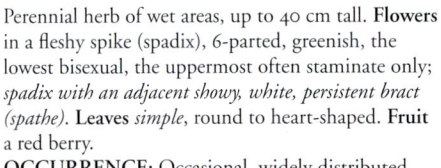

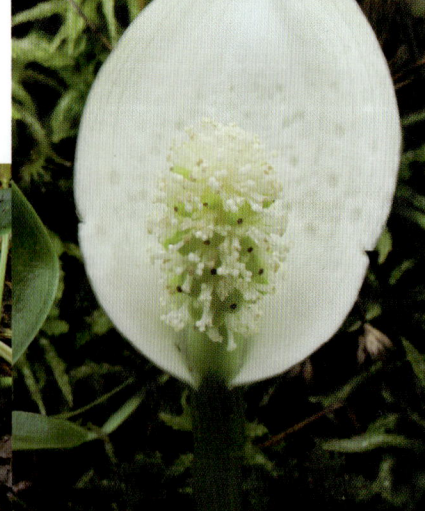

ARACEAE • ARUM FAMILY

Common Duckweed • *Lemna minor*

Perennial, floating herb of lakes, rivers, and pools, up to 1 cm tall. **Flowers** rarely produced. **Leaves** symmetric, 3-veined, 2–6 mm long, often in groups of 2–5. **Roots** up to 15 cm long.
OCCURRENCE: Very rare, documented from only one township in Park.

ASTERACEAE • ASTER FAMILY ▼

Common Yarrow • *Achillea millefolium*

Perennial herb of fields, roadsides, and shorelines, 0.2–1 m tall. **Flower heads** 4–6 mm wide; *rays usually 5, white, 3-toothed at tip*; disc florets white; capitulescence flat-topped, 6–30 cm wide. **Leaves** alternate, *finely dissected*, 3–15 cm long and 1–2.5 cm wide, *aromatic*.
OCCURRENCE: Occasional, widely distributed.
OTHER NAMES: Milfoil, Bloodwort, *Achillea borealis*

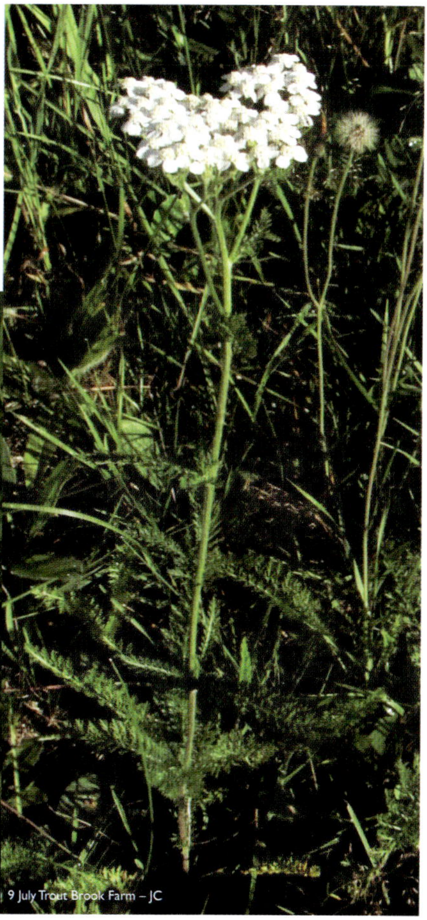

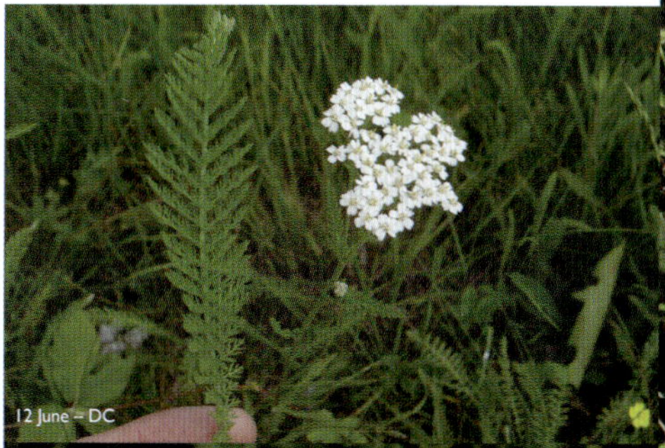

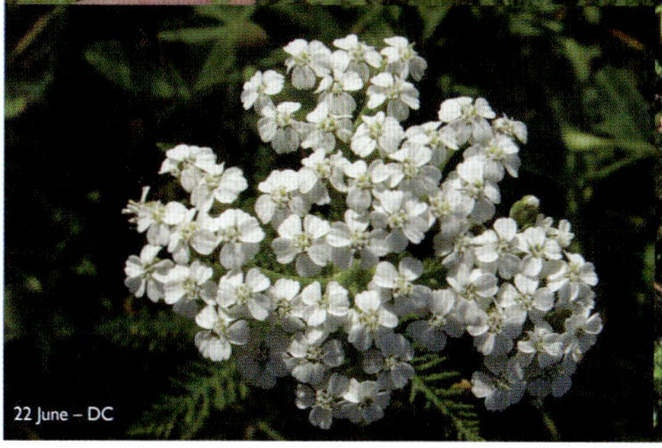

ASTERACEAE • ASTER FAMILY

White Snakeroot • *Ageratina altissima*

Perennial herb of rich deciduous and riparian forests, up to 1.2 m tall. **Flower heads** *bright white*, clustered, *with 9–30 florets*; involucral bracts 3–5 mm long. **Leaves** *opposite*, 4–11 cm long and 2.5–8 cm wide, sharply and doubly toothed; *petioles 1.5–10 cm long*.
OCCURRENCE: Uncommon, with scattered distribution.
NOTES: Poisonous to cattle and humans.

Pearly Everlasting • *Anaphalis margaritacea*

Perennial herb of fields and roadsides, 20–90 cm tall. **Flower heads** numerous, 6–8 mm wide; rays absent; involucral bracts white at apex. **Leaves** alternate, sessile, entire, 3–10 cm long and 0.5–2 cm wide, *variably white to rusty-woolly beneath*. **Stems** *white-woolly*.
OCCURRENCE: Common, widely distributed.

ASTERACEAE • ASTER FAMILY

Small Pussytoes • *Antennaria howellii*

Dioecious perennial herb of grassy or rocky areas, 6–35 cm tall.
Flower heads white-woolly; carpellate heads 7–10 mm long; staminate heads (rare) 4–6 mm long. **Basal leaves** *20–48 mm long and 2–15 mm wide, 1-veined or occasionally with 2 additional veins near margins.* **Cauline leaves** linear, 8–40 mm long, with or without a flat, thin appendage at tip.
OCCURRENCE: Occasional, widely distributed.
NOTES: See *A. neglecta* and *A. parlinii*. *A. howellii* can be difficult to separate from *A. neglecta*, particularly when middle and upper cauline leaves have a flat, curled, scarious appendage at the tip (a character found in some, but not all, specimens of *A. howellii*). When *A. howellii* has these scarious appendages at the tip of cauline leaves, the young basal leaves will be glabrous above; *A. neglecta*, which always has these scarious appendages at the tip of cauline leaves, always has young basal leaves that are woolly above.
OTHER NAMES: Howell's Pussytoes, *Antennaria canadensis*, *Antennaria neodioica*

Field Pussytoes • *Antennaria neglecta*

Dioecious perennial herb of grassy or rocky areas, 4–25 cm tall.
Flower heads white-woolly; carpellate heads 6–10 mm long; staminate heads 3–5 mm long. **Basal leaves** *15–65 mm long and 6–18 mm wide, 1-veined, woolly and dull green on upper surface.* **Cauline leaves** linear, 1.5–25 mm long, *with a flat, thin appendage at tip.*
OCCURRENCE: Rare, with scattered distribution.
NOTES: See *A. howellii* and *A. parlinii*.

ASTERACEAE • ASTER FAMILY

Parlin's Pussytoes • *Antennaria parlinii*

Dioecious perennial herb of grassy or rocky areas, 12–45 cm tall. **Flower heads** white-woolly; carpellate heads 8–13 mm long; staminate heads 6–9 mm long. **Basal leaves** *30–95 mm long and 12–45 mm wide, with or without hairs above, with 3 or 5 prominent veins, veins closest to midrib roughly halfway between midrib and margin.* **Cauline leaves** linear, 3.5–45 mm long, with a flat, thin appendage at tip.
OCCURRENCE: Rare, documented from only one township in Park.
NOTES: See *A. howellii* and *A. neglecta*.

Great Burdock • *Arctium lappa*

Biennial herb of fields and roadsides, up to 3 m tall. **Flower heads** 25–45 mm wide for largest heads, on peduncles typically longer than 2.5 cm, usually in flat-topped or convex clusters. **Leaves** alternate; basal leaves as wide as long, 25–80 cm long and 20–70 cm wide; *petioles of basal leaves solid in cross section near base.*
OCCURRENCE: Uncommon, with scattered distribution.
NOTES: See *A. minus*.

ASTERACEAE • ASTER FAMILY

Common Burdock • *Arctium minus*

Biennial herb of fields and roadsides, up to 1.3 m tall. **Flower heads** *15–25 mm wide for largest heads, on peduncles typically less than 2 cm long, usually in elongated, spire-shaped clusters.* **Leaves** alternate; basal leaves longer than wide, 30–60 cm long and 15–35 cm wide; *petioles of basal leaves hollow near base.*
OCCURRENCE: Uncommon, with scattered distribution.
NOTES: See *A. lappa.*
OTHER NAMES: Lesser Burdock

Lance-leaved Arnica • *Arnica lanceolata*

Perennial herb of alpine ravines, gullies, and middle-elevation streambanks, 20–80 cm tall. **Flower heads** erect to nodding at maturity, *40–60 mm wide; rays yellow, 7–17 per head, 15–20 mm long.* **Leaves** opposite, unlobed, sessile, pubescent, 4–12 cm long and 2–6 cm wide.
OCCURRENCE: Rare, with scattered distribution.
NOTES: Species is listed as Threatened in Maine.
OTHER NAMES: *Arnica mollis*

WILDFLOWERS AND LOW SHRUBS

ASTERACEAE • ASTER FAMILY

Nodding Beggar-ticks • *Bidens cernua*

Annual herb of shorelines and marshes, 0.2–1 m tall. **Flower heads** nodding at maturity, *12–25 mm wide; rays yellow, 6–8 per head, 2–18 mm long.* **Leaves** opposite, simple, unlobed, toothed, sessile, 4–10 cm long and *0.5–2.5 cm wide.* **Fruit** an achene 5–8 mm long, tip with retrorsely barbed awns.
OCCURRENCE: Uncommon, with scattered distribution.
NOTES: See *B. connata* and *B. frondosa*.
OTHER NAMES: Nodding Bur-marigold

Purple-stemmed Beggar-ticks • *Bidens connata*

Annual herb of marshes and shorelines, up to 1.5 m tall. **Flower heads** erect at maturity, *8–20 mm wide; rays usually absent.* **Leaves** opposite, simple, 4–10 cm long and 1–3 cm wide, occasionally with lobes near base.
OCCURRENCE: Very rare, documented from only one township in Park.
NOTES: See *B. cernua* and *B. frondosa*.

ASTERACEAE • ASTER FAMILY

Devil's Beggar-ticks • *Bidens frondosa*

Annual herb of wet soils, 0.1–1.2 m tall. **Flower heads** erect at maturity, *less than 10 mm wide; rays usually absent.* **Leaves** opposite, *1- to 3-times pinnately divided; leaflets 3–5, sharply toothed,* 3.5–6 cm long and 1–2 cm wide.
OCCURRENCE: Occasional, widely distributed.
NOTES: See *B. cernua* and *B. connata*.
OTHER NAMES: Stick-tight

Creeping Thistle • **Cirsium arvense*

Perennial herb of old fields, *colonial*, 0.3–1.2 m tall. **Flower heads** pale pink to purple, 1–2 cm wide near base; *involucral bracts rarely tipped with spines, the spines less than 1 mm long.* **Leaves** alternate, *not extending onto stem as a wing*, with spines on margins, *the largest spines 1–7 mm long.*
OCCURRENCE: Uncommon, with scattered distribution.
NOTES: See *C. muticum* and *C. vulgare*. Species has potential to spread aggressively in some natural habitats.
OTHER NAMES: Canada Thistle

WILDFLOWERS AND LOW SHRUBS 71

ASTERACEAE • ASTER FAMILY

Swamp Thistle • *Cirsium muticum*

Biennial herb of swamps, 0.4–2.3 m tall. **Flower heads** lavender or purple, 1–3 cm wide near base; *involucral bracts usually blunt at tip, rarely tipped with spines, the spines less than 0.5 mm long*. **Leaves** alternate, *not extending onto stem as a wing*, with spines on margins, *the largest spines 2–3 mm long*.
OCCURRENCE: Uncommon, with scattered distribution.
NOTES: See *C. arvense* and *C. vulgare*.

Common Thistle • **Cirsium vulgare*

Biennial herb of fields and roadsides, 0.5–1.5 m tall. **Flower heads** purple, 2–4 cm wide near base; *involucral bracts tipped with spines 2–6 mm long*. **Leaves** alternate, *extending onto stem as a wing*, with spines on margins, *the largest spines 2–10 mm long*.
OCCURRENCE: Rare, with scattered distribution.
NOTES: See *C. arvense* and *C. muticum*. Species has potential to spread aggressively in some natural habitats.
OTHER NAMES: Bull Thistle, Spear Thistle

ASTERACEAE • ASTER FAMILY

Tall White-aster • *Doellingeria umbellata*

Perennial herb of marshes, meadows, and forests, 0.4–2.5 m tall. **Flower heads** numerous, less than 2 cm wide, in a flat-topped cluster; rays white, 6–14 per head, 5–8 mm long; disc florets yellow. **Leaves** alternate, entire, sessile, tapering to both ends, 4–16 cm long and 7–35 mm wide, *the uppermost pinnately veined with prominent veins.*
OCCURRENCE: Common, widely distributed.
OTHER NAMES: Flat-topped White-aster, *Aster umbellatus*

American Burnweed • *Erechtites hieraciifolius*

Annual herb of disturbed sites, 0.1–2 m tall. **Flower heads** numerous, *4–8 mm wide near base, swollen at base; rays absent*; involucral bracts usually in single row except for a few smaller ones at base. **Leaves** alternate, 5–20 cm long and 0.5–8 cm wide; *middle cauline leaves sharply toothed or lobed, margins without stiff hairs.*
OCCURRENCE: Uncommon, with scattered distribution.
OTHER NAMES: Pilewort, *Senecio hieraciifolius*

WILDFLOWERS AND LOW SHRUBS

ASTERACEAE • ASTER FAMILY

Annual Fleabane • *Erigeron annuus*

Annual herb of fields and disturbed sites, *0.6–1.5 m tall*. **Flower heads** 6–12 mm wide near base; rays white, 80–125 per head, 4–10 mm long. **Leaves** alternate; *cauline leaves not clasping, numerous, at least lowest coarsely toothed, the largest usually wider than 1 cm*. **Stems** *with long, spreading pubescence mid-stem*.
OCCURRENCE: Uncommon, with scattered distribution.
OTHER NAMES: Daisy Fleabane, Sweet-scabious

Canada Fleabane • *Erigeron canadensis*

Annual herb of disturbed sites, 8–80 cm tall. **Flower heads** *2–5 mm wide near base; rays white, 25–40 per head, 0.5–1 mm long*. **Leaves** alternate, 1.5–10 cm long and 0.1–2 cm wide; *middle cauline leaves entire, not clasping, margins with stiff hairs*.
OCCURRENCE: Rare, with scattered distribution.
NOTES: Leaves can cause skin rash in some people.
OTHER NAMES: Butter-weed, *Conyza canadensis*

ASTERACEAE • ASTER FAMILY

Philadelphia Fleabane • *Erigeron philadelphicus*

Biennial or short-lived perennial herb of fields, ledges, and shorelines, 0.4–0.8 m tall. **Flower heads** 6–15 mm wide near base; *rays white to pinkish, 150–250 per head, 5–10 mm long*. **Leaves** alternate; *cauline leaves cordate-clasping, the largest usually up to 2.5 cm wide*.
OCCURRENCE: Uncommon, with scattered distribution.

Rough Fleabane • *Erigeron strigosus*

Annual to short-lived perennial herb of fields and disturbed sites, 0.3–0.7 m tall. **Flower heads** 5–8 mm wide near base; *rays white to pinkish, 50–100 per head, 4–6 mm long*. **Leaves** alternate; *cauline leaves not clasping, few, entire or nearly so, the largest usually less than 1 cm wide*. **Stems** *with short, appressed pubescence mid-stem*.
OCCURRENCE: Occasional, widely distributed.
OTHER NAMES: Lesser Daisy Fleabane

WILDFLOWERS AND LOW SHRUBS   75

ASTERACEAE • ASTER FAMILY

Boneset Thoroughwort • *Eupatorium perfoliatum*

Perennial herb of wetlands, 0.4–1.5 m tall. **Flower heads** white, 3.5–6 mm tall, in a crowded, flat-topped capitulescence; disc florets *7–11 per head*. **Leaves** opposite, 5–20 cm long and 1.5–4 cm wide, *rugose above*, pubescent beneath, tapering to tip; *bases of opposite leaves connecting around stem*.
OCCURRENCE: Occasional, widely distributed.
OTHER NAMES: Estuary Boneset

White Wood-aster • *Eurybia divaricata*

Perennial herb of forests, *colonial*, 0.2–1 m tall. **Flower heads** 19–25 mm wide, in a broad, flat-topped cluster; *peduncles without glandular hairs; rays white*, 5–10 per head, 6–12 mm long; outer involucral bracts green at tip, densely ciliate. **Leaves** alternate, 3–20 cm long and 2–10 cm wide, *without any glandular hairs, borne on a petiole, the lowest heart-shaped at base*.
OCCURRENCE: Rare, documented from only northern portion of Park.
OTHER NAMES: *Aster castaneus, Aster divaricatus*

ASTERACEAE • ASTER FAMILY

Large-leaved Wood-aster • *Eurybia macrophylla*

Perennial herb of forests and forest edges, 0.2–1.5 m tall. **Flower heads** 13–16 mm wide; *peduncles with glandular hairs; rays light purple*, 9–20 per head, 7–15 mm long; outer involucral bracts green at tip, densely ciliate. **Leaves** alternate, 4–30 cm long and 3–20 cm wide, *with minute stipitate glands toward tip, borne on a petiole, the lowest heart-shaped at base*.
OCCURRENCE: Common, widely distributed.
OTHER NAMES: Big-leaved Aster, *Aster macrophyllus*

Rough Wood-aster • *Eurybia radula*

Perennial herb of forests and wetlands, 0.1–1.2 m tall. **Flower heads** few, 20–40 mm wide, in a leafy-bracted capitulescence; rays light purple, 15–40 per head; outer involucral bracts green at top, densely ciliate. **Leaves** alternate, 3–10 cm long and 6–30 mm wide, *without any glandular hairs, sessile, sharply toothed*, with rough surfaces. **Rhizomes** present.
OCCURRENCE: Occasional, widely distributed.
OTHER NAMES: Rough-leaved Aster, *Aster radula*

ASTERACEAE • ASTER FAMILY

Common Grass-leaved-goldenrod • *Euthamia graminifolia*

Perennial herb of wetlands, fields, and forests, 0.3–1.5 m tall. **Flower heads** *in clusters of 2 or more on a peduncle, in a flat-topped capitulescence*; rays yellow, 7–35 per head; disc florets yellow, 5–10 per head. **Leaves** *numerous*, alternate, entire, 2.5–15 cm long and *3–12 mm wide*, with 3–5 veins.
OCCURRENCE: Common, widely distributed.
OTHER NAMES: Lance-leaved Goldenrod, Flat-topped Goldenrod, *Solidago graminifolia*

Spotted Joe-Pye Weed • *Eutrochium maculatum*

Perennial herb of marshes, fields, shorelines, and wetlands, 0.6–2.5 m tall. **Flower heads** *purple*, 3.5–7 mm wide near base. **Leaves** *in whorls of 3–6*, 6–30 cm long and 1.5–9 cm wide. **Stems** *usually green with purplish spots or uniformly purple*. **Rhizomes** present.
OCCURRENCE: Common, widely distributed.
OTHER NAMES: *Eupatoriadelphus maculatus, Eupatorium maculatum, Eupatorium purpureum*

78   THE PLANTS OF BAXTER STATE PARK

ASTERACEAE • ASTER FAMILY

Brown Cudweed • *Gnaphalium uliginosum

Annual herb of disturbed sites, 4–30 cm tall. **Flower heads** *woolly*, white to light brown, 2–4 mm long, *in dense, terminal clusters, often overtopped by leaves*. **Leaves** alternate, entire, *white-woolly beneath*. **Stems** white-woolly, branching.
OCCURRENCE: Rare, with scattered distribution.
OTHER NAMES: Low Cudweed

Canada Hawkweed • *Hieracium kalmii*

Perennial herb of fields and forests, 0.1–1.5 m tall. **Flower heads** yellow, ~2.5 cm wide, in a loose, flat-topped cluster; peduncles stout, 2–4 cm long, with few or no stipitate glands. **Leaves** alternate, *evidently toothed, all or mostly cauline and nearly all similar in size, only those at very base and apex reduced in size*, firm to leathery, often pubescent beneath but without any long, bulbous-based hairs. **Stolons** absent.
OCCURRENCE: Very rare, not recently documented.
NOTES: See *H. scabrum*.
OTHER NAMES: Kalm's Hawkweed, *Hieracium canadense*

ASTERACEAE • ASTER FAMILY

Common Hawkweed • *Hieracium lachenalii*

Perennial herb of fields and roadsides, 0.1–1 m tall. **Flower heads** yellow, up to 4 cm wide, 4–12 per stem. **Leaves** alternate, *toothed, well developed during flowering, tapering narrowly to petiole; cauline leaves 4–7 per stem, uniformly green, without red-purple streaks or spots*. **Stolons** absent.
OCCURRENCE: Occasional, widely distributed.
NOTES: See *H. maculatum*.
OTHER NAMES: European Hawkweed, Common Wall Hawkweed, *Hieracium vulgatum*

Spotted Hawkweed • *Hieracium maculatum*

Perennial herb of roadsides, trails, fields, and shorelines, 0.2–0.8 m tall. **Flower heads** yellow, up to 4 cm wide, 4–30 per stem. **Leaves** *usually with coarse teeth, at least the underside suffused with red or purplish streaks or spots*; cauline leaves usually present on upper two-thirds of stem, usually 2–4 per stem. **Stolons** absent.
OCCURRENCE: Common, widely distributed.
NOTES: See *H. lachenalii* and the genus *Pilosella*.

ASTERACEAE • ASTER FAMILY

Panicled Hawkweed • *Hieracium paniculatum*

Perennial herb of dry, rocky forests, 0.3–0.9 m tall. **Flower heads** yellow, very small, 12–50 per stem, ~13 mm wide, on long, thin peduncles. **Leaves** alternate, all or mostly cauline, nearly glabrous, pale-glaucous beneath, thin. **Stolons** *absent*.
OCCURRENCE: Historical record, last documented in 1984, from northern portion of Park.
OTHER NAMES: Allegheny Hawkweed

Savoy Hawkweed • **Hieracium sabaudum*

Perennial herb of disturbed sites and woodland edges, 0.1–1.5 m tall. **Flower heads** yellow, *3–12 per stem*, in a narrow capitulescence. **Leaves** *crowded at base, greatly reduced above, usually glabrous above, with long, bulbous-based hairs beneath*; cauline leaves numerous (±50). **Stolons** *absent*.
OCCURRENCE: Very rare, documented from only one township in Park.
NOTES: See *H. kalmii*.
OTHER NAMES: New England Hawkweed

ASTERACEAE • ASTER FAMILY

Rough Hawkweed • *Hieracium scabrum*

Perennial herb of dry, rocky woods, 20–60 cm tall. **Flower heads** yellow, usually 10–25 per stem in an open capitulescence; *peduncles stout, with abundant stipitate glands.* **Leaves** alternate, all or mostly cauline, *entire or minutely toothed*, thick, reduced in size near top of stem. **Stolons** *absent.*
OCCURRENCE: Uncommon, with scattered distribution.
OTHER NAMES: Sticky Hawkweed

Tall Blue Lettuce • *Lactuca biennis*

Annual or biennial herb of wet forests and openings, *up to 4 m tall.* **Flower heads** blue to white, ~1 cm wide. **Leaves** alternate, usually deeply lobed, with coarse teeth, 10–40 cm long and 4–20 cm wide, *with prickles on midrib.* **Fruit** an achene with a crown of *light brown hairs at apex.*
OCCURRENCE: Occasional, widely distributed.

ASTERACEAE • ASTER FAMILY

Tall Lettuce • *Lactuca canadensis*

Annual or biennial herb of disturbed sites, fields, and forests, up to 2.5 m tall. **Flower heads** yellow, ~6 mm wide. **Leaves** alternate, variable in shape, with margins ranging from nearly entire to deeply lobed, 10–35 cm long and 1.5–12 cm wide, *with prickles on midrib*. **Fruit** an achene with a crown of *white hairs at apex*.
OCCURRENCE: Uncommon, with scattered distribution.
OTHER NAMES: Canada Lettuce, Wild Lettuce

Ox-eye Daisy • **Leucanthemum vulgare*

Perennial herb of fields and disturbed sites, 0.2–1 m tall. **Flower heads** solitary, long-stalked, 2.5–6.5 cm wide; *rays white, 15–35 per head*; disc florets yellow. **Leaves** alternate, 4–15 cm long, usually *lobed or with large teeth, without scent*.
OCCURRENCE: Common, widely distributed.
OTHER NAMES: White Daisy, *Chrysanthemum leucanthemum*

WILDFLOWERS AND LOW SHRUBS   83

ASTERACEAE • ASTER FAMILY

Rayless Chamomile • *Matricaria discoidea*

Annual herb of roadsides and disturbed sites, 5–40 cm tall. **Flower heads** *with a pineapple odor when crushed; rays absent;* disc florets yellow. **Leaves** alternate, 1–5 cm long, 1- to 3-times pinnately divided, *with a pineapple odor when crushed.*
OCCURRENCE: Rare, with scattered distribution
OTHER NAMES: Pineapple-weed, Mayweed, *Matricaria matricarioides*

Tall Rattlesnake-root • *Nabalus altissimus*

Perennial herb of forests and roadsides, up to 2.5 m tall. **Flower heads** nodding, in small axillary and loose terminal clusters; *florets pale yellow to greenish yellow, 4–6 per head; inner involucral bracts 4–6,* longer than outer ones. **Leaves** alternate, thin, variable from unlobed to many-lobed.
OCCURRENCE: Common, widely distributed.
NOTES: See *N. trifoliolatus.*
OTHER NAMES: Tall White Lettuce, *Prenanthes altissima*

ASTERACEAE • ASTER FAMILY

Boott's Rattlesnake-root • *Nabalus boottii*

Perennial herb of *alpine and subalpine zones*, ridges, and ledges, *5–25 cm tall*. **Flower heads** nodding, in small axillary and loose terminal clusters; *florets white, 9–18 per head; inner involucral bracts 7–11*, longer than outer ones. **Leaves** alternate; *lowest triangular, 1.5–5 cm long and 1–3 cm wide*, with weakly toothed margins. **Stems** *decumbent*.
OCCURRENCE: Very rare, documented from only southern portion of Park.
NOTES: See *N. trifoliolatus*. Species is listed as Endangered in Maine.
OTHER NAMES: Alpine Rattlesnake-root, *Prenanthes bootii*

Three-leaved Rattlesnake-root • *Nabalus trifoliolatus*

Perennial herb of rocky summits and moist woods, 0.1–1.5 m tall. **Flower heads** nodding, in small axillary and loose terminal clusters; *florets pale yellow, 8–13 per head; inner involucral bracts 7–14*, longer than outer ones. **Leaves** alternate, highly variable in size and shape, usually 3- to 5-lobed, the lowest usually lobed, uppermost less lobed and smaller.
OCCURRENCE: Occasional, widely distributed.
NOTES: See *N. altissimus* and *N. boottii*.
OTHER NAMES: Gall-of-the-earth, *Prenanthes nana*, *Prenanthes trifoliolata*

ASTERACEAE • ASTER FAMILY

Sharp-toothed Nodding-aster • *Oclemena acuminata*

Perennial herb of *conifer and hardwood forests*, 0.2–1 m tall. **Flower heads** 5–50 per stem, 2.5–4 cm wide; rays white, 9–18 per head. **Leaves** coarsely toothed, *fewer than 20 per stem, the largest wider than 10 mm.* **Stems** *zigzagging on upper portion.*
OCCURRENCE: Common, widely distributed.
NOTES: A hybrid (*Oclemena* ×*blakei*), similar to this species, has been documented at Baxter State Park. It can be recognized by its firmer leaf blades, 0.5–2.4 cm wide, numbering 20–45 per stem, and having small teeth.
OTHER NAMES: Mountain Aster, Whorled Wood Aster, *Aster acuminatus*

Bog Nodding-aster • *Oclemena nemoralis*

Perennial herb of *bogs and peaty shorelines*, 10–90 cm tall. **Flower heads** 1–15 per stem, 2.5–4 cm wide; rays light purple or pink, 13–25 per head. **Leaves** with rough edges, slightly rolled under, often with glands and pubescence beneath, *20 or more per stem, reduced in size toward base of plant, the largest less than 10 mm wide.*
OCCURRENCE: Occasional, widely distributed.
NOTES: See *O. acuminata*.
OTHER NAMES: Bog Aster, *Aster nemoralis*

ASTERACEAE • ASTER FAMILY

Alpine Arctic-cudweed • *Omalotheca supina*

Perennial herb of alpine summits, *2–8 cm tall*. **Flower heads** *1–8 per stem*. **Leaves** *primarily basal*, the largest 12–30 mm long and up to 3 mm wide. **Fruit** an achene with *pappus bristles distinct, the bristles falling separately*.
OCCURRENCE: Very rare, documented from only one township in Park.
NOTES: Species is listed as Endangered in Maine.
OTHER NAMES: *Gnaphalium supinum*

Woodland Arctic-cudweed • *Omalotheca sylvatica*

Perennial herb of fields, roadsides, and shorelines, *10–70 cm tall*. **Flower heads** *10 or more per stem*, in a narrow capitulescence. **Leaves** *alternate*, the largest 20–80 mm long and 2–10 mm wide. **Fruit** an achene with *pappus bristles united at base, the bristles falling together in a ring*.
OCCURRENCE: Very rare, documented from only northern portion of Park.
OTHER NAMES: *Gnaphalium sylvaticum*

WILDFLOWERS AND LOW SHRUBS

ASTERACEAE • ASTER FAMILY

Balsam Groundsel • *Packera paupercula*

Perennial herb of ridges, ledges, and gravel shorelines, 20–50 cm tall.
Flower heads 2–10 per stem; rays yellow, 0–13 per head, 5–10 mm long.
Leaves alternate, progressively reduced above; *basal blades gradually tapering to base*, sharply toothed.
OCCURRENCE: Very rare, documented from only one township in Park.
OTHER NAMES: *Senecio balsamitae, Senecio pauperculus*

New England Groundsel • *Packera schweinitziana*

Perennial herb of wetlands, 0.4–1 m tall. **Flower heads** 8–20 per stem; rays yellow, 8–13 per head, 4–7 mm long. **Leaves** alternate, progressively reduced above; *basal blades truncate or cordate base*, sharply toothed.
OCCURRENCE: Uncommon, with scattered distribution.
OTHER NAMES: Swamp Ragwort, *Senecio robbinsii, Senecio schweinitzianus*

ASTERACEAE • ASTER FAMILY

Northern Sweet-coltsfoot • *Petasites frigidus*

Perennial herb of cedar swamps, 0.1–1 m tall. **Flower heads** 8–13 mm wide, *appearing before leaves*; rays whitish, 8–40 per head, 2–7 mm long. **Leaves** basal only, *deeply cleft, white-woolly beneath*.
OCCURRENCE: Uncommon, with scattered distribution.
OTHER NAMES: Arctic Sweet-coltsfoot, *Petasites palmatus*

Orange King-devil • **Pilosella aurantiaca*

Perennial herb of fields and forests, 15–35 cm tall. **Flower heads** *red-orange*, 3–7 per stem. **Leaves** *with sparse stellate hairs beneath*; cauline leaves absent or much reduced and on only lower third of stem. **Stolons** *numerous*.
OCCURRENCE: Common, widely distributed.
OTHER NAMES: Indian Paintbrush, Orange Paintbrush, *Hieracium aurantiacum*

WILDFLOWERS AND LOW SHRUBS 89

ASTERACEAE • ASTER FAMILY

Yellow King-devil • *Pilosella caespitosa*

Perennial herb of fields and roadsides, 20–70 cm tall. **Flower heads** yellow, 5–25 per stem; involucral bracts bristly with blackish gland-tipped hairs. **Leaves** *bright-green to yellow-green, without a bloom, with sparse stellate hairs beneath (in addition to long hairs); cauline leaves absent or much reduced and on only lower third of stem.* **Stolons** *present, sometimes short and inconspicuous.*
OCCURRENCE: Occasional, widely distributed.
NOTES: See *P. praealta*. A hybrid (*Pilosella* ×*floribunda*), similar to *P. caespitosa*, has been documented at Baxter State Park. It can be recognized by its long rhizomes, tall stem (20–80 cm), and 3–30(–50) flower heads per stem.
OTHER NAMES: Yellow King Devil, Field Hawkweed, *Hieracium caespitosum*, *Hieracium pratense*

Whip King-devil • *Pilosella flagellaris*

Perennial herb of meadows and fields, 5–12 cm tall. **Flower heads** yellow, *mostly 2 or 3 (up to 6) per stem*. **Leaves** pale green, glabrous or nearly so above, *moderately pubescent with stellate hairs beneath*; cauline leaves absent or much reduced and on only lower third of stem. **Stolons** *long, leafy.*
OCCURRENCE: Uncommon, with scattered distribution.
OTHER NAMES: *Hieracium flagellare*

ASTERACEAE • ASTER FAMILY

Mouse-ear King-devil • *Pilosella officinarum*

Perennial herb of fields and lawns, 10–25 cm tall.
Flower heads yellow, *mostly solitary* (rarely 2 per stem), 2.5–3 cm wide. **Leaves** *white-woolly beneath*, bristly above; *cauline leaves absent or much reduced and on only lower third of stem.* **Stolons** *leafy.*
OCCURRENCE: Occasional, widely distributed.
OTHER NAMES: *Hieracium pilosella*

Glaucous King-devil • *Pilosella piloselloides*

Perennial herb of meadows and fields, 10–40 cm tall. **Flower heads** yellow, *11–20 per stem*, on peduncles with few or no branched hairs. **Leaves** *basal only, with few or no hairs above, with few or no minute branched hairs beneath.* **Rhizomes** *present.* **Stolons** *absent.*
OCCURRENCE: Rare, with scattered distribution.
NOTES: See *P. caespitosa* and *P. praealta*.
OTHER NAMES: *Hieracium florentinum, Hieracium piloselloides*

ASTERACEAE • ASTER FAMILY

Tall King-devil • *Pilosella praealta*

Perennial herb of meadows and fields, 25–80 cm tall. **Flower heads** yellow, 5–30 per stem. **Leaves** *dark green, with a thin bloom, usually with minute stellate hairs beneath, with few or no hairs above;* cauline leaves absent or much reduced and on only lower third of stem. **Rhizomes** *present.* **Stolons** *long, leafy.*
OCCURRENCE: Very rare, documented from only one township in Park.
NOTES: See *P. caespitosa* and *P. piloselloides.*
OTHER NAMES: *Hieracium praealtum*

Black-eyed Coneflower • *Rudbeckia hirta*

Biennial or short-lived perennial herb of fields and roadsides, 0.3–1 m tall. **Flower heads** solitary, 5–8 cm wide; *rays yellowish, 8–21 per head; disc florets dark purple-brown.* **Leaves** alternate, entire to finely toothed, *coarsely pubescent.*
OCCURRENCE: Very rare, documented from only one township in Park.
OTHER NAMES: Black-eyed Susan, *Rudbeckia serotina*

ASTERACEAE • ASTER FAMILY

Fall-dandelion • *Scorzoneroides autumnalis*

Perennial herb of disturbed sites, roadsides, and fields, 10–80 cm tall. **Flower heads** yellow, ~2.5 cm wide, gradually tapering at base into peduncles; involucral bracts densely pubescent with black hairs. **Leaves** *basal only, deeply pinnately lobed or toothed*, 4–35 cm long and 0.5–4 cm wide.
OCCURRENCE: Uncommon, with scattered distribution.
OTHER NAMES: *Leontodon autumnalis*

White Goldenrod • *Solidago bicolor*

Perennial herb of dry, rocky, and sandy sites, 10–90 cm tall. **Flower heads** *not restricted to only one side of branches; rays white, 7–9 per head.* **Leaves** alternate, pubescent on both surfaces, *rapidly reduced in size toward top of stem*, the lowest usually toothed, the largest 8–20 cm long and 1.5–6 cm wide. **Stems** *with dense hairs immediately below capitulescence.*
OCCURRENCE: Historical record, last documented in 1984, from northern portion of Park.
NOTES: White rays seem the only sure character for distinguishing this species from *S. hispida*.
OTHER NAMES: Silverrod

ASTERACEAE • ASTER FAMILY

Canada Goldenrod • *Solidago canadensis*

Perennial herb of uplands and wetlands, up to 2 m tall. **Flower heads** *clustered on only one side of branches*; rays yellow, 10–17 per head, 1–3 mm long. **Leaves** alternate, *the lowest falling off by time of flowering*; leaves at mid-stem 8–12 mm wide, sharply toothed, with 3 prominent, pubescent veins. **Stems** pubescent immediately below capitulescence.
OCCURRENCE: Occasional, widely distributed.
OTHER NAMES: Common Goldenrod

Zig-zag Goldenrod • *Solidago flexicaulis*

Perennial herb of rich, hardwood forests, 0.3–1.2 m tall. **Flower heads** *in short axillary clusters, not restricted to only one side of branches*; rays yellow, 3 or 4 per head. **Leaves** alternate, sharply toothed, 7–15 cm long and 3–10 cm wide, narrowing abruptly at base to winged petiole, *gradually reduced in size toward top of stem*. **Stems** zigzagging, pubescent immediately below capitulescence.
OCCURRENCE: Occasional, widely distributed.
OTHER NAMES: *Solidago latifolia*

ASTERACEAE • ASTER FAMILY

Hairy Goldenrod • *Solidago hispida*

Perennial herb of ledges, shorelines, forests, and roadsides, 0.1–1 m tall.
Flower heads *not restricted to only one side of branches*; rays yellow, 6–14 per head. **Leaves** alternate, *softly pubescent, rapidly reduced in size toward top of stem, the lowest usually toothed*; largest 8–20 cm long and 1.5–6 cm wide.
Stems *with dense hairs immediately below capitulescence.*
OCCURRENCE: Uncommon, with scattered distribution.
NOTES: See *S. nemoralis* and *S. puberula*.

Early Goldenrod • *Solidago juncea*

Perennial herb of *dry, upland* sites, 0.3–1.2 m tall. **Flower heads** *clustered on only one side of branches*; rays yellow, 7–12 per head, 4–5 mm long; *flowering early, usually in late July*. **Leaves** alternate, gradually reduced toward top of stem, becoming bract-like, *not sheathing stem, the lowest present and persistent at time of flowering*.
Stems *glabrous immediately below capitulescence.*
OCCURRENCE: Rare, with scattered distribution.
NOTES: See *S. uliginosa*.

ASTERACEAE • ASTER FAMILY

Cutler's Goldenrod • *Solidago leiocarpa*

Perennial herb of *alpine and subalpine zones*, ridges, and ledges, 5–35 cm tall. **Flower heads** in a terminal cluster, *not restricted to only one side of branches*; rays yellow, 6–15 per head. **Leaves** alternate, *basal and lower leaves conspicuously larger than upper ones, decreasing rapidly in size toward top of stem*. **Stems** usually with hairs immediately below capitulescence.
OCCURRENCE: Rare, with scattered distribution.
NOTES: See *S. simplex*. Species is listed as Threatened in Maine.
OTHER NAMES: *Solidago cutleri, Solidago virgaurea*

Large-leaved Goldenrod • *Solidago macrophylla*

Perennial herb of boreal forests to alpine zones, 0.1–1 m tall. **Flower heads** in short axillary clusters, *not restricted to only one side of branches*; rays yellow, 7–12 per head. **Leaves** alternate, sharply toothed, 2–15 cm long and 1–7 cm wide, *gradually reduced in size toward top of stem*. **Stems** straight, pubescent immediately below capitulescence.
OCCURRENCE: Occasional, widely distributed.
NOTES: See *S. flexicaulis*.

ASTERACEAE • ASTER FAMILY

Gray Goldenrod • *Solidago nemoralis*

Perennial herb of dry, open, primarily sandy sites, 0.1–1.3 m tall. **Flower heads** *clustered on only one side of branches, arching or nodding near apex of branches*; rays yellow, 5–9 per head. **Leaves** alternate, *narrow, entire or bluntly toothed, the lowest present and persistent at time of flowering*. **Stems** *with dense hairs immediately below capitulescence*.
OCCURRENCE: Historical record, last documented in 1982, from northern portion of Park.
NOTES: See *S. puberula* and *S. hispida*.

Downy Goldenrod • *Solidago puberula*

Perennial herb of open, rocky, and sandy sites, 0.2–1 m tall. **Flower heads** *not restricted to only one side of branches, not nodding near apex of branches*; rays yellow, 9–16 per head. **Leaves** *alternate, often minutely pubescent, rapidly reduced in size toward top of stem*, the largest 5–15 cm long and 1–3.5 cm wide. **Stems** *often purple, with dense hairs immediately below capitulescence*.
OCCURRENCE: Uncommon, with scattered distribution.
NOTES: See *S. hispida* and *S. nemoralis*.
OTHER NAMES: Dusty Goldenrod

ASTERACEAE • ASTER FAMILY

Rand's Goldenrod • *Solidago randii*

Perennial herb of rocky and sandy sites, up to 90 cm tall. **Flower heads** *not restricted to only one side of branches*; rays yellow, 7–12 per head. **Leaves** alternate, *often glabrous, rapidly reduced in size toward top of stem, the lowest sharply toothed.* **Stems** *glabrous immediately below capitulescence.*
OCCURRENCE: Occasional, widely distributed.
OTHER NAMES: *Solidago simplex*

Common Wrinkle-leaved Goldenrod • *Solidago rugosa*

Perennial herb of various habitats, up to 2.3 m tall. **Flower heads** *clustered on only one side of branches*; rays yellow, 6–11 per head. **Leaves** alternate, *12–22 mm wide at mid-stem, sharply toothed, pubescent beneath on midrib and veins, the lowest deciduous by time of flowering.* **Stems** *with hairs immediately below capitulescence.*
OCCURRENCE: Common, widely distributed.
OTHER NAMES: Rough-stemmed Goldenrod

ASTERACEAE • ASTER FAMILY

Squarrose Goldenrod • *Solidago squarrosa*

Perennial herb of woods, fields, and rocky slopes, 0.3–1.5 m tall. **Flower heads** *not restricted to only one side of branches*; rays yellow, 10–17 per head; *involucral bracts (especially outer) abruptly recurved*. **Leaves** alternate, *rapidly reduced in size toward top of stem, the lowest sharply toothed*. **Stems** *glabrous immediately below capitulescence*.
OCCURRENCE: Occasional, widely distributed.

Bog Goldenrod • *Solidago uliginosa*

Perennial herb of *fens*, 0.4–1.5 m tall. **Flower heads** clustered on only one side of branches or not; rays yellow, 1–8 per head. **Leaves** alternate, glabrous, *the lowest present and persistent at time of flowering; petioles long, nearly or entirely clasping stem at base*. **Stems** *glabrous immediately below capitulescence*.
OCCURRENCE: Rare, with scattered distribution.
NOTES: See *S. juncea*.
OTHER NAMES: Swamp Goldenrod

WILDFLOWERS AND LOW SHRUBS

ASTERACEAE • ASTER FAMILY

Field Sow-thistle • *Sonchus arvensis*

Perennial herb of shorelines, roadsides, and disturbed sites, 0.4–2 m tall. **Flower heads** *bright yellow-orange, 2.5–5 cm wide, with yellow, glandular hairs at base.* **Leaves** alternate, with soft spines on margins, *the uppermost heart-shaped and clasping at base, the lowest deeply lobed*, 6–40 cm long and 2–15 cm wide. **Stems** with milky sap.
OCCURRENCE: Very rare, documented from only northern portion of Park.
OTHER NAMES: Creeping Sow-thistle

Rush American-aster • *Symphyotrichum boreale*

Perennial herb of *fens*, 13–85 cm tall. **Flower heads** with white to light purple rays; *peduncles 0.5–5 cm long; involucral bracts usually appressed.* **Leaves** alternate, toothed, 2–5 mm wide; basal leaves neither heart-shaped nor borne on petioles, not or barely clasping stem. **Rhizomes** long.
OCCURRENCE: Very rare, documented from only one township in Park.
OTHER NAMES: Northern Bog Aster, *Aster borealis, Aster junciformis*

ASTERACEAE • ASTER FAMILY

Heart-leaved American-aster • *Symphyotrichum cordifolium*

Perennial herb of rocky woods, roadsides, and fields, 0.2–1.2 m tall. **Flower heads** with blue to purple rays; peduncles 0.3–2 cm long; *involucral bracts glabrous on outer face*. **Leaves** alternate, toothed; *basal leaves prominently heart-shaped and borne on long petioles; middle and upper stem leaves narrowed or stalked at base on slender petioles, not clasping stem*.
OCCURRENCE: Uncommon, with scattered distribution.
NOTES: See *S. undulatum*.
OTHER NAMES: Heart-leaved Aster, *Aster cordifolius*

Lance-leaved American-aster • *Symphyotrichum lanceolatum*

Perennial herb of moist roadsides, meadows, and thickets, 0.5–1.5 m tall. **Flower heads** *often numerous and clustered on only one side of branches*; rays white; peduncles 0.5–5 cm long; *outer involucral bracts often slightly spreading at tip*. **Leaves** alternate; *basal leaves neither heart-shaped nor borne on petioles*. **Stems** *forming large colonies*. **Rhizomes** *long*.
OCCURRENCE: Rare, with scattered distribution.
OTHER NAMES: *Aster lanceolatus*

ASTERACEAE • ASTER FAMILY

Calico American-aster • *Symphyotrichum lateriflorum*

Perennial herb of forests and other shaded sites, 0.3–1.2 m tall. **Flower heads** with white rays, *clustered on only one side of branches*; peduncles 0–1 cm long; *disc florets with deeply lobed corollas, turning rosy early in flowering; involucral bracts not strongly spreading or recurved at tip.* **Leaves** alternate; *basal leaves neither heart-shaped nor borne on petioles*, the lowest tapering to both ends. **Rhizomes** *short*.
OCCURRENCE: Occasional, widely distributed.
OTHER NAMES: Goblet Aster, Calico Aster, *Aster lateriflorus*

New England American-aster • *Symphyotrichum novae-angliae*

Perennial herb of open sites and forests, 0.3–2 m tall. **Flower heads** with bright purple rays; *peduncles densely glandular*, 0.3–4 cm long; *involucral bracts densely glandular, strongly spreading to recurved at tip, dark green to purple.* **Leaves** alternate, *entire, conspicuously clasping stem*; basal leaves neither heart-shaped nor borne on petioles. **Rhizomes** short.
OCCURRENCE: Occasional, widely distributed.
OTHER NAMES: *Aster novae-angliae*

ASTERACEAE • ASTER FAMILY

New York American-aster • *Symphyotrichum novi-belgii*

Perennial herb of riverbanks and various damp habitats, 0.2–1.4 m tall. **Flower heads** with light purple rays; peduncles densely glandular, 0.4–4 cm long; *involucral bracts glabrous, strongly spreading to recurved at tip, dark green to purple.* **Leaves** alternate, sparsely toothed or entire, *barely clasping stem; basal leaves neither heart-shaped nor borne on petioles.* **Stems** *without stiff, spreading hairs at base.* **Rhizomes** long.
OCCURRENCE: Uncommon, with scattered distribution.
NOTES: See *S. puniceum.*
OTHER NAMES: New York Aster, *Aster novibelgii*

Purple-stemmed American-aster • *Symphyotrichum puniceum*

Perennial herb of marshes and wet ditches, up to 2.5 m tall. **Flower heads** with light purple rays; *peduncles not glandular*, 0.2–3 cm long; *involucral bracts not glandular, strongly spreading to recurved at tip, sometimes tinged with purple.* **Leaves** alternate, *toothed, conspicuously clasping stem; basal leaves neither heart-shaped nor borne on petioles.* **Stems** *with stiff and spreading hairs at base*, rarely glabrous.
OCCURRENCE: Occasional, widely distributed.
NOTES: See *S. novi-belgii.*
OTHER NAMES: *Aster puniceus*

ASTERACEAE • ASTER FAMILY

Tradescant's American-aster • *Symphyotrichum tradescantii*

Perennial herb of rocky and gravelly shorelines, 5–70 cm tall. **Flower heads** *neither crowded nor clustered on only one side of branches, with white rays*; peduncles 0.2–2.5 cm long; *disc florets with lobes strongly spreading or recurved, turning purple; involucral bracts not strongly spreading to recurved at tip.* **Leaves** alternate, glabrous, entire or toothed; *basal leaves not heart-shaped.*
OCCURRENCE: Uncommon, with scattered distribution.
NOTES: See *S. lanceolatum*.
OTHER NAMES: Shore Aster, *Aster saxatilis, Aster tradescantii, Aster vimineus*

Wavy-leaved American-aster • *Symphyotrichum undulatum*

Perennial herb of dry fields, forests, and roadsides, 0.2–1.5 m tall. **Flower heads** with blue to light purple rays; peduncles 0.3–3 cm long; *involucral bracts pubescent on outer face.* **Leaves** alternate; *middle and upper stem leaves on short, broadly winged petioles, clasping stem; basal leaves prominently heart-shaped and borne on long, narrowly winged petioles.*
OCCURRENCE: Historical record, last documented in 1984, from northern portion of Park.
NOTES: See *S. cordifolium*.
OTHER NAMES: *Aster undulatus*

ASTERACEAE • ASTER FAMILY

Red-seeded Dandelion • *Taraxacum laevigatum*

Perennial herb of fields and disturbed sites, 5–30 cm tall. **Flower heads** solitary on stalk from base of plant, ~2.5 cm wide; peduncles unbranched, hollow, with milky sap; *longest involucral bracts with a small, dark, horn-like body at apex.* **Leaves** basal only, *deeply lobed, middle lobes 1–10 mm wide at base.*
OCCURRENCE: Uncommon, with scattered distribution.
OTHER NAMES: *Leontodon erythrospermum, Taraxacum erythrospermum*

Common Dandelion • *Taraxacum officinale*

Perennial herb of fields and disturbed sites, 5–40 cm tall. **Flower heads** solitary on stalk from base of plant, 2–5 cm wide; peduncles unbranched, hollow, with milky sap; *longest involucral bracts without a small, dark, horn-like body at apex.* **Leaves** basal only, *toothed to lobed; middle lobes, when present, usually wider than 10 mm at base.*
OCCURRENCE: Common, widely distributed.
OTHER NAMES: *Leontodon taraxacum*

WILDFLOWERS AND LOW SHRUBS

ASTERACEAE • ASTER FAMILY

Meadow Goat's Beard • *Tragopogon pratensis*

Biennial herb of fields and disturbed sites, 10–80 cm tall. **Flower heads** *solitary, long-stalked, less than 6 cm wide*, opening only in morning, closing by noon; rays yellow, toothed. **Leaves** alternate, entire, linear, *clasping stem*.
OCCURRENCE: Rare, with scattered distribution.
OTHER NAMES: Showy Goatsbeard, Yellow Goatsbeard

Coltsfoot • *Tussilago farfara*

Perennial herb of disturbed sites, 5–50 cm tall. **Flower heads** 1.5–3.5 cm wide, *appearing in early spring before leaves*; rays numerous, small, yellow; main stalk of capitulescence with alternate, scaly bracts. **Leaves** basal only, white-woolly beneath, *hooked at tip*, with long petioles.
OCCURRENCE: Common, widely distributed.
NOTES: Species has potential to spread aggressively in some natural habitats.

BALSAMINACEAE • TOUCH-ME-NOT FAMILY ▼

Spotted Touch-me-not • *Impatiens capensis*

Annual herb of wet areas, 0.5–1.8 m tall. **Flowers** *2–2.5 cm long, deep yellow to orange, abundantly spotted with dark red to brown; flower spur 7–10 mm long, strongly curved underneath and projecting forward.* **Leaves** alternate, short-stalked. **Fruit** a slender capsule, exploding when ripe, dispersing seeds.
OCCURRENCE: Common, widely distributed.
OTHER NAMES: Orange Touch-me-not, Snapweed

BERBERIDACEAE • BARBERRY FAMILY ▼

Blue Cohosh • *Caulophyllum thalictroides*

Perennial herb of floodplains and forests, 20–80 cm tall. **Flowers** *yellow-green to purple, 6-parted, 0.9–1.3 mm wide, in a branching, terminal cluster.* **Leaves** alternate, *large, two or more times divided; leaflets 2- to 5-lobed above middle.*
OCCURRENCE: Historical record, last documented in 1982, from northern portion of Park.

BORAGINACEAE • BORAGE FAMILY ▼

Woodland Forget-me-not • *Myosotis sylvatica*

Perennial herb of forest edges, fields, and disturbed sites, 20–50 cm tall. **Flowers** blue, purple, or white, 5-parted, 5–8 mm wide; sepals densely pubescent with hooked hairs. **Leaves** up to 7 cm long, with simple hairs. **Stems** *with mostly upward-pointing hairs, decumbent at base.*
OCCURRENCE: Uncommon, with scattered distribution.

BRASSICACEAE • MUSTARD FAMILY ▼

Upright Yellow-rocket • *Barbarea stricta*

Biennial or perennial herb of roadsides, fields, and disturbed sites, 0.2–1 m tall. **Flowers** 4-parted; petals yellow, *2.5–4.5 mm long.* **Leaves** alternate; basal leaves petiolate, with 1–4 pairs of lateral lobes and large terminal lobe; uppermost cauline leaves usually toothed or with a few shallow sinuses; *auricles of uppermost leaves with a few marginal hairs.* **Fruit** a silique, 1.2–3 cm long, *usually appressed to rachis when mature; style persistent, beak 0.5–1.6 mm long; fruiting pedicels straight.*
OCCURRENCE: Very rare, documented from only northern portion of Park.
NOTES: See *B. vulgaris.*

BRASSICACEAE • MUSTARD FAMILY

Garden Yellow-rocket • *Barbarea vulgaris*

Biennial or perennial herb of roadsides, fields, and disturbed sites, 0.2–1.2 m tall. **Flowers** 4-parted; petals bright yellow, *6–9 mm long.* **Leaves** alternate; basal leaves petiolate with 1–4 pairs of lateral lobes and large terminal lobe; uppermost cauline leaves toothed, occasionally with a few shallow lobes; *auricles of uppermost leaves without any marginal hairs.* **Fruit** a silique, 1–4 cm long, *usually not appressed to rachis when mature; style persistent; beak 2–3 mm long; fruiting pedicels curved.*
OCCURRENCE: Rare, documented from only northern portion of Park.
NOTES: See *B. stricta*.
OTHER NAMES: Common Winter-cress

Canada Rockcress • *Boechera stricta*

Biennial herb of rocky forests, cliffs, and talus slopes, up to 1 m tall. **Flowers** 4-parted; petals purple to white, *7–10 mm long*; pedicels 10–15 mm long. **Leaves** alternate, the uppermost auricled but usually not clasping, with waxy bloom. **Stems** *with branched hairs near base.* **Fruit** a silique, 4–7 cm long, *closely appressed to rachis.*
OCCURRENCE: Historical record, last documented in 1946, from northern portion of Park.
OTHER NAMES: Drummond's Rockcress, *Arabis drummondii, Boechera drummondii, Turritis drummondii, Turritis stricta*

BRASSICACEAE • MUSTARD FAMILY

Smooth Rockcress • *Borodinia laevigata*

Biennial herb of ledges, balds, rocky slopes, and forests, up to 1 m tall. **Flowers** 4-parted; petals white, *3–6 mm long*; pedicels 7–12 mm long. **Leaves** alternate, the uppermost clasping at base. **Stems** *glaucous, glabrous at base.* **Fruit** a silique, 5–10 cm long, *clearly spreading from rachis.*
OCCURRENCE: Very rare, documented from only northern portion of Park.
NOTES: Species is listed as Threatened in Maine.
OTHER NAMES: *Arabis laevigata, Boechera laevigata, Turritis laevigata*

Shepherd's-purse • **Capsella bursa-pastoris*

Annual herb of open areas, 10–75 cm tall. **Flowers** 4-parted; petals white, less than 3 mm long. **Leaves** in a basal rosette and alternate on stem; *lowest deeply lobed, 5–10 cm long with nearly parallel sides*; uppermost clasping, arrowhead-shaped, entire or with very small teeth and small, ear-shaped appendages at base. **Fruit** *a heart-shaped silicle, 6–13 mm wide, flat, indented on top.*
OCCURRENCE: Historical record, last documented in 1984, from northern portion of Park.
OTHER NAMES: Pick-pocket, *Bursa bursa-pastoris*

BRASSICACEAE • MUSTARD FAMILY

Alpine Bitter-cress • *Cardamine bellidifolia*

Perennial herb of alpine and subalpine zones, *1–8 cm tall*. **Flowers** 4-parted; petals white, *3–5.5 mm long*. **Leaves** *basal only or up to 2 cauline*; basal leaves entire, on long petioles. **Fruit** an erect silique, *15–35 mm long*; beak 1–1.5 mm long; *fruiting pedicels 3–6 mm long*, crowded.
OCCURRENCE: Very rare, documented from only one township in Park.
NOTES: Species is listed as Endangered in Maine.

Two-leaved Toothwort • *Cardamine diphylla*

Perennial herb of floodplains, forests, and rocky slopes, *15–40 cm tall*. **Flowers** 4-parted; petals white to purple, *9–15 mm long*. **Leaves** *palmately divided with 3 leaflets; cauline leaves usually 2, nearly opposite*. **Fruit** a silique, *2–4 cm long*. **Rhizomes** *white, of uniform diameter*.
OCCURRENCE: Uncommon, with scattered distribution.
OTHER NAMES: *Dentaria diphylla, Dentaria incisa*

BRASSICACEAE • MUSTARD FAMILY

Small-flowered Bitter-cress • *Cardamine parviflora*

Annual or biennial herb of dry ledges and woods, 10–40 cm tall. **Flowers** 4-parted; petals white, *2–2.5 mm long*. **Leaves** alternate, divided; *cauline leaves 2–4 cm long, with lateral leaflets 1–3 mm wide*. **Fruit** a silique, *1–2 cm long and 0.6–0.9 mm wide*.
OCCURRENCE: Very rare, documented from only one township in Park.
NOTES: See *C. pensylvanica*.
OTHER NAMES: Dry Land Bitter-cress

Pennsylvania Bitter-cress • *Cardamine pensylvanica*

Biennial or perennial herb of wet woods and swamps, 5–75 cm tall. **Flowers** 4-parted; petals white, 1.5–4 mm long. **Leaves** alternate, divided; *cauline leaves 4–10 cm long, with lateral leaflets 4–25 mm wide*. **Fruit** a silique, 1–3 cm long and *0.8–1.5 mm wide*.
OCCURRENCE: Common, widely distributed.
NOTES: See *C. parviflora*.

112 THE PLANTS OF BAXTER STATE PARK

BRASSICACEAE • MUSTARD FAMILY

Canescent Whitlow-mustard • *Draba cana*

Perennial herb of cliffs, ledges, talus, and rocky slopes on high-pH bedrock, 6–38 cm tall. **Flowers** 4-parted; petals white; *lowest 1–3 flowers subtended by a leafy bract.* **Leaves** *pubescent with stellate hairs;* basal leaves in a dense rosette; cauline leaves few, lanceolate. **Stems** *with branched or stellate hairs.* **Fruit** a silique, *with stellate hairs.*
OCCURRENCE: Very rare, documented from only one township in Park.
NOTES: Species is listed as Endangered in Maine.
OTHER NAMES: Lance-leaved Draba, *Draba breweri, Draba lanceolata*

Poor-man's Pepperweed • *Lepidium virginicum*

Annual herb of fields, roadsides, and disturbed sites, 0.1–1 m tall. **Flowers** 4-parted, in numerous, many-flowered racemes; petals white, 1–2.5 mm long. **Leaves** alternate; *cauline leaves ascending, sharply toothed to entire, not clasping;* basal leaves withered by time of flowering. **Fruit** *a nearly round silique, 2.5–4 mm in diameter, notched at tip.*
OCCURRENCE: Historical record, last documented in 1982, from northern portion of Park.
OTHER NAMES: Wild Peppergrass

BRASSICACEAE • MUSTARD FAMILY

Common Yellow-cress • *Rorippa palustris*

Annual or perennial herb of wetlands, shorelines, and disturbed sites, 0.2–1.3 m tall. **Flowers** 4-parted; petals yellow, 1.8–2.5 mm long. **Leaves** irregularly toothed to deeply cleft or divided, up to 18 cm long. **Stems** wider than 3 mm, branching. **Fruit** a silique, 4–10 mm long and 1.7–3 mm wide.
OCCURRENCE: Rare, documented from only northern portion of Park.

Tumbling Hedge-mustard • **Sisymbrium altissimum*

Annual herb of fields, shorelines, and disturbed sites, 0.6–1.5 m tall. **Flowers** 4-parted; *petals pale yellow*, 6–8 mm long. **Leaves** alternate, *the uppermost deeply lobed*, with lobes up to 2 mm wide. **Stems** branching distally. **Fruit** a silique, 6–9 cm long and 1–2 mm wide; *pedicel nearly as wide as fruit*.
OCCURRENCE: Historical record, last documented in 1982, from northern portion of Park.

Creeping Bellflower • **Campanula rapunculoides*

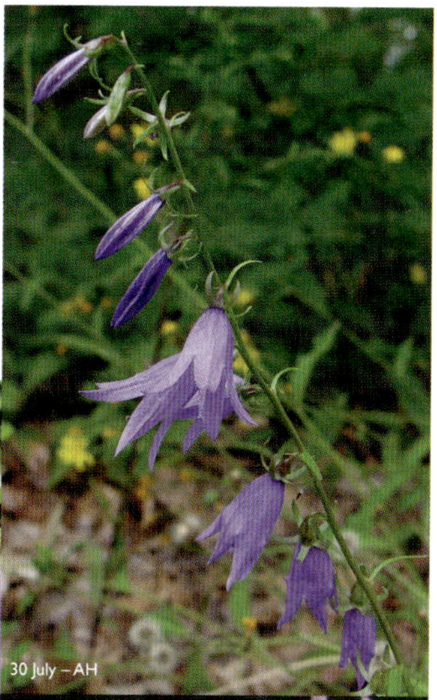
30 July – AH

Perennial herb of roadsides and forest edges, 0.3–1.5 m tall. **Flowers** 5-parted, in a one-sided raceme; petals purple-blue, *fused into a drooping, bell-shaped corolla 2.5–4 cm long.* **Leaves** alternate, toothed; cauline leaves 6–15 cm long and 2.4–6 cm wide.
OCCURRENCE: Very rare, documented from only one township in Park.
NOTES: Historically, a cultivated perennial of flower gardens but now has escaped to roadsides and other disturbed areas.
OTHER NAMES: Roving Bellflower

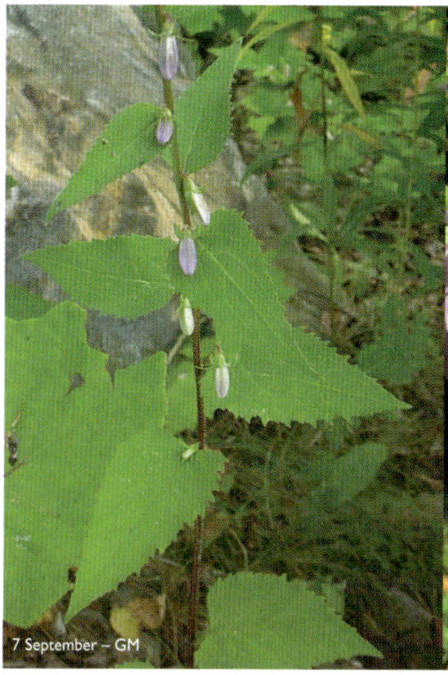
7 September – GM

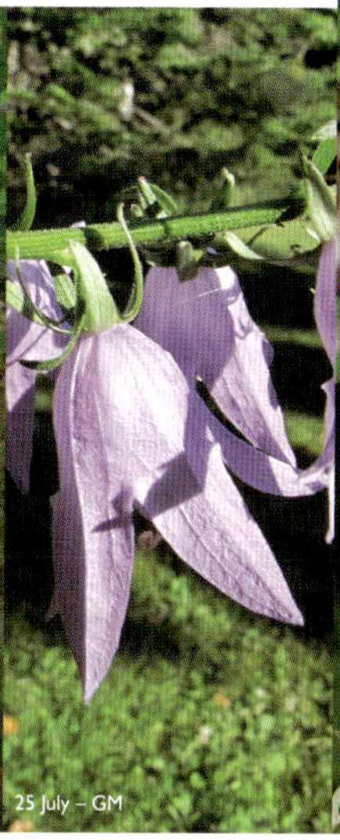
25 July – GM

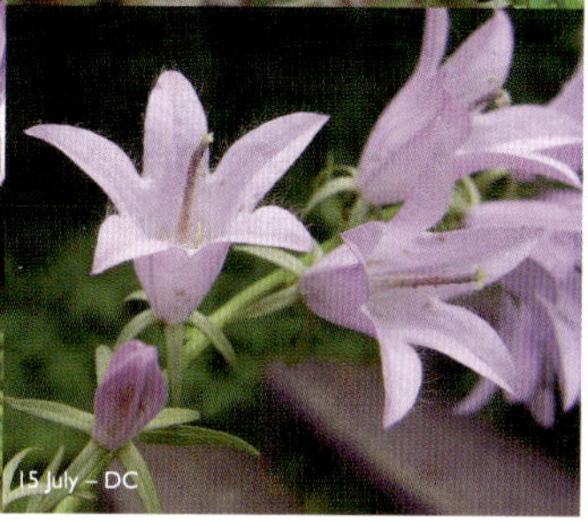
15 July – DC

Harebell • *Campanula rotundifolia*

11 July TSR9 – GM

Slender, perennial herb of cliffs and shorelines, 15–25 cm tall. **Flowers** 5-parted; petals deep blue to purple, *fused into a narrowly bell-shaped, nodding corolla ~2 cm long.* **Leaves** alternate; cauline leaves 1.5–8 cm long and 0.2–0.9 cm wide; basal leaves oval to circular.
OCCURRENCE: Uncommon, with scattered distribution.
OTHER NAMES: Scotch Bellflower

5 July – ML

18 September – ML

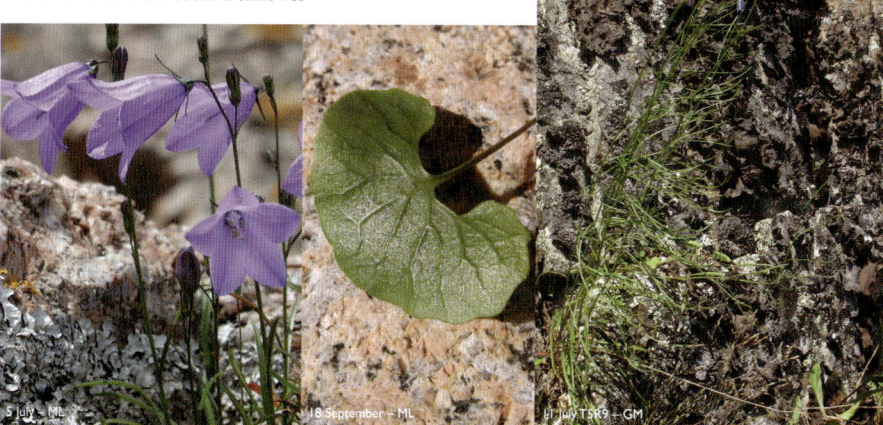

CAMPANULACEAE • BELLFLOWER FAMILY

Cardinal-flower • *Lobelia cardinalis*

Perennial herb of wet meadows and shorelines, up to 1.5 m tall.
Flowers *bright red, showy, 3–4.5 cm long*, with upper lip 2-lobed and lower lip 3-lobed. **Leaves** alternate, irregularly toothed; lower leaves petiolate; upper leaves sessile. **Stems** angled, with milky sap.
OCCURRENCE: Historical record, last documented in 1949, from southern portion of Park.
OTHER NAMES: Red Lobelia

Water Lobelia • *Lobelia dortmanna*

Perennial, *aquatic* herb of lakes and ponds, up to 1 m tall. **Flowers** violet or white, *widely separated*, with upper lip 2-lobed and lower lip 3-lobed. **Leaves** *basal only, in dense rosettes, fleshy, hollow, slightly flattened, slightly curved, blunt at tip*, with a milky sap.
OCCURRENCE: Occasional, widely distributed.
OTHER NAMES: Water-gladiole

CAMPANULACEAE • BELLFLOWER FAMILY

Bladder-pod Lobelia • *Lobelia inflata*

Annual herb of fields, open woods, and roadsides, up to 1 m tall. **Flowers** 5-parted, *6–8 mm long*; petals light blue, fused into tubular, 2-lipped corolla; *lower lip of corolla pubescent near base; bracteoles minute, attached at base of pedicel.* **Leaves** alternate, toothed, *10–35 mm wide*, reduced toward top of stem. **Stems** much-branched, *pubescent at base*.
OCCURRENCE: Uncommon, with scattered distribution.
NOTES: See *L. kalmii*.
OTHER NAMES: Indian-tobacco

Brook Lobelia • *Lobelia kalmii*

Perennial herb of fens, stream and river shorelines, and seeps, often on high-pH bedrock, 20–40 cm tall. **Flowers** 5-parted; petals violet or white; upper lip 2-lobed and lower lip 3-lobed; *lower lip of corolla glabrous; bracteoles minute, attached near middle of pedicel.* **Leaves** alternate; *cauline leaves 1–4 mm wide*. **Stems** *glabrous at base*.
OCCURRENCE: Very rare, documented from only one township in Park.
NOTES: See *L. inflata*.

CAPRIFOLIACEAE • HONEYSUCKLE FAMILY ▼

Bush-honeysuckle • *Diervilla lonicera*

Deciduous shrub of dry woods, forest edges, and rocky slopes, *in dense colonies*, 0.3–1.2 m tall. **Flowers** *5-parted, funnel-shaped*, usually in groups of 3; *petals yellow (turning reddish)*, 12–20 mm long. **Leaves** opposite, toothed, 5–15 cm long, *tapering to a long, drawn-out tip*. **Twigs** woody at base, round in cross section. **Fruit** a slender, 2-valved capsule, 8–15 mm long, with many seeds.
OCCURRENCE: Common, widely distributed.
OTHER NAMES: Northern Bush-honeysuckle, *Diervilla diervilla*

American Twinflower • *Linnaea borealis*

Evergreen, perennial herb of wet woods, up to 40 cm tall. **Flowers** white with rose-purple markings, 5-parted, funnel-shaped, fragrant, *borne in nodding pairs* from long, erect peduncles; petals 10–15 mm long. **Leaves** opposite, leathery, pubescent, 1–2 cm long.
OCCURRENCE: Common, widely distributed.
NOTES: This species, the only one in the genus, was named for celebrated Swedish botanist Linnaeus (1707–1778), who developed the binomial classification system to name plants and animals still used today.
OTHER NAMES: Pink-bells, *Linnaea americana*

CAPRIFOLIACEAE • HONEYSUCKLE FAMILY

American Honeysuckle • *Lonicera canadensis*

Deciduous shrub of wet woods, up to 1.5 m tall. **Flowers** yellow-green or stray-colored, growing in pairs from leaf axils, *radially symmetric or nearly so; corolla (tube and lobes) 1.5–2.2 cm long; peduncles glabrous, longer than 1 cm; bractlets subtending flowers 1.2–3 mm long.* **Leaves** opposite, *widest below middle, glabrous above, with cilia.* **Fruit** a red-black berry, *in pairs* diverging in opposite directions.
OCCURRENCE: Common, widely distributed.
NOTES: See *L. villosa*.
OTHER NAMES: Fly Honeysuckle

Swamp Honeysuckle • *Lonicera oblongifolia*

Deciduous shrub of fens and swamps (often ones dominated by cedar), up to 2 m tall. **Flowers** yellow, growing in pairs from leaf axils; *corolla lobes spreading, giving 2-lipped appearance*; corolla (tube and lobes) 0.6–1.4 cm long; *peduncles longer than 1 cm, minutely pubescent; bractlets subtending flowers absent.* **Leaves** opposite, *widest at or above middle, pubescent above, with a tangled mat of soft hairs beneath; margins minutely pubescent but with no long, stiff cilia.* **Fruit** a *red or purplish red* berry.
OCCURRENCE: Very rare, documented from only one township in Park.
NOTES: See *L. villosa*. Species is listed as Special Concern in Maine.

CAPRIFOLIACEAE • HONEYSUCKLE FAMILY

Mountain Honeysuckle • *Lonicera villosa*

Deciduous shrub of bogs, swamps, and wet thickets, up to 1 m tall. **Flowers** creamy yellow, *radially symmetric*, growing in pairs from leaf axils; corolla (tube and lobes) 0.6–1.4 cm long; *peduncles less than 1 cm long, pubescent*. **Leaves** *opposite*, widest at or above middle, pubescent above, with straight, stiff hairs beneath; margins with long, stiff cilia. **Fruit** a *dark blue*, glaucous berry.
OCCURRENCE: Uncommon, with scattered distribution.
NOTES: See *L. canadensis* and *L. oblongifolia*.
OTHER NAMES: Mountain Fly Honeysuckle

Common Snowberry • **Symphoricarpos albus*

Deciduous shrub of open areas, up to 2 m tall. **Flowers** white or pink, 4- or 5-parted; *petals fused into a bell shape*. **Leaves** *opposite, entire*, 1.5–4 cm long and 1–3 cm wide. **Twigs** *pubescent when young, hollow*. **Fruit** a *white* berry, 1–1.5 cm long.
OCCURRENCE: Historical record, last documented in 1985, from one township in Park.
OTHER NAMES: Waxberry

CARYOPHYLLACEAE • PINK FAMILY ▼

Mouse-ear Chickweed • *Cerastium fontanum*

Short-lived perennial herb of gardens and open areas, up to 65 cm tall. **Flowers** 5-parted, ~0.5 cm wide; petals white, cleft so deeply there appear to be 10, *about as long as sepals*. **Leaves** opposite, entire, sessile, *1–2 cm long and 3–12 mm wide*, densely pubescent.
OCCURRENCE: Occasional, widely distributed.
OTHER NAMES: *Cerastium vulgatum*

Low Baby's-breath • *Gypsophila muralis*

Annual herb of fields and disturbed sites, 4–40 cm tall. **Flowers** 5-parted; petals pink, *6–10 mm long; pedicels and sepals without glands or hairs*. **Leaves** opposite, acute, 3–32 mm long and 0.2–2 mm wide; *lowest internodes minutely pubescent*. **Stems** much branched throughout.
OCCURRENCE: Very rare, documented from only one township in Park.
NOTES: See *Spergularia rubra*.
OTHER NAMES: *Psammophiliella muralis*

WILDFLOWERS AND LOW SHRUBS 121

CARYOPHYLLACEAE • PINK FAMILY

Blunt-leaved Grove-sandwort • *Moehringia lateriflora*

Perennial herb of open areas, 5–40 cm tall. **Flowers** 5-parted, 0.5–1.5 cm wide, solitary or several in a cluster; petals white, entire; sepals 2–3 mm long. **Leaves** opposite, *oval, with blunt tip, 1- to 3-veined*, 2–15 mm wide, with underside of midrib pubescent, *not tufted at base*. **Rhizomes** extensive, thin.
OCCURRENCE: Very rare, documented from only one township in Park.
OTHER NAMES: Blunt-leaved Sandwort, *Arenaria lateriflora*.

Mountain Sandwort • *Mononeuria groenlandica*

Perennial herb of mountain summits and rocky outcrops, often forming mats, 5–13 cm tall. **Flowers** 5-parted, 1–1.5 cm wide; *borne in clusters of 3–15*; petals white, with small notch at tip; sepals 3–5.5 mm long. **Leaves** opposite, 4–12 mm long and 0.3–0.8 mm wide. **Stems** *with sterile leafy shoots at base*.
OCCURRENCE: Uncommon, with scattered distribution.
NOTES: Species is listed as Special Concern in Maine.
OTHER NAMES: Smooth Sandwort, Mountain Sandplant, *Arenaria groenlandica, Minuartia groenlandica*

CARYOPHYLLACEAE • PINK FAMILY

Bird's-eye Pearlwort • *Sagina procumbens*

Perennial herb of open ground and roadsides, growing in tufts, up to 10 cm tall. **Flowers** 4- or 5-parted; petals white, shorter than sepals or absent; sepals 1.5–2.5 mm long; styles 4 or 5, less than 0.5 mm long. **Leaves** *3–15 mm long and less than 1 mm wide.*
OCCURRENCE: Very rare, documented from only one township in Park.
OTHER NAMES: Birds-eye

Common Soapwort • *Saponaria officinalis*

Perennial herb of roadsides and disturbed sites, 40–80 cm tall. **Flowers** 5-parted, 2–3 cm wide, often in pairs; *petals pink or white, with small notch at tip and an awl-shaped basal appendage.* **Leaves** opposite, ovate, with 3–5 prominent ribs.
OCCURRENCE: Very rare, documented from only one township in Park.
OTHER NAMES: Bouncing Bet, *Lychnis saponaria*

CARYOPHYLLACEAE • PINK FAMILY

Moss Campion • *Silene acaulis*

Perennial, *matted, cushion-forming* herb of alpine and subalpine zones, 3–6 cm tall. **Flowers** *5-parted*, borne singly on peduncles 2–2.5 cm long; *petals bright pink, 8–12 mm long*. **Leaves** crowded at base, evergreen, overlapping, widest at or near base, 4–10 mm long and 0.8–1.5 mm wide, margins scabrous or ciliate.
OCCURRENCE: Historical record, last documented in 1847, from southern portion of Park.
NOTES: Species is listed as historical in Maine.
OTHER NAMES: Cushion Pink

Bladder Campion • *Silene vulgaris*

Perennial herb of roadsides, fields, and shorelines, up to 1 m tall. **Flowers** 1–2 cm wide, 5-parted; petals white, cleft so deeply there appear to be 10; *sepals swollen and bladder-like, prominently veined*. **Leaves** opposite, entire. **Stems** glabrous.
OCCURRENCE: Occasional, widely distributed.
NOTES: Blooms spring into summer.
OTHER NAMES: Maiden's-tears, *Silene cucubalus*, *Silene inflata*

CARYOPHYLLACEAE • PINK FAMILY

Red Sand-spurry • *Spergularia rubra*

Annual or short-lived perennial herb of dry and sandy soils, 4–25 cm tall. **Flowers** 5-parted; petals pink, *less than 5 mm long; sepals with stalked glands.* **Leaves** opposite, 3–35 mm long and 0.4–1.2 mm wide, *with a mucro at tip; tufts of smaller leaves in leaf axils;* stipules silvery, much longer than wide.
OCCURRENCE: Rare, documented from only northern portion of Park.
NOTES: See *Gypsophila muralis.*
OTHER NAMES: Roadside Sand-spurrey, *Arenaria rubra, Tissa rubra*

Boreal Stitchwort • *Stellaria borealis*

Perennial herb of alpine and subalpine zones, fields, shorelines, and wetlands, 25–50 cm tall. **Flowers** solitary in forks of stem or in branched cymes; petals white, *occasionally absent;* sepals 5, 1.5–3.5 mm long. **Leaves** opposite, sessile, sparsely ciliate at least toward base. **Stems** branched, angled, weak.
OCCURRENCE: Uncommon, with scattered distribution.
OTHER NAMES: *Alsine borealis*

WILDFLOWERS AND LOW SHRUBS 125

CARYOPHYLLACEAE • PINK FAMILY

Grass-leaved Stitchwort • *Stellaria graminea*

Perennial herb of fields and meadows, 20–50 cm tall. **Flowers** 5-parted, 5–18 mm wide; petals white, cleft so deeply there appear to be 10, *about as long as sepals; sepals prominently 3-veined, 4.5–5.5 mm long.* **Leaves** opposite, *sessile*, 1.5–4 cm long, *widest below middle.* **Stems** *smooth*, weak.
OCCURRENCE: Uncommon, with scattered distribution.
NOTES: See *S. longifolia*.
OTHER NAMES: Lesser Stitchwort, *Alsine graminea*

Long-leaved Stitchwort • *Stellaria longifolia*

Perennial herb of wetlands, shorelines, and fields, nearly always in wet areas, 15–45 cm tall. **Flowers** 5-parted, 5–10 mm wide; petals white, cleft so deeply there appear to be 10, *longer than sepals; sepals obscurely 3-veined, 3.5–4.5 mm long.* **Leaves** opposite, *sessile*, 1–6 cm long, *widest at or above middle.* **Stems** *scabrous on angles.*
OCCURRENCE: Rare, documented from only northern portion of Park.
NOTES: See *S. graminea*.
OTHER NAMES: *Alsine longifolia*

126 THE PLANTS OF BAXTER STATE PARK

CARYOPHYLLACEAE • PINK FAMILY

Common Stitchwort • *Stellaria media*

Annual, weak-stemmed herb of disturbed areas, up to 40 cm tall. **Flowers** 1–3 mm long, *inconspicuous*, 5-parted; petals white, cleft so deeply there appear to be 10. **Leaves** opposite, cordate, 0.5–4 cm long, *the lowest with distinct petioles*. **Stems** branched, with 1 or 2 lines of hairs.
OCCURRENCE: Rare, with scattered distribution.
OTHER NAMES: *Alsine media*

CISTACEAE • ROCK-ROSE FAMILY ▼

Round-fruited Pinweed • *Lechea intermedia*

Slender, erect, perennial herb of dry, open areas, up to 60 cm tall. **Flowers** *small, with 3 inconspicuous, red petals*, rarely open, in a panicle about one-third to one-half height of plant. **Leaves** green, the uppermost 2–5.3 mm wide; *basal rosettes forming in late summer and persisting through winter*.
OCCURRENCE: Rare, with scattered distribution.
NOTES: Blooms late summer to fall.
OTHER NAMES: Large-pod Pinweed, Intermediate Pinweed

COLCHICACEAE • COLCHICUM FAMILY ▼

Sessile-leaved Bellwort • *Uvularia sessilifolia*

Perennial, spring herb of forests, up to 40 cm tall.
Flowers *light yellow, drooping, with 6 tepals, the tepals 1.3–2.5 cm long.* **Leaves** alternate, *sessile, pale on underside, scabrous on margins.* **Fruit** a green, triangular capsule.
OCCURRENCE: Common, widely distributed.
OTHER NAMES: Wild Oats, Little Merrybells

COMANDRACEAE • BASTARD-TOADFLAX FAMILY ▼

False Toadflax • *Geocaulon lividum*

Perennial herb of alpine and subalpine bogs, 10–30 cm tall.
Flowers *with 5 petal-like greenish to purple sepals; typically in groups of 3 from axils of leaves; center flower carpellate and outer flowers staminate.* **Leaves** alternate, entire, 1–4 cm long, blunt or rounded at tip; veins becoming yellow with age. **Fruit** *a yellow-orange to bright red drupe, 5–12 mm wide.*
OCCURRENCE: Very rare, with scattered distribution.
NOTES: Species is listed as Special Concern in Maine. Easy to overlook when in bloom.
OTHER NAMES: Northern Comandra, *Comandra lividum*

CONVOLVULACEAE • MORNING-GLORY FAMILY ▼

Hedge False Bindweed • *Calystegia sepium*

Perennial, herb of fields, roadsides, and shorelines, up to 3 m long. **Flowers** pink to white, *funnel-shaped, 5–7 cm long*, with 2 green bracts partly enclosing base. **Leaves** entire, usually square-lobed at base. **Stems** twining or trailing.
OCCURRENCE: Very rare, documented from only one township in Park.
OTHER NAMES: Wild Morning-glory, Hedge Bindweed, *Convolvulus sepium*

CORNACEAE • DOGWOOD FAMILY ▼

Bunchberry • *Chamaepericlymenum canadense*

Perennial herb of forests and clearings, 5–30 cm tall. **Flowers** in small cymes subtended by *4 petal-like, white, expanded bracts*. **Leaves** in a cluster of 4 or 6, opposite, but appearing whorled, 2–9 cm long, entire, *with curving, parallel veins*, occasionally with a pair of smaller, opposite leaves below whorl. **Fruit** *a cluster of red drupes*.
OCCURRENCE: Common, widely distributed.
OTHER NAMES: Canada Dwarf-dogwood, Dwarf Cornel, Canada Bunchberry, Crackerberry, Crunchberry, *Cornus canadensis*

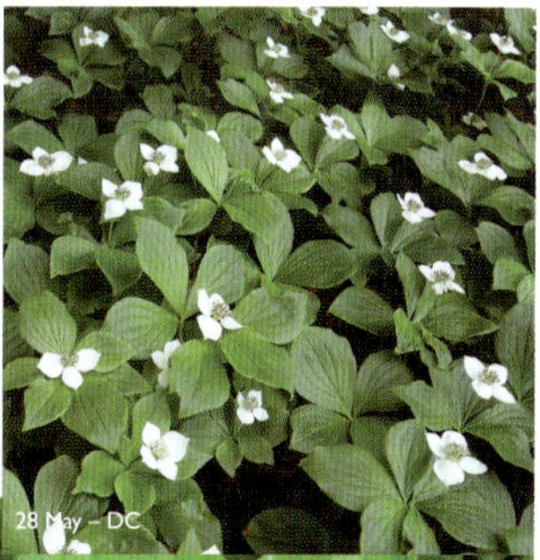

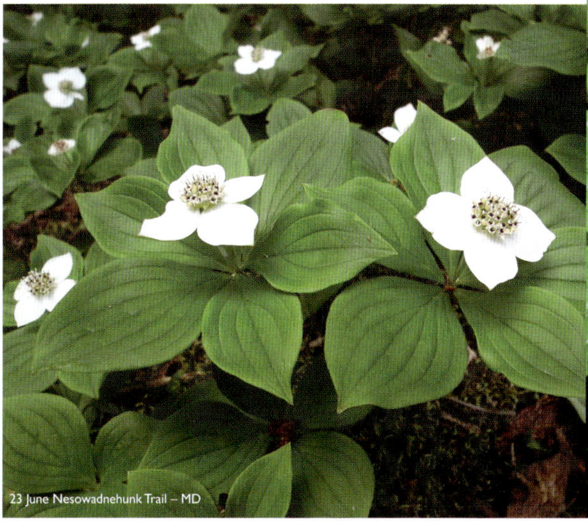

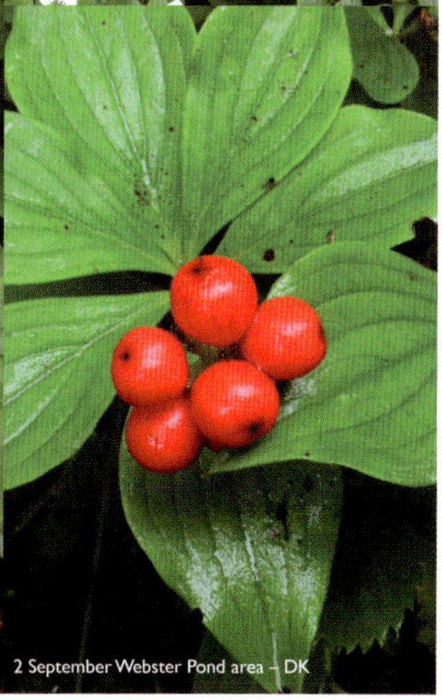

DIAPENSIACEAE • DIAPENSIA FAMILY ▼

Cushion-plant • *Diapensia lapponica*

Evergreen shrub of alpine and sub-alpine zones, *forming dense, rounded tussocks*, 3–8 cm tall. **Flowers** *5-parted; petals white*, turning pink with age, *fused at base*. **Leaves** *opposite, in dense basal rosettes, 7–15 mm long and 1.3–2.3 mm wide, entire.*
OCCURRENCE: Uncommon, with scattered distribution.
NOTES: Species is listed as Special Concern in Maine. This species is very sensitive to high temperatures.

DROSERACEAE • SUNDEW FAMILY ▼

Spatulate-leaved Sundew • *Drosera intermedia*

Perennial herb of bogs and swamps, up to 25 cm tall. **Flowers** white, usually 5-parted, in a one-sided raceme. **Leaves** basal only, with numerous large, red, glandular hairs on blade; *blades spoon-shaped, longer than wide; petioles glabrous.*
OCCURRENCE: Occasional, widely distributed.
NOTES: The glandular hairs exude a sticky substance that traps insects.
OTHER NAMES: Narrow-leaved Sundew

DROSERACEAE • SUNDEW FAMILY

Round-leaved Sundew • *Drosera rotundifolia*

Perennial herb of bogs and swamps, up to 30 cm tall. **Flowers** white, usually 5-parted, in a one-sided raceme. **Leaves** basal only, with numerous large, red, glandular hairs on blade and petiole; *blades round or slightly wider than long*.
OCCURRENCE: Common, widely distributed.
NOTES: The glandular hairs exude a sticky substance that traps insects.

ELATINACEAE • WATERWORT FAMILY ▼

Small Waterwort • *Elatine minima*

Small, aquatic, annual herb of lakes and ponds, in mats up to 10 cm wide. **Flowers** *axillary*, with 2 greenish petals and 2 sepals. **Leaves** *tiny, opposite, rounded, 0.7–5 mm long*; branchlets ascending, 0.2–5 cm long. **Stems** sometimes rooting at nodes.
OCCURRENCE: Rare, documented from only one township in Park.

ERICACEAE • HEATH FAMILY ▼

Bog-rosemary • *Andromeda polifolia*

Evergreen shrub of bogs, 0.1–0.7 m tall. **Flowers** pink or white, 5-parted, bell-shaped, nodding. **Leaves** *alternate, 2–8 mm wide*, pale blue-green above when young, turning darker when mature, *white-pubescent beneath*; margins rolled toward lower surface.
OCCURRENCE: Common, widely distributed.
NOTES: See *Kalmia polifolia*.
OTHER NAMES: *Andromeda glaucophylla*

Red Bearberry • *Arctostaphylos uva-ursi*

Matted, evergreen shrub of rocky ledges, up to 50 cm tall. **Flowers** white or pink, urn-shaped, in clusters of 5–12. **Leaves** alternate, *entire*, 5–15 mm wide, *without impressed veins above, growing obliquely (at roughly 45 degrees) to stem; petiole unwinged*. **Branchlets** *pubescent*. **Fruit** berry-like, *red, mealy*.
OCCURRENCE: Rare, documented from only northern portion of Park.
NOTES: See *Vaccinium vitis-idaea* and *Arctous alpina*.
OTHER NAMES: Common Bearberry, Kinnikinnick

ERICACEAE • HEATH FAMILY

Alpine-bearberry • *Arctous alpina*

Deciduous shrub of alpine and subalpine zones, 3–20 cm tall. **Flowers** white to yellowish, urn-shaped, usually in clusters of 2 or 3, *opening before leaves unfold.* **Leaves** alternate, *toothed*, usually 1–8 mm wide, *with conspicuous impressed veins above, turning bright red in fall; petiole winged.* **Branchlets** *glabrous.* **Fruit** berry-like, *purple to black, juicy.*
OCCURRENCE: Rare, with scattered distribution.
NOTES: See *Arctostaphylos uva-ursi.* Species is listed as Threatened in Maine.
OTHER NAMES: Mountain Bearberry, *Arctostaphylos alpina*

Leatherleaf • *Chamaedaphne calyculata*

Evergreen shrub of bogs and wetlands, up to 1.5 m tall. **Flowers** white, bell-shaped, *in one-sided, leafy racemes at branch tips.* **Leaves** alternate, *leathery, dull, with rusty scales on underside*, usually 10–15 mm wide.
OCCURRENCE: Common, widely distributed.
OTHER NAMES: Cassandra, *Cassandra calyculata*

ERICACEAE • HEATH FAMILY

Noble Prince's-pine • *Chimaphila umbellata*

Evergreen, woody-based herb of upland woods, 10–30 cm tall. **Flowers** white or pink, 10–15 mm wide, nodding. **Leaves** opposite or whorled, toothed, shiny, *widest above middle, not marked with white along veins.*
OCCURRENCE: Common, widely distributed.
OTHER NAMES: Pipsissewa

ERICACEAE • HEATH FAMILY

Red Crowberry • *Empetrum atropurpureum*

Evergreen, mat-forming shrub of alpine and subalpine areas, up to 80 cm long. **Flowers** inconspicuous. **Leaves** alternate to nearly whorled, narrow, 4–7 mm long; margins pubescent but without glandular hairs. **Twigs** *with dense white hairs toward tip, without any glandular hairs.* **Fruit** a *purple to reddish purple,* fleshy drupe.
OCCURRENCE: Uncommon, with scattered distribution.
NOTES: See *E. nigrum.*
OTHER NAMES: *Empetrum eamesii* ssp. *atropurpureum*

Black Crowberry • *Empetrum nigrum*

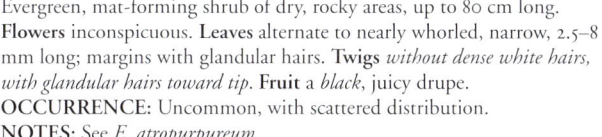

Evergreen, mat-forming shrub of dry, rocky areas, up to 80 cm long. **Flowers** inconspicuous. **Leaves** alternate to nearly whorled, narrow, 2.5–8 mm long; margins with glandular hairs. **Twigs** *without dense white hairs, with glandular hairs toward tip.* **Fruit** a *black,* juicy drupe.
OCCURRENCE: Uncommon, with scattered distribution.
NOTES: See *E. atropurpureum.*
OTHER NAMES: Curlewberry, *Empetrum eamesii* ssp. *hermaphroditum*

WILDFLOWERS AND LOW SHRUBS 135

ERICACEAE • HEATH FAMILY

Trailing-arbutus • *Epigaea repens*

Evergreen, trailing shrub of dry openings and wooded edges, up to 80 cm long. **Flowers** white or pink, tubular, 5-parted, *with a spicy fragrance, in compact, terminal clusters of 1–8.* **Leaves** alternate, entire, *leathery, with bristly hairs on margins; petioles densely pubescent.*
OCCURRENCE: Occasional, widely distributed.
OTHER NAMES: Mayflower

Creeping Snowberry • *Gaultheria hispidula*

Evergreen, mat-forming, woody shrub of wet woods and bogs, up to 20 cm tall. **Flowers** white, *4-parted.* **Leaves** alternate, *3–10 mm long, bristly beneath.* **Stems** *trailing or prostrate.* **Fruit** a *white* berry, often hidden under leaves and stems, wintergreen-flavored.
OCCURRENCE: Common, widely distributed.
NOTES: See *Vaccinium macrocarpon* and *Linnaea borealis*.
OTHER NAMES: Creeping Spicy-wintergreen, *Chiogenes hispidula, Chiogenes serpyllifolia*

ERICACEAE • HEATH FAMILY

Eastern Spicy-wintergreen • *Gaultheria procumbens*

Evergreen shrub of woods and open areas, 5–20 cm tall. **Flowers** white, *5-parted*. **Leaves** alternate, *1.5–5 cm long, glabrous or with bristles beneath*, ciliate. **Stems** *erect*. **Fruit** a *red*, wintergreen-flavored berry-like capsule.
OCCURRENCE: Common, widely distributed.
OTHER NAMES: Checkerberry, Teaberry

Black Huckleberry • *Gaylussacia baccata*

Deciduous shrub of woods, thickets, and rocky shorelines, up to 2 m tall. **Flowers** *reddish*, 5-parted, urn-shaped, with resin dots on bracts, sepals, and pedicels; pedicels less than 7 mm long. **Leaves** alternate, entire, 3–5.5 cm long, *with small, yellow resin dots beneath* visible with hand lens. **Fruit** a sweet drupe, black when ripe, *with 10 hard nutlets*.
OCCURRENCE: Occasional, widely distributed.
NOTES: See *Vaccinium angustifolium*.

ERICACEAE • HEATH FAMILY

Moss-plant • *Harrimanella hypnoides*

Evergreen, moss-like, matted shrub of alpine and subalpine gullies, ravines, and open slopes, 2–20 cm tall. **Flowers** *white, cup-shaped, nodding,* fused at base with lobes distinct for half their length, tips not recurved; *pedicels deep red.* **Leaves** alternate, *densely crowded, loosely appressed to stem, blades 1–3 mm long and 0.2–0.8 mm wide,* entire to toothed, leaving small pegs on stem when fallen off; petioles absent. **Fruit** a dehiscent capsule, 1.5–2 mm long and 1.5–2 mm wide.
OCCURRENCE: Very rare, documented from only one township in Park.
NOTES: Species is listed as Threatened in Maine.
OTHER NAMES: Moss Bell-heather, *Cassiope hypnoides*

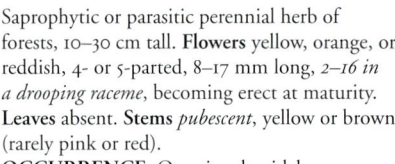

Yellow Pine-sap • *Hypopitys monotropa*

Saprophytic or parasitic perennial herb of forests, 10–30 cm tall. **Flowers** yellow, orange, or reddish, 4- or 5-parted, 8–17 mm long, *2–16 in a drooping raceme,* becoming erect at maturity. **Leaves** absent. **Stems** *pubescent,* yellow or brown (rarely pink or red).
OCCURRENCE: Occasional, widely distributed.
NOTES: See *Monotropa uniflora.*
OTHER NAMES: False Beech-drops, *Monotropa hypopithys*

ERICACEAE • HEATH FAMILY

Sheep American-laurel • *Kalmia angustifolia*

Evergreen shrub of dry openings and wetlands, up to 1 m tall. **Flowers** pink, 5-parted, 0.6–1.2 cm wide, *borne in clusters in axils of previous year's leaves*; pedicels glandular. **Leaves** usually in whorls of 3, 1.5–8 cm long and 0.5–2.5 cm wide, flat, with a distinct petiole, light green beneath, *without purple hairs on midrib beneath*. **Branchlets** *round in cross section on new growth.*
OCCURRENCE: Common, widely distributed.
NOTES: See *K. polifolia*.
OTHER NAMES: Sheep Laurel

Bog American-laurel • *Kalmia polifolia*

Evergreen shrub of bogs, up to 1 m tall. **Flowers** pink, 5-parted, 1–1.6 cm wide, borne in clusters at branch tips; pedicels without glands. **Leaves** *opposite*, 1.5–4.5 cm long and 0.3–1.5 cm wide, sessile, white beneath, *with purple hairs on midrib beneath*; margins distinctly rolled toward lower surface. **Branchlets** *more or less flat in cross section on new growth.*
OCCURRENCE: Occasional, widely distributed.
NOTES: See *Andromeda polifolia* and *K. angustifolia*.
OTHER NAMES: Pale Laurel, Bog Laurel

WILDFLOWERS AND LOW SHRUBS

ERICACEAE • HEATH FAMILY

Alpine-azalea • *Kalmia procumbens*

Evergreen, mat-forming, dwarf shrub of *alpine and subalpine zones*, rarely over 10 cm tall. **Flowers** *pink*, sometimes whitish, 5-parted; petals fused at base, cleft for half their length, lobes spreading. **Leaves** *opposite, crowded, 3–8 mm long and 2–4 mm wide, margins rolled under*, tomentose beneath but without purple hairs on midrib; *petioles absent or 0.1–2 mm long, often flattened against stem*.
OCCURRENCE: Very rare, documented from only one township in Park.
NOTES: Species is listed as Threatened in Maine.
OTHER NAMES: *Azalea procumbens, Loiseleuria procumbens*

One-flowered-shinleaf • *Moneses uniflora*

Evergreen herb of bogs and mossy woods, 3–14 cm tall. **Flowers** white or very light pink, *4- or 5-parted, solitary, terminal, nodding, fragrant*, 12–25 mm wide. **Leaves** *clustered at base*, finely toothed; blades rounded, 6–22 mm long and 5–20 mm wide; petioles 5–10 mm long. **Fruit** a capsule, *erect when mature*.
OCCURRENCE: Uncommon, widely distributed.
NOTES: See *Pyrola minor* and *Orthilia secunda*.
OTHER NAMES: One-flowered Pyrola, Single Delight

ERICACEAE • HEATH FAMILY

One-flowered Indian-pipe • *Monotropa uniflora*

Saprophytic or parasitic perennial herb of rich woods, 5–30 cm tall. **Flowers** *white* or pinkish, *solitary, 8–19 mm long, nodding, becoming erect in fruit*. **Leaves** absent. **Stems** glabrous, white, rarely pink or red, blackening late in season.
OCCURRENCE: Occasional, widely distributed.
NOTES: See *Hypopitys monotropa*.
OTHER NAMES: Ghost-flower

One-sided-shinleaf • *Orthilia secunda*

Evergreen herb of wet woods and bogs, 10–20 cm tall. **Flowers** pale yellow to white, ~5 mm long; *style straight, 2.5–5 mm long, clearly protruding beyond corolla at maturity; sepals finely toothed; inflorescence turned more or less horizontal at flowering time in a one-sided raceme*, becoming erect and less obviously one-sided in fruit; bracts at base of lower pedicels 4–9 mm long and *0.4–1.8 mm wide*; bracts at base of peduncle (among basal leaves) 2–7 per peduncle, 3–9 mm long and *1–2 mm wide*. **Leaves** basal only, shiny above, dull beneath; blades 2–6 cm long and 0.7–3 cm wide, usually longer than petiole.
OCCURRENCE: Common, widely distributed.
NOTES: See *Pyrola chlorantha*, *Pyrola minor*, and *Moneses uniflora*.
OTHER NAMES: One-sided Pyrola, *Pyrola secunda*

ERICACEAE • HEATH FAMILY

Purple Mountain-heath • *Phyllodoce caerulea*

Dwarf, mat-forming, evergreen shrub of *alpine and subalpine* zones, 5–15 cm tall. **Flowers** *purple, 5-parted, urn-shaped, nodding*; petals fused at base, constricted at throat, 1 cm long; pedicels minutely stipitate-glandular, erect. **Leaves** *alternate, needle-shaped, crowded, finely toothed*, 4–10 mm long and 1–1.3 mm wide.
OCCURRENCE: Very rare, documented from only one township in Park.
NOTES: Species is listed as Threatened in Maine.
OTHER NAMES: *Andromeda caerulea*

American Shinleaf • *Pyrola americana*

Perennial herb of forests, 15–30 cm tall. **Flowers** *white*, bilaterally symmetric, borne in a spirally arranged raceme; styles 6–9 mm long, curving down then upturned at end; *sepals 2–4.3 mm long and 1.4–2 mm wide*; bracts at base of lower pedicels 4–12 mm long and *1.2–3 mm wide*; bracts at base of peduncle (among basal leaves) 1–5 per peduncle, 5.5–13 mm long and 2–4 mm wide. **Leaves** basal only, dark green and shiny above; blades round, 16–73 mm long and 16–70 mm wide, *usually same length as petiole*.
OCCURRENCE: Uncommon, widely distributed.
NOTES: See *P. asarifolia* and *P. elliptica*.
OTHER NAMES: Round-leaved Pyrola, *Pyrola rotundifolia*

ERICACEAE • HEATH FAMILY

Pink Shinleaf • *Pyrola asarifolia*

Perennial herb of forests and shorelines, 8–64 cm tall. **Flowers** *pink to reddish purple*, bilaterally symmetric, borne in a spirally arranged raceme; styles 7–10 mm long, curving down then upturned at end; *sepals 1.4–5.5 mm long and 1.3–2.7 mm wide, triangular, overlapping at base*; bracts at base of lower pedicels 3–17 mm long and 1–3.6 mm wide; bracts at base of peduncle (among basal leaves) 1–3 per peduncle, 7–16 mm long and 2.5–5 mm wide. **Leaves** basal only, dark green and shiny above; blades round to wider than long, 24–71 mm long and 13–49 mm wide, toothed or entire, *usually same length as petiole*.
OCCURRENCE: Rare, documented from only northern portion of the Park.
NOTES: See *P. americana* and *P. elliptica*.

Green-flowered Shinleaf • *Pyrola chlorantha*

Perennial herb of woods, 10–25 cm tall. **Flowers** *green-white*, bilaterally symmetric, borne in a spirally arranged raceme; styles 5–7 mm long, curving down then upturned at end; *sepals 1.2–1.7 mm long and 1.3–1.9 mm wide, about as long as wide*; bracts at base of lower pedicels 3–5 mm long and 0.5–0.8 mm wide; bracts at base of peduncle (among basal leaves) absent or 1 per peduncle, 2.8–5 mm long and 0.5–0.8 mm wide. **Leaves** basal only, dark green and shiny above; blades 18–28 mm long and 10–30 mm wide, *usually shorter than petiole*.
OCCURRENCE: Rare, documented from only one township in Park.
NOTES: See *Orthilia secunda*.
OTHER NAMES: Green-flowered Pyrola, *Pyrola virens*

ERICACEAE • HEATH FAMILY

Elliptic-leaved Shinleaf • *Pyrola elliptica*

Perennial herb of rich forests, 15–30 cm tall. **Flowers** white to greenish white, bilaterally symmetric, borne in a spirally arranged raceme; styles 5–7 mm long, curving down then upturned at end; *sepals 1.2–2.1 mm long and 1.2–1.9 mm wide*; bracts at base of lower pedicels 2.5–10 mm long and *0.3–1 mm wide*; bracts at base of peduncle (among basal leaves) 1 or 2 per peduncle, 3.5–10 mm long and *0.5–1.2 mm wide*. **Leaves** basal only, green above, blades 12–80 mm long and 11–57 mm wide, *usually longer than petiole*.
OCCURRENCE: Uncommon, widely distributed.
NOTES: See *P. americana* and *P. asarifolia*.

Little Shinleaf • *Pyrola minor*

Perennial herb of boreal and subalpine forests, 12–21 cm tall. **Flowers** white or pink, *radially symmetric, borne in a spirally arranged raceme, not one-sided*; style straight, 0.5–1.5 mm long, scarcely protruding beyond corolla; sepals *1.3–1.8 mm long and 1.3–1.8 mm wide*; bracts at base of lower pedicels 4–5 mm long and *0.5–1.4 mm wide*; bracts at base of peduncle (among basal leaves) absent or 1 or 2 per peduncle, 4–6.5 mm long and *0.7–2 mm wide*. **Leaves** dark green and shiny above, blades 20–30 mm long and 14–27 mm wide, rounded at apex, *usually longer than petiole*.
OCCURRENCE: Very rare, documented from only one township in Park.
NOTES: See *Moneses uniflora*. Species is listed as Special Concern in Maine.
OTHER NAMES: Mountain Pyrola

ERICACEAE • HEATH FAMILY

Rhodora • *Rhododendron canadense*

Deciduous shrub of bogs and wet thickets, up to 1 m tall.
Flowers *bright magenta, bilaterally symmetric*, fragrant, *emerging before leaves*. **Leaves** alternate, thin, entire, gray-green, 1–8 cm long, with fine hairs beneath; margins usually rolled toward lower surface.
OCCURRENCE: Common, widely distributed.
OTHER NAMES: *Rhodora canadensis*

Labrador-tea • *Rhododendron groenlandicum*

Evergreen shrub of bogs, up to 1.5 m tall. **Flowers** white, *5-parted, radially symmetric*, not fragrant, in terminal clusters; pedicels 1.2–2.5 cm long. **Leaves** alternate, thick, entire, fragrant when crushed, 2–5 cm long, *with dense white to rusty-woolly hairs beneath*; margins rolled toward lower surface.
OCCURRENCE: Common, widely distributed.
OTHER NAMES: *Ledum groenlandicum*

ERICACEAE • HEATH FAMILY

Lapland Rosebay • *Rhododendron lapponicum*

Evergreen shrub of *alpine and subalpine areas,* up to 50 cm tall. **Flowers** *rose to purple, fragrant*; pedicels 2.5–14 mm long, with numerous scales. **Leaves** alternate, thick, 0.4–2 cm long and 0.2–0.7 cm wide, *surfaces with numerous straw-colored or golden scales.*
OCCURRENCE: Very rare, documented from only one township in Park.
NOTES: Species is listed as Threatened in Maine.
OTHER NAMES: *Azalea lapponica*

Common Lowbush Blueberry • *Vaccinium angustifolium*

Deciduous, *colonial* shrub of dry open areas, *0.1–1 m tall.* **Flowers** white to pink, 5-parted, *urn-shaped, 4–7 mm long*; anthers without projecting horns. **Leaves** alternate, thin, not shiny, *finely and regularly toothed with gland-tipped teeth*; mature blades 15–32 mm long and 6–22 mm wide. **Branchlets** with rounded bumps on surface, glabrous or with short hairs. **Fruit** a sweet, blue berry, *4–9 mm wide.*
OCCURRENCE: Common, widely distributed.
NOTES: See *V. boreale* and *V. myrtilloides*.
OTHER NAMES: Low Sweet Blueberry

ERICACEAE • HEATH FAMILY

Northern Blueberry • *Vaccinium boreale*

Tiny, deciduous, *colonial* shrub of high-elevation open areas and bogs, *less than 9 cm tall*. **Flowers** white to pink, *cylindric, 3–4 mm long*; anthers without projecting horns. **Leaves** alternate, thin, not shiny, *finely toothed*; mature blades 8–21 mm long and 2–6 mm wide. **Branchlets** with rounded bumps on surface, glabrous or with short hairs. **Fruit** a blue berry, *3–5 mm wide*.
OCCURRENCE: Rare, with scattered distribution.
NOTES: See *V. angustifolium*. Species is listed as Special Concern in Maine.
OTHER NAMES: Sweet Hurts, Alpine Blueberry

Dwarf Blueberry • *Vaccinium cespitosum*

Deciduous shrub of alpine and subalpine ridges and plateaus, forming dense mats, up to 60 cm tall. **Flowers** pink to white, 5-parted, nodding, urn-shaped, 4–7 mm long, *solitary in leaf axils; calyx continuous with pedicel, lobes absent; anthers with projecting horns*. **Leaves** alternate, thin, not shiny, *with gland-tipped teeth, bright green*; blades 10–30 mm long and 3–12 mm wide. **Branchlets** lacking rounded bumps on surface, with short, fine hairs. **Fruit** a blue to dull black berry, 5–9 mm wide.
OCCURRENCE: Rare, with scattered distribution.

ERICACEAE • HEATH FAMILY

Large Cranberry • *Vaccinium macrocarpon*

Woody, evergreen shrub of bogs, wet areas, and sandy areas, up to 30 cm tall. **Flowers** white to pink, 4-parted; *pedicels 20–30 mm long, with green, leaf-like bracteoles emerging above middle of pedicel, 2–4 mm long and 1–2 mm wide.* **Leaves** alternate, leathery, shiny, 5–17 mm long and 2–8 mm wide, without dark glands beneath, *rounded at tip; margins flat or slightly rolled toward lower surface.* **Fruit** a red berry.
OCCURRENCE: Uncommon, widely distributed.
NOTES: See *V. oxycoccos*.
OTHER NAMES: American Cranberry

Velvet-leaved Blueberry • *Vaccinium myrtilloides*

Deciduous, *colonial* shrub of forests and wet areas, 0.2–1 m tall. **Flowers** pale to mostly white, urn-shaped, 4–6.5 mm long; anthers without projecting horns. **Leaves** alternate, thin, not shiny, *with long hairs, entire or nearly so,* 1–5 cm long; mature blades 8–45 mm long. **Branchlets** with rounded bumps on surface, *densely and evenly covered with long hairs.* **Fruit** a blue berry, 7–10 mm wide.
OCCURRENCE: Common, widely distributed.
NOTES: See *V. angustifolium*.
OTHER NAMES: Sourtop Blueberry

ERICACEAE • HEATH FAMILY

Small Cranberry • *Vaccinium oxycoccos*

Woody, evergreen shrub of bogs, up to 30 cm tall. **Flowers** white to pink, 4-parted; *pedicels 20–30 mm long, with red, scale-like bracteoles at or below middle of pedicel, 0.5–2 mm long and less than 1 mm wide*. **Leaves** alternate, leathery, shiny, 2–10 mm long and 1–3 mm wide, without dark glands beneath, *acute at tip; margins distinctly rolled toward lower surface*. **Fruit** a red berry.
OCCURRENCE: Occasional, widely distributed.
NOTES: See *V. macrocarpon*.
OTHER NAMES: Bog Cranberry

Alpine Bilberry • *Vaccinium uliginosum*

Deciduous shrub of rocky barrens and mountain slopes, forming dense mats, up to 20 cm tall. **Flowers** white to pink, *4-parted*, urn-shaped, *in clusters of 1–4 at tip of branches*; pedicels 0.1–7 mm long; *anthers with projecting horns*. **Leaves** alternate, not shiny, entire, *leathery, blue-green, oval, blunt at tip*; blades 8–14 mm long and 3–7 mm wide. **Branchlets** lacking rounded bumps on surface. **Fruit** a sweet, blue-black berry.
OCCURRENCE: Uncommon, with scattered distribution.
OTHER NAMES: Alpine Blueberry, Bog Blueberry

ERICACEAE • HEATH FAMILY

Mountain Cranberry • *Vaccinium vitis-idaea*

Low, barely woody, evergreen shrub, up to 20 cm tall. **Flowers** pink to reddish, *4-parted, bell-shaped*, in short, drooping racemes; *pedicels less than 5 mm long*. **Leaves** alternate, leathery, shiny, 5–18 mm long and 4–9 mm wide, *with dark glands beneath, held at right angle to stem; margins rolled under*. **Fruit** a *red* berry.
OCCURRENCE: Occasional, widely distributed.
NOTES: Known and prized as lingonberries in Scandinavia. See *Arctostaphylos uva-ursi*.
OTHER NAMES: Rock Cranberry, Cowberry, Lingonberry

ERIOCAULACEAE • PIPEWORT FAMILY ▼

Seven-angled Pipewort • *Eriocaulon aquaticum*

Submerged to emergent, perennial herb of pond and lake shallows, up to 2 m tall. **Flowers** *white to gray, borne in flat-topped, globular clusters* with blackish bracts below. **Leaves** basal only, slender, tapering to tips, 2–8 mm long. **Roots** *conspicuously septate*.
OCCURRENCE: Common, widely distributed.
OTHER NAMES: White-buttons, *Eriocaulon septangulare*

FABACEAE • LEGUME FAMILY ▼

Garden Bird's-foot-trefoil • *Lotus corniculatus*

Perennial herb of disturbed areas, up to 1 m tall. **Flowers** *yellow*, often marked with red, 10–16 mm long, usually 4–8 per umbel. **Leaves** alternate, pinnately divided; leaflets 5, 5–15 mm long, *the lowest 2 resembling stipules*. **Stems** glabrous. **Fruit** a legume, 1.5–3.5 cm long.
OCCURRENCE: Uncommon, with scattered distribution.

Rabbit-foot Clover • *Trifolium arvense*

Annual herb of roadsides and disturbed sites, 10–40 cm tall. **Flowers** *grayish pink or grayish white; inflorescence oblong, appearing furry*. **Leaves** alternate, finely toothed at tip, divided, *with silky hairs; stipules longer than petioles*. **Stems** *soft-pubescent*, branched.
OCCURRENCE: Uncommon, with scattered distribution.
OTHER NAMES: Stone Clover

FABACEAE • LEGUME FAMILY

Palmate Hop Clover • *Trifolium aureum*

Biennial or annual herb of fields and roadsides, 20–50 cm tall. **Flowers** bright yellow, *5–7 mm long*, the uppermost petal usually flat; *heads 10–13 mm wide*. **Leaves** alternate, finely toothed, divided; *terminal leaflet sessile or petiolule approximately same length as minute petiolules of lateral leaflets*.
OCCURRENCE: Occasional, widely distributed.
NOTES: See *T. dubium*.
OTHER NAMES: *Trifolium agrarium*

Lesser Hop Clover • *Trifolium dubium*

Annual herb of fields and disturbed sites, 5–40 cm tall. **Flowers** bright yellow, *2.5–4 mm long*, the uppermost petal usually folded down middle like a roof ridge; *heads less than 10 mm wide*. **Leaves** alternate, finely toothed, divided; *terminal leaflet with a petiolule up to 1 mm long, longer than petiolules of lateral leaflets*.
OCCURRENCE: Rare, documented from only one township in Park.
NOTES: See *T. aureum*.

FABACEAE • LEGUME FAMILY

Alsike Clover • *Trifolium hybridum*

Perennial herb of fields and roadsides, 25–60 cm tall. **Flowers** *bicolored, pink and white*, sweet smelling, 8–10 mm long, *pedicillate*, in dense heads 2–3.5 cm wide. **Leaves** alternate, finely toothed, divided; *stipules lance-ovate with free tips*. **Stems** *erect, not rooting at nodes*.
OCCURRENCE: Uncommon, with scattered distribution.
NOTES: See *T. repens*.

Red Clover • *Trifolium pratense*

Short-lived perennial of fields and roadsides, 15–60 cm tall. **Flowers** *pink-purple*, 12–18 mm long, *sessile*, in *globose heads*. **Leaves** alternate, divided; leaflets finely toothed, blunt, widest near middle, *often blotched with a white "V"*; stipules triangular, bristle-pointed. **Stems** appressed-pubescent.
OCCURRENCE: Occasional, widely distributed.

FABACEAE • LEGUME FAMILY

White Clover • *Trifolium repens*

Perennial herb of lawns and fields, up to 40 cm tall. **Flowers** *uniformly white*, 8–10 mm long, in rounded heads 1–3 cm wide. **Leaves** basal only, divided; *stipules connate, forming sheath with tips separated.* **Stems** *creeping, rooting at nodes.*
OCCURRENCE: Common, widely distributed.
NOTES: See *T. hybridum*.

Bird Vetch • *Vicia cracca*

Perennial vine of fields and roadsides, up to 2 m long. **Flowers** 20–50 per cluster, blue to purple, 8–15 mm long, *in dense, one-sided racemes.* **Leaves** alternate, pinnately divided; *leaflets in 5–11 pairs,* linear to narrowly oblong, 1–3 cm long; stipules entire. **Stems** pubescent. **Fruit** a legume, 2–3 cm long and 5–7 mm wide.
OCCURRENCE: Occasional, widely distributed.
OTHER NAMES: Tufted Vetch, Cow Vetch

GENTIANACEAE • GENTIAN FAMILY ▼

American Spurred-gentian • *Halenia deflexa*

Annual herb of forests, fields, and shorelines, 0.2–0.9 m tall. **Flowers** *purplish or greenish, 4-parted, with conspicuous basal spurs in larger flowers.* **Leaves** *opposite,* entire.
OCCURRENCE: Rare, documented from only northern portion of Park.
OTHER NAMES: *Halenia heterantha, Swertia deflexa, Tetragonanthus deflexus*

GROSSULARIACEAE • GOOSEBERRY FAMILY ▼

Eastern Black Currant • *Ribes americanum*

Deciduous shrub of forests, swamps, and rocky slopes, up to 1.5 m tall. **Flowers** greenish white, 5-parted, usually in clusters of 5 or more in spreading to drooping racemes. **Leaves** alternate, 3- to 5-lobed, *with bright yellow resinous dots on both surfaces.* **Stems** *without bristles or prickles, with bright yellow resinous dots.* **Fruit** a *black,* sweet berry, hanging in drooping clusters, shed without pedicel attached.
OCCURRENCE: Rare, documented from only northern portion of Park.
NOTES: See *R. rubrum* and *R. triste*.

WILDFLOWERS AND LOW SHRUBS 155

GROSSULARIACEAE • GOOSEBERRY FAMILY

Skunk Currant • *Ribes glandulosum*

Deciduous shrub of rocky woods and shorelines, up to 1 m tall. **Flowers** white to pinkish, usually in clusters of 5 or more *in ascending racemes*; ovary with gland-tipped hairs. **Leaves** alternate, 3- to 5-lobed, *with a skunk-like odor when crushed*, without resinous dots beneath. **Stems** *without bristles or prickles*. **Fruit** *a dark red berry with gland-tipped hairs, shed without pedicel attached, in ascending racemes.*
OCCURRENCE: Common, widely distributed.
NOTES: See *R. lacustre*.

Hairy-stemmed Gooseberry • *Ribes hirtellum*

Deciduous shrub of rocky woods and shorelines, up to 1.5 m tall. **Flowers** white or mostly white, 5-parted, *solitary or in clusters of 2–4*. **Leaves** alternate, 3-lobed. **Stems** *usually without spines at nodes; internodes usually with sparse bristles*. **Fruit** *a purple-black berry, glabrous, shed with entire pedicel attached.*
OCCURRENCE: Uncommon, with scattered distribution.
OTHER NAMES: Smooth Gooseberry, Canada Gooseberry

GROSSULARIACEAE • GOOSEBERRY FAMILY

Bristly Swamp Currant • *Ribes lacustre*

Deciduous shrub of wet woods and swamps, up to 2 m tall. **Flowers** mostly orange, wide and flat, usually in clusters of 5 or more *in spreading to drooping racemes; ovary with red, gland-tipped hairs.* **Leaves** alternate, *with 3–5 relatively deep lobes.* **Stems** with spines at nodes and *dense prickles along internodes.* **Fruit** *a black to dark purple berry with red gland-tipped hairs,* shed without pedicel attached, in spreading to drooping racemes.
OCCURRENCE: Common, widely distributed.
NOTES: See *R. glandulosum*.
OTHER NAMES: Bristly Black Currant, Spiny Swamp Currant

Garden Red Currant • **Ribes rubrum*

Deciduous shrub of fields, roadsides, and shorelines, up to 1.5 m tall. **Flowers** *greenish yellow,* usually in clusters of 5 or more, in spreading to drooping racemes; *axis of raceme and pedicels glabrous and without glands.* **Leaves** alternate, *without resinous dots beneath; terminal lobe rounded in outline.* **Stems** *ascending to erect,* without bristles or prickles. **Fruit** a red berry, glabrous, shed without pedicel attached, in spreading to drooping racemes.
OCCURRENCE: Rare, with scattered distribution.
NOTES: See *R. americanum* and *R. triste*.

GROSSULARIACEAE • GOOSEBERRY FAMILY

Swamp Red Currant • *Ribes triste*

Deciduous shrub of forests, talus slopes, and moist habitats, up to 1 m tall. **Flowers** green-purple, usually in clusters of 5 or more, in spreading to drooping racemes; *axis of raceme and pedicels usually with a few short glands, loosely pubescent*. **Leaves** alternate, *without resinous dots beneath; terminal lobe triangular, with more or less straight sides*. **Stems** straggling to ascending, without bristles or prickles. **Fruit** a bright red berry, glabrous, shed without pedicel attached, in spreading to drooping racemes.
OCCURRENCE: Uncommon, with scattered distribution.
NOTES: See *R. americanum* and *R. glandulosum*.

HALORAGACEAE • WATER-MILFOIL FAMILY ▼

Alternate-flowered Water-milfoil • *Myriophyllum alterniflorum*

Perennial, aquatic herb of clear lakes and rivers, often forming dense stands, up to 2.5 m long. **Flowers** with 8 stamens; *distal flowers in emergent spikes in axils of alternate, entire bracts*. **Submerged leaves** finely divided like a feather, in whorls of 3 or 4(–5), *usually less than 1 cm long when mature*, usually with 6–16 narrow segments. **Stems** very slender, much branched. **Winter buds** (**turions**) *absent*. **Fruit** a schizocarp, rarely formed, without wings or ridges.
OCCURRENCE: Uncommon, with scattered distribution.
NOTES: See *M. sibiricum* and *M. verticillatum*.

HALORAGACEAE • WATER-MILFOIL FAMILY ▼

Farwell's Water-milfoil • *Myriophyllum farwellii*

Perennial, aquatic herb of clear lakes and rivers, often forming dense stands, up to 1 m long. **Flowers** with 4 stamens; *distal flowers in axils of submerged leaves.* **Submerged leaves** finely divided like a feather, *usually in whorls of 3(–4) but occasionally alternate or opposite*, 1–2 cm long when mature, with 10–14 narrow segments, very limp when removed from water. **Stems** slender, very delicate, *never emergent.* **Winter buds** (**turions**) *present, dark green, elongate*; developing turions often appearing as reduced dark green to black leaves at base of new shoots. **Fruit** a schizocarp, *1.5–2.5 mm long, with evident rough ridges.*
OCCURRENCE: Rare, with scattered distribution.
NOTES: See *M. humile.*

Northern Water-milfoil • *Myriophyllum sibiricum*

Perennial, aquatic herb of shallow lakes and rivers, often forming dense stands, up to 6 m long. **Flowers** with 8 stamens; *distal flowers in emergent spikes in axils of whorled, entire or toothed bracts.* **Submerged leaves** finely divided like a feather, in whorls of (3–)4, 1.3–3.2 cm long when mature, usually with 6–18 narrow segments. **Stems** whitish or tan, fairly stout. **Winter buds** (**turions**) *present, dark green, cylindrical or tapering toward apex*; developing turions often appearing as reduced black leaves at base of new shoots. **Fruit** a schizocarp, *1.5–2.7 mm long, without wings or ridges.*
OCCURRENCE: Very rare, documented from only one township in Park.
NOTES: See *M. verticillatum* and *M. alterniflorum*

HALORAGACEAE • WATER-MILFOIL FAMILY

Slender Water-milfoil • *Myriophyllum tenellum*

Perennial, aquatic herb of lakes and rivers, often forming dense mats, up to 0.7 m long. **Flowers** with 4 stamens; *distal flowers in emergent spikes in axils of minute, scale-like bracts*. **Submerged leaves** *alternate, scale-like, 0.3–1 mm long*. **Stems** usually simple, unbranched.
OCCURRENCE: Occasional, with scattered distribution.

Whorled Water-milfoil • *Myriophyllum verticillatum*

Perennial, aquatic herb of lakes and rivers, often forming dense stands, up to 3 m long. **Flowers** with 8 stamens; *distal flowers in emergent spikes in axils of whorled, deeply dissected bracts*. **Submerged leaves** finely divided like a feather, in whorls of (3–)4, 1.2–3 cm long when mature, usually with 12–22 narrow segments. **Winter buds (turions)** *present, wider at apex, reddish brown*. **Fruit** a schizocarp, 2–2.7 mm long.
OCCURRENCE: Very rare, documented from only one township in Park.
NOTES: See *M. alterniflorum* and *M. sibiricum*.

HEMEROCALLIDACEAE • DAY-LILY FAMILY ▼

Orange Day-lily • *Hemerocallis fulva*

Perennial herb of fields and roadsides, up to 1.5 m tall.
Flowers orange, funnel-shaped, not fragrant, in terminal clusters. **Leaves** basal only, entire, yellowish green, 2.5–3 cm wide. **Rhizomes** present.
OCCURRENCE: Very rare, documented from only one township in Park.

HYDROCHARITACAE • FROG'S-BIT FAMILY ▼

Free-flowered Waterweed • *Elodea nuttallii*

Perennial, aquatic herb of lakes and streams, up to 1 m long.
Flowers white to purple, 3-parted, on long, thread-like stalks.
Leaves whorled, 3(–4) per node, usually less than 1.7 mm wide, pointed at tip, often with edges folded under.
OCCURRENCE: Uncommon, with scattered distribution.
OTHER NAMES: *Anacharis nuttallii*

HYDROCHARITACAE • FROG'S-BIT FAMILY

Wavy Waternymph • *Najas flexilis*

Annual, aquatic herb of lakes and ponds, up to 1 m long. **Flowers** tiny, developing in leaf axils. **Leaves** opposite, occasionally crowded, narrow, with a broad base where attached to stem, *minutely toothed, up to 3 cm long and 0.2–0.6 mm wide*. **Stems** often heavily branched.
OCCURRENCE: Uncommon, with scattered distribution.
OTHER NAMES: Slender Naiad, Northern Waternymph, *Najas caespitosa*

Tape-grass • *Vallisneria americana*

Perennial, aquatic herb of slow-moving water, up to 2 m long. **Carpellate flowers** solitary, white, 3.5–6.5 mm wide, *raised to surface of water on long, often spiraled, peduncles*. **Leaves** *basal only, thin, ribbon-like*, up to 2 m long and 3–10 mm wide, *with prominent central stripe*.
OCCURRENCE: Historical record, last documented in 1985, from northern portion of Park.
NOTES: After flowers are fertilized, the flower stalk coils downwards and pulls the fruit underwater.
OTHER NAMES: Water-celery, *Vallisneria spiralis*

HYPERICACEAE • ST. JOHN'S-WORT FAMILY ▼

Northern St. John's-wort • *Hypericum boreale*

Perennial herb of bogs and wet meadows, 10–50 cm tall. **Flowers** yellow; petals 2–3 mm long, without black spots; *sepals widest near middle*; stamens 8–15; styles usually 3, distinct nearly to base; inflorescence with opposite branches, *with leaf-like bracts*. **Leaves** opposite, 3–20 mm long and 1–7 mm wide, palmately veined with 3 or 5 longitudinal veins from base. **Fruit** a capsule, *rounded at apex*; styles usually not persistent on capsule; *sepals shorter than capsule*.
OCCURRENCE: Uncommon, with scattered distribution.
NOTES: See *H. canadense, H. majus,* and *H. mutilum.*

Lesser St. John's-wort • *Hypericum canadense*

Annual or perennial herb of moist ground and low thickets, 10–70 cm tall. **Flowers** yellow; *petals 2–3 mm long*, without black spots; *sepals widest below middle*; stamens 12–22; styles 3, distinct nearly to base; *inflorescence with opposite branches, with awl-shaped bracts*. **Leaves** opposite, 10–40 mm long and 0.5–4 mm wide, palmately veined with 1 or 3 longitudinal veins from base. **Stems** *unscented when crushed*. **Fruit** a capsule, *pointed at apex*; styles usually not persistent on capsule; *sepals shorter than capsule*.
OCCURRENCE: Uncommon, with scattered distribution.
NOTES: See *H. boreale, H. gentianoides, H. majus*

HYPERICACEAE • ST. JOHN'S-WORT FAMILY

Pale St. John's-wort • *Hypericum ellipticum*

Perennial herb of marshes, bogs, and river shorelines, 20–50 cm tall. **Flowers** bright yellow; *petals 5–10 mm long*, without black spots; stamens 20–40; *styles connate*; bracts of inflorescence lance-shaped. **Leaves** opposite, 10–40 mm long and 3–20 mm wide, *pinnately veined with one central longitudinal vein and lateral veins clearly visible*. **Fruit** a capsule; *styles persistent in fruit, connate, appearing as a single, straight beak at apex.*
OCCURRENCE: Occasional, widely distributed.
OTHER NAMES: Elliptic St. John's-wort

Fraser's St. John's-wort • *Hypericum fraseri*

Perennial herb of bogs, swamps, and wet woods, 20–80 cm tall. **Flowers** pink; *petals 5–8 mm long; sepals 3–5 mm long when mature; styles 0.5–1 mm long*. **Leaves** opposite, blue-green. **Fruit** a capsule; styles on capsule 0.5–1 mm long.
OCCURRENCE: Common, widely distributed.
NOTES: See *H. virginicum*.
OTHER NAMES: *Hypericum virginicum* var. *fraseri*, *Triadenum fraseri*

HYPERICACEAE • ST. JOHN'S-WORT FAMILY

Orange-grass St. John's-wort • *Hypericum gentianoides*

Annual herb of dry, sandy or rocky soils, 5–30 cm tall. **Flowers** yellow; petals 1.5–4 mm long; sepals 1.5–2.5 mm long; stamens usually 5–10; styles 3, distinct nearly to base; *inflorescence with alternate branches* and minute bracts. **Leaves** opposite, minute, scale-like, 1–3 mm long and *less than 1 mm wide*, 1-veined, appressed to stem. **Stems** *with a citrus smell when crushed*. **Fruit** a capsule, pointed at apex; styles usually not persistent on capsule.
OCCURRENCE: Very rare, documented from only one township in Park.
NOTES: See *H. canadense*.
OTHER NAMES: Pineweed

Greater Canada St. John's-wort • *Hypericum majus*

Perennial herb of wet fields, shorelines, and ditches, 0.1–0.8 m tall. **Flowers** yellow; *petals 3.5–4 mm long*, without black spots; *sepals widest below middle*; stamens 14–21; styles 3, distinct nearly to base; bracts of inflorescence awl-shaped. **Leaves** opposite, 1.5–4 cm long and 0.3–0.9 cm wide, palmately veined with 5 or 7 longitudinal veins from base. **Fruit** a capsule, *pointed at apex*; styles usually not persistent on capsule.
OCCURRENCE: Very rare, documented from only one township in Park.
NOTES: See *H. boreale, H. canadense,* and *H. mutilum*.

HYPERICACEAE • ST. JOHN'S-WORT FAMILY

Dwarf St. John's-wort • *Hypericum mutilum*

Perennial herb of wet fields, swamps, and ditches, 0.1–0.8 m tall. **Flowers** yellow; petals 2–3 mm long, without black spots; *sepals widest near middle*; stamens 5–16; styles 3, distinct nearly to base; *bracts of inflorescence tiny, much smaller than leaves*. **Leaves** opposite, 0.3–3.5 cm long and 0.3–1.5 cm wide, palmately veined with 3 or 5 longitudinal veins from base. **Fruit** a capsule, *rounded at apex*; styles usually not persistent on capsule; *sepals approximately as long as capsule*.
OCCURRENCE: Very rare, documented from only one township in Park.
NOTES: See *H. boreale*, *H. canadense*, and *H. majus*.

Common St. John's-wort • **Hypericum perforatum*

Perennial herb of fields, roadsides, and disturbed sites, 0.3–0.9 m tall. **Flowers** deep yellow; petals 6–10 mm long, *with black dots only along margins*; sepals with abundant black dots and streaks; stamens 20–40 per flower; styles 3, distinct nearly to base; bracts of inflorescence leaf-like. **Leaves** opposite, 0.8–2.6 cm long and 0.1–1 cm wide, with black dots beneath. **Stems** woody at base, *with 2 opposite, raised ridges, extending down from base of nodes*. **Fruit** a capsule; styles usually not persistent on capsule.
OCCURRENCE: Occasional, widely distributed.
NOTES: See *H. punctatum*.

HYPERICACEAE • ST. JOHN'S-WORT FAMILY

Shrubby St. John's-wort • *Hypericum prolificum

Deciduous shrub of low fields, swamps, and thickets, up to 2.5 m tall. **Flowers** yellow; petals 4.5–10 mm long; *stamens numerous, more than 100, sometimes obscuring petals*; bracts of inflorescence leaf-like. **Leaves** opposite, blue-green, the largest 3–7 cm long and 0.4–1.4 cm wide. **Stems** *sharply 2-edged, woody at base, much branched*. **Fruit** a slender capsule, 6–14 mm long and 3–5 mm wide, containing numerous seeds.
OCCURRENCE: Very rare, documented from only one township in Park.
OTHER NAMES: *Hypericum spathulatum*

Spotted St. John's-wort • *Hypericum punctatum*

Perennial herb of fields and roadsides, 0.3–1.2 m tall. **Flowers** pale yellow; petals 4–8 mm long, *with abundant, conspicuous black dots and streaks; sepals with few or no black dots and streaks*; stamens 20–40 per flower; styles 3, distinct nearly to base; bracts of inflorescence leaf-like. **Leaves** opposite, *1–7 cm long and 0.4–2 cm wide, with black dots beneath*. **Stems** woody at base, *round in cross section*. **Fruit** a capsule; styles usually not persistent on capsule.
OCCURRENCE: Historical record, last documented in 1984, from northern portion of Park.
NOTES: See *H. perforatum*.
OTHER NAMES: *Hypericum subpetiolatum*

HYPERICACEAE • ST. JOHN'S-WORT FAMILY

Virginia St. John's-wort • *Hypericum virginicum*

Perennial herb of bogs, swamps, and wet woods, 20–80 cm tall. **Flowers** pink; petals 8–10 mm long; sepals 5–8 mm long when mature; styles 1.8–3 mm long. **Leaves** opposite, blue-green. **Fruit** a capsule; styles on capsule 1.8–3 mm long.
OCCURRENCE: Occasional, with scattered distribution.
NOTES: See *H. fraseri*.
OTHER NAMES: *Triadenum virginicum*

IRIDACEAE • IRIS FAMILY ▼

Blue Iris • *Iris versicolor*

Perennial herb of wetlands and shorelines, up to 1 m tall. **Flowers** purple to blue, with white and yellow at base, ~10 cm wide, with 3 longer, spreading, recurving, outer tepals and 3 distinct inner tepals 2–5 cm long. **Leaves** alternate, *flat, entire, 1–3 cm wide, shorter than flowering stems.* **Fruit** a blunt, 3-lobed capsule, 3.5–5.5 cm long.
OCCURRENCE: Common, widely distributed.
OTHER NAMES: Northern Blue Flag

IRIDACEAE • IRIS FAMILY

Strict Blue-eyed-grass • *Sisyrinchium montanum*

Perennial herb of fields and roadsides, 15–45 cm tall. **Flowers** violet-blue, with 6 tepals, overtopped by a pointed bract; *outer bract green to bronze, rarely suffused with purple, with edges fused for 2–5 mm at base.* **Leaves** basal only, very slender, *the largest 2–3 mm wide.* **Stems** *distinctly winged, 2–3.7 mm wide.* **Fruit** a round capsule, 5–7 mm long.
OCCURRENCE: Occasional, widely distributed.
NOTES: See *S. mucronatum.*
OTHER NAMES: Common Blue-eyed-grass

Needle-tipped Blue-eyed-grass • *Sisyrinchium mucronatum*

Perennial herb of forest edges, fields, and shorelines, 10–50 cm tall. **Flowers** violet-blue, with 6 tepals, overtopped by a pointed bract; *outer bract usually suffused with purple, with edges fused for 1–2.7 mm at base.* **Leaves** basal only, very slender, *the largest 1–2 mm wide.* **Stems** *0.9–2 mm wide, with barely visible wing margins.* **Fruit** a round capsule, 2.5–4 mm long.
OCCURRENCE: Historical record, last documented in 1985, from one township in Park.
NOTES: See *S. montanum.*
OTHER NAMES: *Sisyrinchium intermedium*

LAMIACEAE • MINT FAMILY ▼

Split-lipped Hemp-nettle • *Galeopsis bifida*

Annual herb of fields and disturbed sites, 0.2–1 m tall. **Flowers** pink, 13–15 mm long, *middle lobe of lower lip notched at tip; sepals with teeth 7.5–11 mm long*. **Leaves** opposite, coarsely toothed, *usually wedge-shaped at base*, 3–10 cm long and 1–5 cm wide. **Stems** bristly, soft, swollen below joints, branching.
OCCURRENCE: Occasional, widely distributed.
NOTES: See *G. tetrahit*.
OTHER NAMES: *Galeopsis tetrahit* var. *bifida*

Brittle-stemmed Hemp-nettle • *Galeopsis tetrahit*

Annual herb of fields and disturbed sites, 20–75 cm tall. **Flowers** white or pink, 15–24 mm long, *middle lobe of lower lip without a notch at tip; sepals with teeth 5–8 mm long*. **Leaves** opposite, coarsely toothed, *usually rounded at base*, 3–10 cm long and 1–5 cm wide. **Stems** bristly, soft, swollen below joints, branching.
OCCURRENCE: Rare, with scattered distribution.
NOTES: See *G. bifida*.

LAMIACEAE • MINT FAMILY

Gill-over-the-ground • *Glechoma hederacea*

Perennial herb of meadows, and fields, up to 45 cm tall. **Flowers** blue-purple, 1–2.2 cm long; *upper lobe small, notched at apex.* **Leaves** *opposite, 1–2.5 cm long and 1.5–5 cm wide.* **Stems** *creeping, rooting at nodes.*
OCCURRENCE: Very rare, documented from only southern portion of Park.
OTHER NAMES: *Nepeta hederacea*

American Water-horehound • *Lycopus americanus*

Perennial herb of wetlands, 0.1–1 m tall. **Flowers** white, 4-parted; *sepals 4-lobed, lobes longer than 1 mm and more than 2-times as long as wide.* **Leaves** *opposite, 3–10 cm long; lowest blades toothed to lobed.* **Fruit** *a nutlet, 1–1.5 mm long and 0.6–1 mm wide; persistent sepals distinctly longer than mature nutlets.* **Tubers** *absent.*
OCCURRENCE: Occasional, widely distributed.
NOTES: See *L. uniflorus* and *L. virginicus*.
OTHER NAMES: Cut-leaved Water-horehound

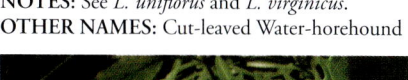

WILDFLOWERS AND LOW SHRUBS

LAMIACEAE • MINT FAMILY

Northern Water-horehound • *Lycopus uniflorus*

Perennial herb of wetlands, 0.1–1 m tall. **Flowers** white, 5-parted; sepals 5-lobed, lobes usually less than 1 mm long and less than 2-times as long as wide. **Leaves** opposite, 2–11 cm long; *lowest blades coarsely toothed but not lobed*. **Fruit** a nutlet, 1.1–1.8 mm long and up to 1.3 mm wide; *persistent sepals shorter than or equaling mature nutlets*. **Tubers** present, white.
OCCURRENCE: Common, widely distributed.
NOTES: See *L. americanus* and *L. virginicus*.
OTHER NAMES: Bugleweed

Virginia Water-horehound • *Lycopus virginicus*

Perennial herb of wetlands, 0.2–1.2 m tall. **Flowers** white, 4-parted; sepals 4-lobed, lobes less than 1 mm long and less than 2-times as long as wide. **Leaves** opposite, 3–12 cm long, occasionally with a reddish tinge; *lowest blades coarsely toothed but not lobed*. **Fruit** a nutlet, 1.3–2 mm long and 0.7–1.3 mm wide; *persistent sepals shorter than or equaling mature nutlets*. **Tubers** occasionally present.
OCCURRENCE: Historical record, last documented in 1984, from northern portion of Park.
NOTES: See *L. americanus* and *L. uniflorus*.

LAMIACEAE • MINT FAMILY

American Wild Mint • *Mentha canadensis*

Perennial herb of wetlands and shorelines, *with a strong peppermint odor*, 0.1–1.2 m tall. **Flowers** white to lavender, 4- or 5-parted. **Leaves** opposite, toothed, 1.5–8.5 cm long and 0.6–4 cm wide; *upper blades tapering to base in a wedge shape; blades gradually decreasing in size from base to apex of stem.*
OCCURRENCE: Occasional, widely distributed.
OTHER NAMES: *Mentha arvensis* ssp. *canadensis*

Common Selfheal • *Prunella vulgaris*

Perennial herb of disturbed sites, 10–80 cm tall. **Flowers** blue or purple, borne in uninterrupted spikes longer than wide; *upper lip 3-lobed; lower lip deeply cleft into 2 narrow segments; middle lobe of lower lip fringed.* **Leaves** opposite, obscurely toothed, 2–9 cm long and 0.7–4 cm wide, distinctly longer than wide.
OCCURRENCE: Common, widely distributed.
OTHER NAMES: Heal-all

LAMIACEAE • MINT FAMILY

Hooded Skullcap • *Scutellaria galericulata*

Perennial herb of shorelines and wet thickets, *usually in open, sunny areas*, 0.1–1 m tall. **Flowers** blue to purple, pubescent, *12–32 mm long, borne in pairs from axils of leaves*. **Leaves** opposite, toothed, 1–8 cm long and 0.3–3 cm wide; *petioles less than 4 mm long*. **Stems** *with downward-curving hairs.*
OCCURRENCE: Occasional, widely distributed.
NOTES: See *S. lateriflora*.
OTHER NAMES: Marsh Skullcap, *Scutellaria epilobiifolia*

Mad Dog Skullcap • **Scutellaria lateriflora*

Perennial herb of shorelines and wet thickets, *usually in shaded areas*, 15–75 cm tall. **Flowers** light blue, *5–7 mm long, borne in lateral, elongate racemes from leaf axils*. **Leaves** opposite, coarsely toothed, 2–12 cm long and 0.6–6 cm wide; *petioles longer than 4 mm*. **Stems** *with hairs curving upward.*
OCCURRENCE: Occasional, widely distributed.
NOTES: See *S. galericulata*.
OTHER NAMES: Blue Skullcap

LAMIACEAE • MINT FAMILY

Marsh Hedge-nettle • *Stachys palustris

Perennial herb of fields and disturbed sites, 0.2–1.2 m tall. **Flowers** *purple or dark pink; sepals with both glandular hairs and non-glandular hairs; non-glandular hairs usually shorter than 1 mm and of similar length as glandular hairs.* **Leaves** opposite, regularly toothed, 3.5–9 cm long and 1–4 cm wide. **Stems** pubescent below inflorescence; *hairs stiff, mostly pointing downwards.*
OCCURRENCE: Very rare, documented from only one township in Park.
NOTES: See *S. pilosa*.

Hairy Hedge-nettle • *Stachys pilosa*

Perennial herb of fields, wetlands, shorelines, and disturbed sites, 0.3–1 m tall. **Flowers** *light pink; sepals with both glandular hairs and non-glandular hairs; non-glandular hairs usually 1.5–3 mm long and conspicuously longer than glandular hairs.* **Leaves** opposite, regularly toothed, 3.5–9 cm long and 1–4 cm wide. **Stems** pubescent below inflorescence; *hairs soft, mostly spreading.*
OCCURRENCE: Rare, with scattered distribution.
NOTES: See *S. palustris*.
OTHER NAMES: *Stachys borealis*

LAMIACEAE • MINT FAMILY

American Germander • *Teucrium canadense*

Perennial herb of shorelines and wet meadows, 0.2–1 m tall. **Flowers** pink-purple, 1–2 cm long, whorled in terminal, cylindric spikes; *upper lip apparently absent*; lower lip appearing 5-lobed; middle lobe of lower lip much larger than others; *stamens distinctly pointed upward, protruding from flower base.* **Leaves** opposite, toothed, *thick*, 5–12 cm long and 1.5–3 cm wide, with crooked hairs on underside; petioles 5–15 mm long. **Stems** pubescent.
OCCURRENCE: Rare, with scattered distribution.
OTHER NAMES: Canada Germander

LENTIBULARIACEAE • BLADDERWORT FAMILY ▼

Horned Bladderwort • *Utricularia cornuta*

Perennial, semi-aquatic herb of wet soils, flowering racemes 12–38 cm tall. **Flowers** bright yellow, 15–25 mm long; *spur 7–12 mm long, longer than lower lip*; scapes with 3–5 flowers. **Leaves** (modified leaf-like branches) alternate, *not often seen, embedded in substrate and difficult to extract*, simple, undivided; bladders embedded in substrate.
OCCURRENCE: Occasional, widely distributed.
NOTES: See other species of *Utricularia*.

LENTIBULARIACEAE • BLADDERWORT FAMILY

Mixed Bladderwort • *Utricularia geminiscapa*

Perennial, aquatic herb of *bogs*, ponds, and sluggish streams, flowering racemes 5–25 cm tall. **Flowers** *bright yellow, not streaked with red*; spur 2–3 mm long; *scapes with 2–5 flowers; submerged stems occasionally with small, solitary flowers without petals*. **Leaves** (modified leaf-like branches) alternate, very numerous, *delicate, limp, 10–20 mm long*, divided into filiform, slightly flattened segments; uppermost segments often bristle-tipped at apex and usually with lateral bristles; *bladders 1–2 mm long*. **Winter buds** *green, 2–5 mm wide*.
OCCURRENCE: Rare, with scattered distribution.
NOTES: See other species of *Utricularia*, particularly *U. vulgaris*.

Creeping Bladderwort • *Utricularia gibba*

Aquatic, annual or perennial herb of ponds, swamps, and sluggish streams, sometimes creeping over wet ground, flowering racemes 2–10 cm tall. **Flowers** yellow, streaked with red; spur 3–4 mm long, shorter than to equal in length to lower lip; *scapes with 1 flower*. **Leaves** (modified leaf-like branches) alternate, very delicate, 5–15 mm long, divided into filiform segments; *uppermost segments without any marginal or apical bristles*; bladders 1–1.6 mm long. **Winter buds** *minute, 1 mm wide*.
OCCURRENCE: Rare, with scattered distribution.
NOTES: See other species of *Utricularia*, particularly *U. minor* and *U. vulgaris*.
OTHER NAMES: Humped Bladderwort

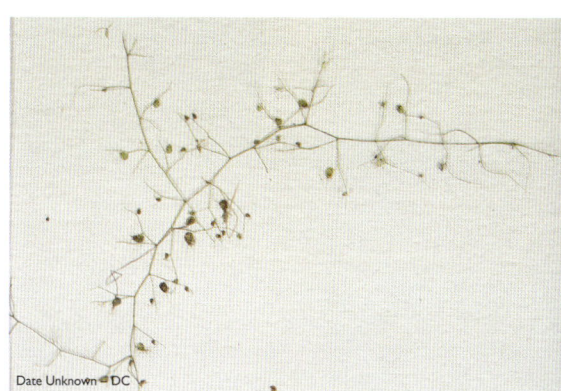

LENTIBULARIACEAE • BLADDERWORT FAMILY

Flat-leaved Bladderwort • *Utricularia intermedia*

Perennial, aquatic herb of bogs and ponds, flowering racemes 9–20 cm tall. **Flowers** *bright yellow, 8–18 mm long*; spur 4–7 mm long, *nearly as long as lower lip and more or less appressed to lower lip*; scapes with 3–5 flowers. **Leaves** (modified leaf-like branches) alternate, 10–20 mm long, *divided into flat segments*; uppermost segments with marginal and apical bristles; branches dimorphic, green ones above substrate with no bladders and white ones buried in substrate with abundant bladders; bladders 2.5–5.5 mm long. **Winter buds** 7–11 mm wide.
OCCURRENCE: Occasional, widely distributed.
NOTES: See other species of *Utricularia*, particularly *U. minor* and *U. vulgaris*.
OTHER NAMES: Northern Bladderwort

Lesser Bladderwort • *Utricularia minor*

Perennial, aquatic herb of shallow water along shorelines, bogs, and meadows, flowering racemes 5–24 cm tall. **Flowers** *pale yellow*, with a purplish tinge or stripes on lower lip, *3.5–8 mm long*; spur 1.5–3.2 mm long, *about half as long as lower lip and projecting at a right angle to it*; scapes with 2–6 flowers. **Leaves** (modified leaf-like branches) alternate, *2–8 mm long, divided into flat segments*; uppermost segments occasionally with marginal and apical bristles; branches dimorphic, green ones above substrate with fewer bladders and white ones buried in substrate with abundant bladders; bladders 0.8–1.5 mm long, *present on green branches*. **Winter buds** 3–4 mm wide.
OCCURRENCE: Very rare, documented from only one township in Park.
NOTES: See other species of *Utricularia*, particularly *U. gibba* and *U. intermedia*.

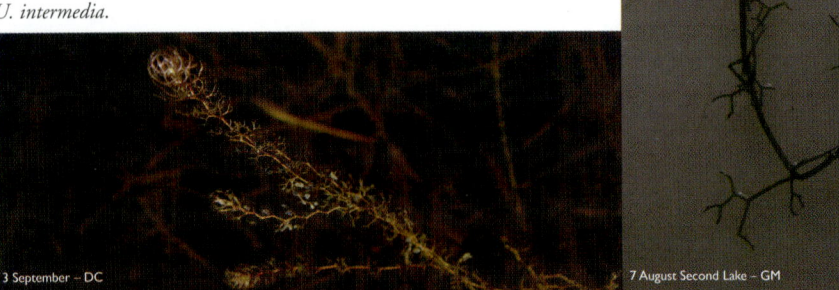

LENTIBULARIACEAE • BLADDERWORT FAMILY

Eastern Purple Bladderwort • *Utricularia purpurea*

Perennial, aquatic herb of ponds, swamps, and sluggish streams, flowering racemes 2.5–10 cm tall. **Flowers** *purple or pink*, with a yellow spot on lower lip; spur 4–6 mm long; *scapes with 2 or 3 flowers*. **Leaves** (modified leaf-like branches) *whorled*, divided into filiform segments, *conspicuously inrolled at tips when young*; bladders borne only at tip of leaf segments, 1.5–2.5 mm long. **Winter buds** poorly developed.
OCCURRENCE: Occasional, with scattered distribution.
NOTES: See other species of *Utricularia*.
OTHER NAMES: Greater Purple Bladderwort

Floating Bladderwort • *Utricularia radiata*

Aquatic, annual or perennial herb of ponds and sluggish streams, flowering racemes 3.5–11 cm tall. **Flowers** *dull yellow*; spur 4.6–6 mm long, shorter than lower lip; scapes with 3 or 4 flowers, *on a peduncle held above water by a whorl of inflated branches*. **Leaves** (modified leaf-like branches) alternate to whorled, divided into filiform segments; bladders 1.5–2 mm long, scattered along branches. **Winter buds** *minute*, up to 1 mm wide.
OCCURRENCE: Rare, with scattered distribution.
NOTES: See other species of *Utricularia*.
OTHER NAMES: Inflated Bladderwort, *Utricularia inflata*

LENTIBULARIACEAE • BLADDERWORT FAMILY

Resupinate Bladderwort • *Utricularia resupinata*

Perennial, aquatic herb of shallow water of ponds, pools, and shorelines, often forming mats, flowering scape 4–20 cm tall. **Flowers** *light purple to rose-pink;* spur 3.5–6 mm long, half as long as lower lip; *scapes with 1 flower.* **Leaves** (modified leaf-like branches) alternate, arising from horizontal stolons, entirely or mostly embedded in substrate and difficult to find, 4–11 cm long; bladders numerous, 0.8–1 mm long.
OCCURRENCE: Very rare, documented from only one township in Park.
NOTES: See other species of *Utricularia*.

Greater Bladderwort • *Utricularia vulgaris*

Perennial, aquatic herb of ponds, swamps, and sluggish streams, flowering racemes 10–40 cm tall. **Flowers** *bright yellow, streaked with red;* spur 4–9 mm long, as long as lower lip; *scapes with 6–15 flowers.* **Leaves** (modified leaf-like branches) alternate, very numerous, *15–90 mm long*, divided into filiform, slightly flattened segments; uppermost segments bristle-tipped at apex and usually with lateral bristles; *bladders 1–5 mm long.* **Winter buds** *grayish, 10–20 mm wide.*
OCCURRENCE: Occasional, widely distributed.
NOTES: See other species of *Utricularia*, particularly *U. geminiscapa* and *U. gibba*.
OTHER NAMES: *Utricularia macrorhiza*

LILIACEAE • LILY FAMILY ▼

Yellow Blue-bead Lily • *Clintonia borealis*

Perennial herb of northern forests, up to 70 cm tall. **Flowers** *light yellow to greenish, ~2 cm wide, 3 to 6 in a drooping cluster*. **Leaves** *basal only*, oval to lance-shaped, 10–30 cm long, *usually 3 or 4 per plant*. **Fruit** *a blue berry*.
OCCURRENCE: Common, widely distributed.
OTHER NAMES: Bluebead-lily

American Trout-lily • *Erythronium americanum*

Perennial herb of hardwood forests and rich slopes, 10–20 cm tall. **Flowers** *yellow, solitary, nodding*. **Leaves** *basal only, green, irregularly mottled*, entire, 8–23 cm long and 1–6 cm wide. **Stems** slender. **Fruit** a 3-valved capsule, with numerous seeds.
OCCURRENCE: Very rare, documented from only one township in Park.

WILDFLOWERS AND LOW SHRUBS 181

LILIACEAE • LILY FAMILY

Canada Lily • *Lilium canadense*

Perennial herb of riparian forests, wet fields, and wetlands, 0.4–2 m tall. **Flowers** yellow or orange-yellow, *nodding, not fragrant, with 6 tepals 5–8 cm long.* **Leaves** *in 6–10 whorls, with 3–12 leaves per whorl,* 4–17 cm long and 1–3.6 cm wide. **Fruit** an erect capsule, with numerous, packed, flat seeds in 2 rows; 3–5.2 cm long and 1.5–2.3 cm wide.
OCCURRENCE: Rare, with scattered distribution.

Indian Cucumber Root • *Medeola virginiana*

Perennial herb of hardwood and mixed forests, 20–90 cm tall. **Flowers** *green to yellow, hanging downward from upper whorl of leaves.* **Leaves** borne in 2 whorls: one with 5–9 leaves at middle stem and another with fewer leaves at top of stem, becoming purple-tinged at base in fruit. **Fruit** an erect cluster of dark purple berries.
OCCURRENCE: Common, widely distributed.

182 THE PLANTS OF BAXTER STATE PARK

LILIACEAE • LILY FAMILY

Clasping-leaved Twistedstalk • *Streptopus amplexifolius*

Perennial herb of forests, shorelines, and alpine areas, 0.5–1.2 m tall. **Flowers** green-yellow, bell-shaped, with 6 tepals; *pedicels glabrous*. **Leaves** alternate, *with 0–6 cilia per cm, the uppermost clasping stem*. **Stems** freely branched, *with glabrous nodes*. **Fruit** a red berry.
OCCURRENCE: Occasional, widely distributed.
NOTES: See *S. lanceolatus*.
OTHER NAMES: *Uvularia amplexifolia*

Lance-leaved Twistedstalk • *Streptopus lanceolatus*

Perennial herb of hardwood and mixed forests, 15–40 cm tall. **Flowers** pink with darker reddish stripes, bell-shaped, with 6 recurving tepals; *pedicels usually pubescent*. **Leaves** alternate, *with 30–50 cilia per cm, the uppermost not clasping stem*. **Stems** simple or occasionally branched, *with pubescent nodes*. **Fruit** a red berry.
OCCURRENCE: Common, widely distributed.
NOTES: See *S. amplexifolius* and *Polygonatum pubescens* (Ruscaceae). The hybrid between *S. amplexifolius* and *S. lanceolatus*, called *S.* ×*oreopolis*, has been found in Baxter, primarily in alpine and subalpine areas. It has 22–40 cilia on the leaves and red-purple to purple tepals.
OTHER NAMES: Rose Twisted-stalk, Rose Mandarin, Rosybells, *Streptopus roseus*

WILDFLOWERS AND LOW SHRUBS

LYTHRACEAE • LOOSESTRIFE FAMILY ▼

Purple Loosestrife • *Lythrum salicaria*

Perennial herb of wetlands and shores, 0.5–1.5 m tall. **Flowers** *magenta, in an elongate, dense spike*; petals 5 or 6, *crinkled*, 7–12 mm long. **Leaves** *opposite or sometimes 3 in a whorl*, 3–10 cm long, somewhat clasping. **Stems** erect, *square*.
OCCURRENCE: Very rare, documented from only one township in Park.
NOTES: See *Chamaenerion angustifolium* (Onagraceae). Species has potential to spread aggressively in some natural habitats.

MALVACEAE • MALLOW FAMILY ▼

Musk Mallow • *Malva moschata*

Perennial herb of disturbed sites, 0.4–1 m tall. **Flowers** *white or pink, 4–5 cm wide*, borne in axils of upper leaves; petals with small indentation at tip. **Leaves** alternate, *deeply 3- to 5-lobed or divided, pubescent beneath*. **Stems** pubescent with simple hairs.
OCCURRENCE: Very rare, documented from only one township in Park.

MELANTHIACEAE • BUNCHFLOWER FAMILY ▼

Red Trillium • *Trillium erectum*

Perennial herb of floodplains, forests, talus and rocky slopes, 20–60 cm tall. **Flowers** *dark red to red-purple*, with 3 petals 2–4.5 cm long and 1–2 cm wide. **Leaves** *sessile*, tapering to a point, in a whorl of 3, often wider than long.
OCCURRENCE: Common, widely distributed.
NOTES: See *T. undulatum*.
OTHER NAMES: Stinking Benjamin, Red Wakerobin, *Trillium purpureum*

Painted Trillium • *Trillium undulatum*

Perennial herb of wet woods and streambanks, 10–50 cm tall. **Flowers** *white with crimson veins near base*, with 3 petals. **Leaves** *on a short petiole*, tapering to a point, in a whorl of 3, often longer than wide.
OCCURRENCE: Common, widely distributed.
NOTES: See *T. erectum*.
OTHER NAMES: Painted Wakerobin, *Trillium erythrocarpum*

MELANTHIACEAE • BUNCHFLOWER FAMILY

American False Hellebore • *Veratrum viride*

Perennial herb of riparian forests, marshes, and swamps, up to 2 m tall. **Flowers** *green to yellowish green, pubescent, with 6 tepals 8–13 mm long and 3–5 mm wide.* **Leaves** alternate, *sessile or nearly so, somewhat clasping, 15–35 cm long and 8–20 cm wide.* **Stems** leafy to tip.
OCCURRENCE: Occasional, widely distributed.
NOTES: All parts of plant are poisonous.

MELASTOMATACEAE • MELASTOMA FAMILY ▼

Virginia Meadow-beauty • *Rhexia virginica*

Perennial herb of damp, sandy, or gravelly meadows and shorelines, 15–60 cm tall. **Flowers** *bright magenta to purple, 4-parted, 2–3 cm wide*; anthers on twisted stalks. **Leaves** opposite, 2–7 cm long, *strongly 3-veined*, finely and sharply toothed, often with bristles on upper surface. **Stems** simple or branched, distinctly square at base.
OCCURRENCE: Very rare, documented from only one township in Park.

MENYANTHACEAE • BUCKBEAN FAMILY ▼

Buck-bean • *Menyanthes trifoliata*

Emergent, perennial herb of peatland margins, 10–50 cm tall. **Flowers** white, *1.5–3 cm wide*, in erect clusters; *petals with prominent white hairs on inner surface*. **Leaves** basal only, *divided, with 3 oval leaflets*. **Fruit** a many-seeded capsule.
OCCURRENCE: Uncommon, widely distributed.
OTHER NAMES: Bogbean

Little Floating-heart • *Nymphoides cordata*

Perennial, aquatic herb of quiet ponds. **Flowers** white, 5-parted, *~1 cm wide*, with tubers growing near or among flowers. **Leaves** basal only, *simple, heart-shaped, 1.5–5 cm wide, floating*, 1 per stem; margins entire or slightly scalloped. **Fruit** an oval-shaped capsule, 3–5 mm long.
OCCURRENCE: Very rare, documented from only one township in Park.

MYRICACEAE • BAYBERRY FAMILY ▼

Sweet-fern • *Comptonia peregrina*

Deciduous shrub of dry, disturbed sites, often on bare, mineral soil, up to 1.5 m tall. **Staminate flowers** borne in hanging catkins. **Carpellate flowers** borne in round, bur-like clusters. **Leaves** alternate, *very aromatic when crushed*, 3–15.5 cm long and 0.3–2.9 cm wide, *pinnately lobed*. **Fruit** a barrel-shaped, hard, smooth nut, *within a globular bur formed from longer persistent bracts*.
OCCURRENCE: Very rare, documented from only one township in Park.
NOTES: This flowering plant is sometimes mistaken for a fern because of its fern-like leaves.
OTHER NAMES: *Myrica peregrina*

Small Bayberry • *Morella caroliniensis*

Deciduous shrub of swamps, bogs, and shorelines, 0.3–2 m tall. **Flowers** borne on old wood mostly below leaf bases. **Leaves** alternate, *dark and shiny green above*, pale beneath, aromatic when crushed, 4–8 cm long, mostly entire, sometimes sparsely toothed on distal half of blade. **Twigs** with *red*, round, bluntly pointed winter buds. **Fruit** *an achene, nearly round, covered with a thick layer of blue-white or grayish wax*, at maturity 3.5–4.5 mm wide.
OCCURRENCE: Rare, with scattered distribution.
NOTES: See *Myrica gale*. Wax on fruit can be used as an additive when making bayberry-scented candles.
OTHER NAMES: Candleberry, *Myrica pensylvanica*

MYRICACEAE • BAYBERRY FAMILY

Sweet Gale • *Myrica gale*

Deciduous shrub of mineral-rich wetlands, 0.3–1.5 m tall. **Carpellate flowers** borne at tip of previous year's twigs. **Leaves** alternate, *dull green above*, 2–6 cm long, wider above middle, *sharply toothed with 1–4 pairs of teeth usually restricted to distal third of blade*, tapering to base. **Twigs** with *dark brown*, oval, winter buds. **Fruit** *a flat achene covered by 2 conspicuous bracts, not encrusted with wax*.
OCCURRENCE: Common, widely distributed.
NOTES: See *Morella caroliniensis*.

MYRSINACEAE • MARLBERRY FAMILY ▼

Starflower • *Lysimachia borealis*

Perennial herb of forests, 10–20 cm tall. **Flowers** *solitary, white*, usually 7-parted, 8–14 mm wide; pedicels 2–5 cm long. **Leaves** *in a solitary whorl at apex of stem*, 4–10 cm long, occasionally with lower scale-like leaves alternating on stem.
OCCURRENCE: Common, widely distributed.
OTHER NAMES: *Trientalis borealis*

WILDFLOWERS AND LOW SHRUBS

MYRSINACEAE • MARLBERRY FAMILY

Fringed Yellow-loosestrife • *Lysimachia ciliata*

Perennial herb of roadsides, wet meadows, wet woods, and shorelines, 0.4–1.3 m tall. **Flowers** yellow, 5-parted, 1.7–2.5 cm wide, *solitary in upper axils*. **Leaves** opposite, 4–15 cm long and 1.5–6.5 cm wide; *petioles 0.5–6 cm long, usually ciliate for entire length*. **Stems** erect. **Fruit** a capsule, 5–7 mm long.
OCCURRENCE: Rare, with scattered distribution.

Creeping Yellow-loosestrife • **Lysimachia nummularia*

Perennial, *often mat-forming*, herb of floodplains, forests, fields, and disturbed sites, up to 60 cm tall. **Flowers** *yellow, 5-parted, solitary in axils*. **Leaves** opposite, semi-evergreen, *with orange to black glandular dots beneath; blades orbicular or nearly so, 1–2.5 cm long; petioles short, distinct, glabrous*. **Stems** *extensively creeping, rooting at nodes*.
OCCURRENCE: Historical record, last documented in 1985.
NOTES: See *Chrysosplenium americanum* (Saxifragaceae). Species has potential to spread aggressively in some natural habitats (particularly floodplains and rich woods).

MYRSINACEAE • MARLBERRY FAMILY

Swamp Yellow-loosestrife • *Lysimachia terrestris*

Perennial herb of wetlands, 0.2–1 m tall. **Flowers** *yellow with black or red streaks, 5-parted, in a terminal raceme 5–30 cm long.* **Leaves** *usually opposite*, with minute translucent or colored dots, 3.5–10 cm long and 4–16 mm wide, *often with small bulbils emerging from axils later in season*.
OCCURRENCE: Common, widely distributed.
NOTES: See *L. thyrsiflora*.
OTHER NAMES: Swamp Candles

Tufted Yellow-loosestrife • *Lysimachia thyrsiflora*

Perennial herb of marshes and bogs, 0.2–1.7 m tall. **Flowers** *yellow, 6- or 7-parted, in short, dense, axillary racemes from middle and lower leaves; petals entire, narrow, 3–5 mm long.* **Leaves** opposite or whorled, with minute colored dots, sometimes pubescent along midvein beneath, 5–15 cm long and 1–5.5 cm wide.
OCCURRENCE: Rare, with scattered distribution.
NOTES: See *L. terrestris*.

WILDFLOWERS AND LOW SHRUBS

NYMPHAEACEAE • WATER-LILY FAMILY ▼

Water-shield • *Brasenia schreberi*

Perennial, aquatic herb of quiet waters. **Flowers** *dull red-purple; petals 12–16 mm long.* **Leaves** *oval*, floating, dark green, 3.5–13.5 cm long and 2–8 cm wide; *petioles attached to center of blade.*
OCCURRENCE: Rare, with scattered distribution.
NOTES: Underwater portions of plant covered with a gelatinous film.
OTHER NAMES: Purple Wen-dock

Small-leaved Pond-lily • *Nuphar microphylla*

Perennial, aquatic herb of lakes, rivers, and streams. **Flowers** yellow, *1–2 cm wide; sepals usually 5 or 6 per flower; anthers 1–3 mm long, shorter than filaments.* **Leaves** floating, *sinus usually at least two-thirds as long as midrib*; petioles flattened on upper side to filiform, *not winged*. **Fruit** a many-seeded, leathery berry, yellow, green, brown, or rarely purple, *deeply constricted below stigmatic disk; stigmatic disk red, 2.5–7 mm in diameter.*
OCCURRENCE: Uncommon, with scattered distribution.
NOTES: See *N. variegata*. A hybrid (*Nuphar ×rubrodisca*), similar to *N. microphylla*, has been documented at Baxter State Park. It can be recognized by having poor fruit set, anthers 3–6 mm long, and a leaf sinus about half as long as midrib.

NYMPHAEACEAE • WATER-LILY FAMILY

Bullhead Pond-lily • *Nuphar variegata*

Perennial herb of lakes and ponds. **Flowers** yellow, *2.5–5 cm wide; sepals usually 6–12 per flower; anthers 4–6 mm long, longer than filaments.* **Leaves** floating, *sinus rarely more than half as long as midrib*; petioles flat, upper portion *often winged.* **Fruit** a many-seeded, leathery berry, usually purple-tinged, *slightly constricted below stigmatic disk; stigmatic disk green, 2–20 mm in diameter.*
OCCURRENCE: Common, widely distributed.
NOTES: See *N. microphylla.*
OTHER NAMES: Spatterdock

17 June Kidney Pond – HC

11 July South Branch Pond – ML

11 July South Branch Pond – ML

White Water-lily • *Nymphaea odorata*

Perennial, aquatic herb of lakes and ponds. **Flowers** *white, sometimes pinkish, 7–20 cm wide, strongly fragrant,* with numerous petals tapering to tip. **Leaves** nearly round, with a narrow cleft, long-stalked, *purplish on underside.*
OCCURRENCE: Uncommon, with scattered distribution.
OTHER NAMES: Fragrant Water-lily

16 August – DC

10 September T2R9 – GM

WILDFLOWERS AND LOW SHRUBS   193

ONAGRACEAE • EVENING-PRIMROSE FAMILY ▼

Narrow-leaved Fireweed • *Chamaenerion angustifolium*

Perennial herb of meadows and fields, commonly becoming established after fire or other disturbance, up to 3 m tall. **Flowers** *magenta, 4-parted, in a crowded, terminal raceme; petals 1–2 cm long*, longer than sepals. **Leaves** alternate, thin, green above, net-veined beneath, the largest 3–20 cm long, the lowest scale-like. **Stems** glabrous. **Fruit** a capsule, 3–8 cm long.
OCCURRENCE: Common, widely distributed.
NOTES: See *Lythrum salicaria* (Lythraceae).
OTHER NAMES: Great Willow-herb, Wickup, *Chamerion angustifolium, Epilobium angustifolium*

Small Enchanter's-nightshade • *Circaea alpina*

Perennial herb of rich forests, 5–25 cm tall. **Flowers** *white, 2-parted; petals deeply cleft, 0.6–2 mm long; pedicels glabrous, erect to ascending*. **Leaves** *opposite*, thin, *2–6 cm long*, sharply and coarsely wavy-margined or toothed. **Stems** weak, *unbranched below flowers*. **Fruit** a capsule, 1.6–2.6 mm long and 0.9–1.2 mm wide, bristly with hooked hairs, *not ridged or grooved*.
OCCURRENCE: Common, widely distributed.
NOTES: See *C. canadensis*. Named for Circe, a sorceress in Homer's Odyssey.
OTHER NAMES: Dwarf Enchanter's-nightshade

ONAGRACEAE • EVENING-PRIMROSE FAMILY

Broad-leaved Enchanter's-nightshade • *Circaea canadensis*

Perennial herb of floodplains and forests, 15–70 cm tall. **Flowers** *white, 2-parted; petals deeply cleft,* 1.6–3.9 mm long; *pedicels pubescent, widely spreading.* **Leaves** *opposite,* 6–12 cm long, shallowly wavy-margined or toothed. **Fruit** a capsule, 2.8–4.5 mm long and 2–2.8 mm wide, *with several strong ridges and grooves.*
OCCURRENCE: Very rare, with scattered distribution.
NOTES: See *C. alpina*.

Pimpernel Willow-herb • *Epilobium anagallidifolium*

Perennial, *mat-forming* herb of alpine and subalpine areas, ridges, and ledges, 2–10 cm tall. **Flowers** *pale purple,* 4-parted; petals notched at tip, *3–4.5 mm long.* **Leaves** at least lowest opposite, *10–25 mm long and 3–7 mm wide; margins entire or weakly toothed.* **Stems** unbranched above base, with hairs in *decurrent lines from leaf bases.* **Fruit** a dehiscent capsule, 23–40 mm long.
OCCURRENCE: Very rare, with scattered distribution.
NOTES: See *E. hornemannii* and *E. lactiflorum*. Species is listed as Endangered in Maine.
OTHER NAMES: *Epilobium alpinum*

ONAGRACEAE • EVENING-PRIMROSE FAMILY

Fringed Willow-herb • *Epilobium ciliatum*

Perennial herb of wet sites, 0.1–1 m tall. **Flowers** pale pink, numerous, 4-parted; petals notched at tip, 1.5–5 mm long. **Leaves** at least lowest opposite, thin, pale green, *scarcely or not veiny on lower surface*, 1–12 cm long and 5–35 mm wide; *margins weakly toothed, flat*. **Stems** 4-angled, *usually sparingly branched above base, with hairs in decurrent lines from leaf bases*. **Fruit** a capsule, 4–10 cm long. **Seeds** *with a whitish tuft of hair*.
OCCURRENCE: Occasional, widely distributed.
NOTES: See. *E. coloratum*.
OTHER NAMES: Glandular Willow-herb, *Epilobium glandulosum*

Eastern Willow-herb • *Epilobium coloratum*

Perennial herb of wet sites, 0.3–1 m tall. **Flowers** pink, numerous, 4-parted; petals notched at tip, 3–5 mm long. **Leaves** at least lowest opposite, *very veiny on lower surface*, 4–15 cm long and 5–25 mm wide; *margins sharply and irregularly toothed, flat*. **Stems** 4-angled, *usually bushy-branched above base, with hairs in decurrent lines from leaf bases*. **Fruit** a capsule, 3–5 cm long. **Seeds** *with a red-brown tuft of hair*.
OCCURRENCE: Uncommon, with scattered distribution.
NOTES: See *E. ciliatum*.
OTHER NAMES: Purple-veined Willow-herb

ONAGRACEAE • EVENING-PRIMROSE FAMILY

Hornemann's Willow-herb • *Epilobium hornemannii*

Perennial, *mat-forming* herb of alpine and subalpine areas, ridges, and ledges, 6–40 cm tall. **Flowers** *pale purple*, 4-parted; petals notched at tip, *4–8 mm long; inflorescence nodding before anthesis.* **Leaves** at least lowest opposite, *19–40 mm long and 8–15 mm wide, margins weakly toothed, flat.* **Stems** *unbranched above base, with hairs in decurrent lines from leaf bases.* **Fruit** a dehiscent capsule, 35–75 mm long.
OCCURRENCE: Rare, with scattered distribution.
NOTES: See *E. anagallidifolium* and *E. lactiflorum*. Species is listed as Endangered in Maine.

White-flowered Willow-herb • *Epilobium lactiflorum*

Perennial, *mat-forming* herb of alpine and subalpine areas, ridges, and ledges, up to 50 cm tall. **Flowers** *white*, 4-parted; petals notched at tip, *2.5–4 mm long.* **Leaves** at least lowest opposite, *19–40 mm long and 8–15 mm wide; margins weakly toothed, flat.* **Stems** *unbranched above base, with hairs in decurrent lines from leaf bases.* **Fruit** a dehiscent capsule, 35–75 mm long.
OCCURRENCE: Historical record, last documented in 1929, from one township in Park.
NOTES: See *E. anagallidifolium* and *E. hornemannii*. Species is listed as historical in Maine.

ONAGRACEAE • EVENING-PRIMROSE FAMILY

Bog Willow-herb • *Epilobium leptophyllum*

Perennial herb of wet sites, 0.2–1 m tall. **Flowers** pink, 4-parted; petals notched at tip, 4–6.5 mm long; inflorescence erect to arching in bud. **Leaves** opposite, 1.5–6 cm long and 1–5 mm wide, *with abundant, minute hairs on upper surface; margins entire or undulate, not toothed,* rolled toward lower surface. **Stems** round, *with hairs not in decurrent lines from leaf bases; hairs appressed.* **Fruit** a pubescent capsule up to 5 cm long.
OCCURRENCE: Uncommon, with scattered distribution.
NOTES: See *E. palustre* and *E. strictum*.
OTHER NAMES: Narrow-leaved Willow-herb

Marsh Willow-herb • *Epilobium palustre*

Perennial herb of bogs and other wetlands, 10–80 cm tall. **Flowers** pink or white, 4-parted; petals notched at tip, 4–8 mm long; inflorescence nodding in bud. **Leaves** opposite, 2–7 cm long and 2–15 mm wide, *with very few or no hairs on upper surface; margins entire or undulate, not toothed,* rolled toward lower surface. **Stems** round, *with hairs not in decurrent lines from leaf bases; hairs appressed.* **Fruit** a capsule, 3–9 cm long.
OCCURRENCE: Rare, with scattered distribution.
NOTES: See *E. leptophyllum* and *E. strictum*.
OTHER NAMES: Swamp Willow-herb, *Epilobium lineare*

ONAGRACEAE • EVENING-PRIMROSE FAMILY

Downy Willow-herb • *Epilobium strictum*

Perennial herb of bogs, fens, marshes, and shorelines, 30–60 cm tall. **Flowers** pink, 4-parted; petals notched at tip, 5–8 mm long. **Leaves** at least lowest opposite, 2–4 cm long and 3–8 mm wide, *with dense, soft pubescence on upper surface; margins entire or undulate, not toothed,* rolled toward lower surface. **Stems** *with hairs not in decurrent lines from leaf bases; hairs dense, divergent, not appressed.*
OCCURRENCE: Very rare, documented from only one township in Park.
NOTES: See *E. leptophyllum* and *E. palustre*.

Common Evening-primrose • *Oenothera biennis*

Biennial or short-lived perennial herb of dry, open, sometimes sandy, soils, up to 2 m tall. **Flowers** yellow, 4-parted; *petals 12–25 mm long and 14–27 mm wide; sepals without a minute, knob-like appendage just below tip.* **Leaves** alternate, *thin,* unevenly small-toothed. **Fruit** a capsule, 1–4 cm long and *3–5 mm wide.*
OCCURRENCE: Rare, with scattered distribution.
NOTES: See *O. parviflora*.
OTHER NAMES: Yellow Evening-primrose

Small-flowered Evening-primrose • *Oenothera parviflora*

Biennial or short-lived perennial herb of gravelly or sandy soils, 10–80 cm tall. **Flowers** yellow, 4-parted; *petals 8–15 mm long and 9–20 mm wide; sepals with a minute, knob-like appendage just below tip*. **Leaves** alternate, *thick,* minutely toothed to nearly entire. **Fruit** a capsule, 1.5–4 cm long and *6–10 mm wide.*

OCCURRENCE: Rare, with scattered distribution.

NOTES: See. *O. biennis.*

OTHER NAMES: Northern Evening-primrose, *Oenothera cruciata*

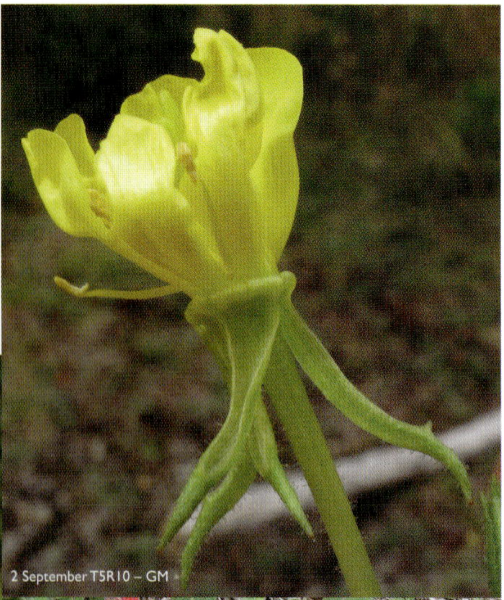
2 September T5R10 – GM

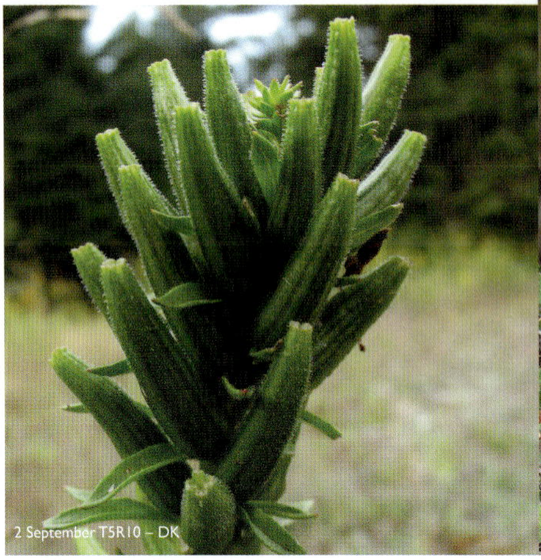
2 September T5R10 – DK

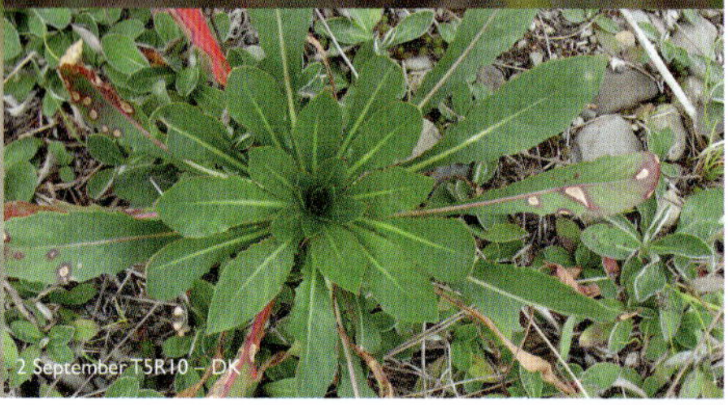
2 September T5R10 – DK

Little Evening-primrose • *Oenothera perennis*

Perennial herb of fields and roadsides, 20–80 cm tall. **Flowers** yellow, 4-parted; petals notched at tip, *3–10 mm long; tip of inflorescence nodding in bud.* **Leaves** alternate, reduced toward top of stem, 1–6 cm long. **Fruit** a 4-winged or quadrangular capsule, *tapering to base*, 8–11 mm long and 3–3.5 mm wide.

OCCURRENCE: Uncommon, widely distributed.

OTHER NAMES: Small Sundrops

26 June T5R10 – GM

29 July – AH

26 June T5R10 – GM

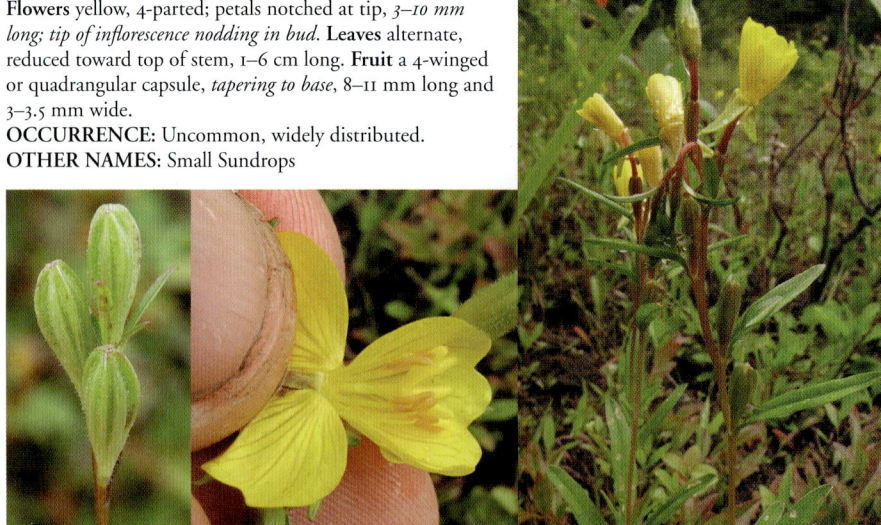

ORCHIDACEAE • ORCHID FAMILY ▼

Dragon's-mouth • *Arethusa bulbosa*

Perennial herb of bogs, 2–40 cm tall. **Flowers** *pink to magenta, terminal, one per stem; floral bracts inconspicuous.* **Leaves** basal only, solitary, 5–23 cm long and 3–12 mm wide, *developing after plant blooms.* **Fruit** an erect capsule, 15–25 mm long.
OCCURRENCE: Very rare, with scattered distribution.
NOTES: See *Calopogon tuberosus* and *Pogonia ophioglossoides*.

Tuberous Grass-pink • *Calopogon tuberosus*

Perennial herb of bogs and wet meadows, 12–85 cm tall. **Flowers** *magenta-pink, ~2.5 cm wide, with a yellow crest on top lip, 2–4 per stem.* **Leaves** basal only, slender, 2–50 cm long and 2–35 mm wide, shorter than height of plant. **Fruit** a capsule, 13–30 mm long and 5–10 mm wide.
OCCURRENCE: Rare, with scattered distribution.
NOTES: See *Arethusa bulbosa* and *Pogonia ophioglossioides*.
OTHER NAMES: *Calopogon pulchellus*

WILDFLOWERS AND LOW SHRUBS 201

ORCHIDACEAE • ORCHID FAMILY

Fairy-slipper • *Calypso bulbosa*

Perennial herb of cedar swamps, 5–22 cm tall. **Flowers** pink to magenta, solitary. **Leaves** *solitary, basal only, entire, leathery, strongly ribbed, with a long petiole, 1.2–6.2 cm long and 3–5.4 cm wide,* produced in autumn, withering by spring shortly after flowering. **Fruit** an erect capsule, 2–3 cm long and 1–1.5 cm wide.
OCCURRENCE: Historical record, last documented in 1984, from northern portion of Park.
OTHER NAMES: Venus' Slipper

Long-bracted Green Orchid • *Coeloglossum viride*

Perennial herb of forested wetlands, fens, meadows, and subalpine ravines, 6–80 cm tall. **Flowers** greenish, often suffused with red or brown, especially in plants of exposed areas, with an inconspicuous spur; *bracts subtending flowers widely spreading, 2 to 6 times as long as flowers.* **Leaves** *alternate, fleshy, 2 or 3 per stem, the largest 2–18 cm long and 1–7 cm wide.* **Fruit** a capsule, 7–14 mm long and 4–5 mm wide.
OCCURRENCE: Historical record, last documented in 1856, from southern portion of Park.
NOTES: See *Platanthera aquilonis.*
OTHER NAMES: Frog Orchid, *Coeloglossum bracteatum, Habenaria bracteata*

ORCHIDACEAE • ORCHID FAMILY

Spotted Coral-root • *Corallorhiza maculata*

Perennial herb of forests, 15–60 cm tall, *plants typically flowering July through August.* **Flowers** tan to yellowish, often red distally, usually spotted with purple; lip white, usually spotted with purple; *sepals with 3 veins; spur present.* **Leaves** absent at flowering. **Stems** usually purple to brown, lacking chlorophyll. **Fruit** a capsule, 9–24 mm long and 5–9 mm wide.
OCCURRENCE: Very rare, with scattered distribution.
NOTES: See *C. trifida*.

Early Coral-root • *Corallorhiza trifida*

Saprophytic, perennial herb of thickets, bogs, and forests, 5–30 cm tall, *plants typically flowering May though June.* **Flowers** light yellow-green, often spotted with purple; lip white, often spotted with purple; *sepals with 1 vein; spur absent.* **Leaves** absent at flowering. **Stems** bright greenish yellow. **Fruit** a capsule, 4.5–15 mm long and 4.3–6 mm wide.
OCCURRENCE: Rare, with scattered distribution.
NOTES: See *C. maculata*.

ORCHIDACEAE • ORCHID FAMILY

Pink Lady's-slipper • *Cypripedium acaule*

Perennial herb of dry woods, 15–60 cm tall. **Flowers** pink, sometimes white, solitary; *lip 3–6 cm long, forming an inflated pouch with a cleft in center.* **Leaves** 2, *basal only, creased at veins forming subtle ridges,* 9–30 cm long and 2.5–15 cm wide.
OCCURRENCE: Occasional, widely distributed.
OTHER NAMES: Moccasin Flower

Yellow Lady's-slipper • *Cypripedium parviflorum*

Perennial herb of evergreen swamps, hardwood forests, and shorelines, 0.7–2 m tall. **Flowers** *pale to deep yellow; lip 15–54 mm long, forming an inflated pouch.* **Leaves** *alternate, 3–5 per stem,* 5–18 cm long and 2–10 cm wide.
OCCURRENCE: Very rare, documented from only one township in Park.
NOTES: The variety documented in Baxter State Park was *C. parviflorum* var. *pubescens*.

ORCHIDACEAE • ORCHID FAMILY

Broad-leaved Helleborine • *Epipactis helleborine*

Perennial herb of dry or wet forests and shorelines, 10–80 cm tall. **Flowers** green, pink, purple, or yellowish, *1–3 cm wide, 15–50 per inflorescence; petals 9–11 mm long and 4–6 mm wide.* **Leaves** *alternate, 3–10 per stem, 4–18 cm long and 1.5–8.5 cm wide, clasping stem.* **Fruit** a capsule, 9–14 mm long.
OCCURRENCE: Occasional, widely distributed.
NOTES: This species is Maine's only nonnative orchid.

Dwarf Rattlesnake-plantain • *Goodyera repens*

Perennial herb of coniferous forests and rich woods, often forming dense mats, 5–20 cm tall. **Flowers** white, pubescent, in a one-sided spike-like raceme or loosely spiraled; *lip with a deep pouch about as deep as long.* **Leaves** basal only, *deep green with distinct, netted, white veins, usually with 5 veins, the largest blades 1.1–3.2 cm long and 0.5–1.8 cm wide.*
OCCURRENCE: Rare, with scattered distribution.
NOTES: See *G. tesselata*.
OTHER NAMES: Lesser Rattlesnake-plantain

WILDFLOWERS AND LOW SHRUBS   205

ORCHIDACEAE • ORCHID FAMILY

Checkered Rattlesnake-plantain • *Goodyera tesselata*

Perennial herb of dry or wet forests, often forming dense mats, 13–32 cm tall. **Flowers** white, pubescent, loosely spiraled on a spike-like raceme; *lip with a shallow pouch longer than deep*. **Leaves** basal only, *dull green with 5–9 pale white veins, the largest blades 2–8 cm long and 1–2.5 cm wide*.
OCCURRENCE: Uncommon, with scattered distribution.
NOTES: See *G. repens*.

Green Adder's-mouth • *Malaxis unifolia*

Perennial herb of swamps, bogs, and shorelines, 3–50 cm tall. **Flowers** *tiny, greenish, many in a compact raceme 1–13 cm long; flower lip 3-toothed at tip*. **Leaves** solitary, entire, bright green, glossy, *borne at center of stem*, 1.6–10 cm long and 0.5–5 cm wide. **Stems** swollen at base.
OCCURRENCE: Rare, with scattered distribution.

ORCHIDACEAE • ORCHID FAMILY

Auricled Twayblade • *Neottia auriculata*

Perennial herb of streambanks and riparian forests, 5–25 cm tall. **Flowers** pale green to blue-green; *lip shallowly cleft into 2 rounded lobes; axis of inflorescence with glandular hairs; pedicels stout, 2.5–5 mm long.* **Leaves** opposite, in a solitary pair on stem, 2.5–6 cm long and 1.4–4.2 cm wide.
OCCURRENCE: Very rare, with scattered distribution.
NOTES: See *N. convallarioides* and *N. cordata*. Species is listed as Threatened in Maine.
OTHER NAMES: *Listera auriculata*

Broad-leaved Twayblade • *Neottia convallarioides*

Perennial herb of floodplains and swamps, 10–30 cm tall. **Flowers** yellowish green, sometimes faintly tinged with purple; *lip shallowly cleft into 2 rounded lobes; axis of inflorescence with glandular hairs; pedicels slender, 2.5–7 mm long.* **Leaves** opposite, in a solitary pair on stem, 2–7 cm long and 1.5–5.8 cm wide.
OCCURRENCE: Historical record, last documented in 1901, from southern portion of Park.
NOTES: See *N. auriculata* and *N. cordata*.
OTHER NAMES: *Listera convallarioides*

WILDFLOWERS AND LOW SHRUBS

ORCHIDACEAE • ORCHID FAMILY

Heart-leaved Twayblade • *Neottia cordata*

Perennial herb of swamps, forests, and glacial basins, 10–25 cm tall. **Flowers** yellow-green, green, or reddish purple; *lip deeply cleft about halfway to its base into 2 sharp-pointed, narrow lobes; axis of inflorescence glabrous; pedicels slender, 2–3 mm long.* **Leaves** opposite, in a solitary pair on stem, 0.9–2 cm long and 0.7–2 cm wide.
OCCURRENCE: Historical record, last documented in 1984, in the Park.
NOTES: See *N. auriculata* and *N. convallarioides*.
OTHER NAMES: *Listera cordata*

North Wind Bog-orchid • *Platanthera aquilonis*

Perennial herb of fens, meadows, forests, and wetlands, 5–60 cm tall. **Flowers** *scentless; lower lip green to yellow-green, 2.5–6 mm long; spur 2–5 mm long.* **Leaves** alternate, 2.7–23 cm long and *0.4–3.7 cm wide.*
OCCURRENCE: Uncommon, with scattered distribution.
NOTES: See *P. dilatata* and *P. huronensis*.

ORCHIDACEAE • ORCHID FAMILY

White-fringed Bog-orchid • *Platanthera blephariglottis*

Perennial herb of bogs, fens, and fields, 0.4–1 m tall. **Flowers** *white; lower lip with a fringed margin; spur 1.5–2.5 cm long.* **Leaves** *alternate, 5–35 cm long and 1–5 cm wide.*
OCCURRENCE: Rare, with scattered distribution.
OTHER NAMES: *Habenaria blephariglottis*

Little Club-spur Bog-orchid • *Platanthera clavellata*

Perennial herb of wet sands and shorelines, 8–47 cm tall. **Flowers** *greenish white, twisted to one side; lower lip obscurely 3-lobed, 3–7 mm long and 3–4 mm wide; spur 7–13 mm long, slender, swollen at tip, curving.* **Leaves** *alternate, entire, with one well developed at or below middle of stem, others much reduced, 3–19 cm long and 0.8–3.5 cm wide.*
OCCURRENCE: Occasional, widely distributed.
OTHER NAMES: Green Woodland Orchis, *Habenaria clavellata*

WILDFLOWERS AND LOW SHRUBS

ORCHIDACEAE • ORCHID FAMILY

White Northern Bog-orchid • *Platanthera dilatata*

Perennial herb of alpine and subalpine areas, shorelines, ledges, and wetlands, up to 1 m tall. **Flowers** with a strong spicy frangrance, suggestive of cloves; lower lip bright white, 4–11 mm long and 2–5 mm wide. **Leaves** alternate, 3.5–32 cm long and *0.3–7 cm wide*.
OCCURRENCE: Uncommon, with scattered distribution.
NOTES: See *P. aquilonis* and *P. huronensis*.
OTHER NAMES: Bog-candle, *Habenaria dilatata*

Greater Purple-fringed Bog-orchid • *Platanthera grandiflora*

Perennial herb of rich forests and wet meadows, 0.5–1.2 m tall. **Flowers** *magenta-pink*, in a loose raceme 3–9 cm wide; lower lip 3-lobed, 10–25 mm long; margin fringed more than one-third deep; *spur 1.5–3.5 cm long, projecting from under lip; opening to spur circular.* **Leaves** alternate, entire, 13–24 cm long and 2.5–9 cm wide.
OCCURRENCE: Rare, with scattered distribution.
NOTES: See *P. psycodes*.
OTHER NAMES: Large Purple-fringed Orchis, *Habenaria fimbriata*, *Habenaria grandiflora*

ORCHIDACEAE • ORCHID FAMILY

Hooker's Bog-orchid • *Platanthera hookeri*

Perennial herb of forests, 18–45 cm tall. **Flowers** *yellowish green; lower lip 8–23 mm long and 1–6 mm wide, tending to turn upward near tip, with entire margins; spur 11–27 mm long, tapering to a rounded apex; scape naked, only rarely with a bract.* **Leaves** *in a basal pair, 5–17 cm long and 4–13 cm wide.*
OCCURRENCE: Rare, with scattered distribution.
NOTES: See *P. obtusata* and *P. orbiculata*.
OTHER NAMES: *Habenaria hookeri*

Lake Huron Green Bog-orchid • *Platanthera huronensis*

Perennial herb of wet meadows, marshes, and shorelines, up to 1 m tall. **Flowers** *with a sweet fragrance; lower lip greenish white, 5–12 mm long; spur 4–12 mm long.* **Leaves** *alternate, 5–30 cm long and 0.6–6 cm wide.*
OCCURRENCE: Very rare, with scattered distribution.
NOTES: See *P. aquilonis* and *P. dilatata*.
OTHER NAMES: *Habenaria huronensis*

WILDFLOWERS AND LOW SHRUBS

ORCHIDACEAE • ORCHID FAMILY

Blunt-leaved Bog-orchid • *Platanthera obtusata*

Perennial herb of fens, forests, fields, and swamps, 10–30 cm tall. **Flowers** greenish white to yellowish green; lower lip unlobed, 2.5–8 mm long and less than 2 mm wide; spur 5–9 mm long, tapering to tip. **Leaves** *basal only, solitary or in a pair, ascending*; 3.5–15 cm long and 0.8–5 cm wide.
OCCURRENCE: Very rare, with scattered distribution.
NOTES: See *P. hookeri* and *P. orbiculata*.
OTHER NAMES: *Habenaria obtusata*

Round-leaved Bog-orchid • *Platanthera orbiculata*

Perennial herb of forests and swamps, 17–60 cm tall. **Flowers** *greenish white; lower lip 7–17 mm long and 1–2.5 mm wide, tending to turn downward near tip, with entire margins; spur 14–27 mm long, parallel-sided to club-shaped at tip; scape with 1–6 bracts between basal leaves and inflorescence.* **Leaves** in a basal pair, 5–21 cm long and 3–22 cm wide.
OCCURRENCE: Rare, with scattered distribution.
NOTES: See *P. hookeri* and *P. obtusata*.
OTHER NAMES: *Habenaria orbiculata*

ORCHIDACEAE • ORCHID FAMILY

Lesser Purple-fringed Bog-orchid • *Platanthera psycodes*

Perennial herb of wet woods, meadows, and shorelines, 0.2–1 m tall. **Flowers** *magenta-pink, ~2 cm long*, in a loose, terminal raceme 2.5–4 cm wide; lower lip 3-lobed, 5–13 mm long; margin fringed less than one-third deep; *spur 1.2–1.8 cm long, projecting from under lip; opening to spur oblong*. **Leaves** alternate, *2–6 per stem, wider nearer tip than middle*, 5–22 cm long and 1.5–7 cm wide.
OCCURRENCE: Uncommon, with scattered distribution.
NOTES: See *P. grandiflora*.
OTHER NAMES: Small Purple-fringed Orchis, *Habenaria psycodes*

Rose Pogonia • *Pogonia ophioglossoides*

Perennial herb of bogs, 8–35 cm tall. **Flowers** *pink, terminal, usually solitary, ~2 cm long, with a yellow, bearded lip; floral bract conspicuous, leaf-like*, 7–37 mm long and 2–12 mm wide. **Leaves** *solitary*, 1.4–12 cm long and 4–32 mm wide, clasping middle of stem, *present at flowering time*. **Fruit** an erect capsule, 14–30 mm long and 4–8 mm wide.
OCCURRENCE: Uncommon, widely distributed.
NOTES: See *Arethusa bulbosa* and *Calopogon tuberosus*.
OTHER NAMES: Snakemouth

ORCHIDACEAE • ORCHID FAMILY

Nodding Ladies'-tresses • *Spiranthes cernua*

Perennial herb of wet to dry meadows, marshes, fens, and shorelines, 10–50 cm tall, *flowering late summer and fall.* **Flowers** *whitish with green-yellow markings on lip; lower lip oblong, not constricted below tip; lateral sepals free, not united with dorsal sepal and lateral petals; rachis moderately to densely pubescent.* **Leaves** alternate, the largest up to 26 cm long and 2 cm wide.
OCCURRENCE: Very rare, documented from only northern portion of Park.
NOTES: See *S. romanzoffiana*.

Hooded Ladies'-tresses • *Spiranthes romanzoffiana*

Perennial herb of wet meadows, marshes, fens, and shorelines, 8–55 cm tall, *flowering early summer through late summer.* **Flowers** *white to yellowish white, very fragrant; lower lip 4.8–10.2 mm long and 1.6–6.8 mm wide, strongly constricted below expanded tip; lateral sepals united for at least half their length with dorsal sepal and lateral petals; rachis glabrous to very sparsely pubescent.* **Leaves** alternate, the largest up to 26 cm long and 3 cm wide.
OCCURRENCE: Very rare, documented from only one township in Park.
NOTES: See *S. cernua*.

OROBANCHACEAE • BROOM-RAPE FAMILY ▼

Slender-leaved Agalinis • *Agalinis tenuifolia*

Annual herb of dry fields and roadsides, 20–80 cm tall. **Flowers** *light purple to pink, 10–15 mm long, in a terminal raceme; petals glabrous on inside surface; pedicels 0.7–2.7 cm long.* **Leaves** *opposite,* 20–50 mm long and 1–6 mm wide. **Stems** bushy branched. **Fruit** a capsule, 2–5.5 mm long.
OCCURRENCE: Very rare, documented from only one township in Park.
OTHER NAMES: *Gerardia tenuifolia*

Northern Painted-cup • *Castilleja septentrionalis*

Perennial herb of alpine ravines, ledges, floodplains, and shorelines, 20–60 cm tall. **Flowers** *irregular, 2-lipped, yellowish, surrounded by cream to purplish bracts; inflorescence villous.* **Leaves** alternate, entire, 3–10 cm long.
OCCURRENCE: Rare, with scattered distribution.
NOTES: Species is listed as Special Concern in Maine.

OROBANCHACEAE • BROOM-RAPE FAMILY

Beech-drops • *Epifagus virginiana*

Perennial parasitic herb of dry beech forests, 15–45 cm tall. **Flowers** *whitish with pink to purple-brown blotches or stripes, 2.2–3.2 mm long; pedicels minutely glandular-pubescent on side near stem.* **Leaves** alternate, without chlorophyll, *reduced to scales, 2–4 mm long.* **Stems** *reddish purple to yellow-brown with fine purple lines, persisting to next season.*
OCCURRENCE: Uncommon, with scattered distribution.

Common Eyebright • *Euphrasia nemorosa*

Annual herb of fields and roadsides, 10–45 cm tall. **Flowers** white with purple lines, *5–8 mm long; subtending leafy bracts pubescent, rounded to truncate at base, ascending to spreading, with bristle-tipped, spreading teeth; sepals densely pubescent throughout, with somewhat bristle-tipped lobes.* **Leaves** at least lowest opposite, *pubescent,* coarsely toothed.
OCCURRENCE: Rare, with scattered distribution.
NOTES: See *E. stricta*.
OTHER NAMES: *Euphrasia americana*

OROBANCHACEAE • BROOM-RAPE FAMILY

Oakes' Eyebright • *Euphrasia oakesii*

Annual herb of *alpine and subalpine zones*, ridges, and ledges, 3–10 cm tall. **Flowers** *pale violet with purple lines, 2.5–4.5 mm long; lower lip 3-lobed; inflorescence densely crowded*. **Leaves** at least lowest opposite, glabrous to pubescent. **Stems** retrorse-pubescent.
OCCURRENCE: Very rare, documented from only one township in Park.
NOTES: Species is listed as Endangered in Maine.
OTHER NAMES: *Euphrasia williamsii*

Strict Eyebright • **Euphrasia stricta*

Annual herb of fields and roadsides, 15–40 cm tall. **Flowers** white with purple lines, 5–8 mm long; *subtending leafy bracts glabrous or at most scabrous on veins, tapering to base, erect to ascending, with bristle-tipped, ascending teeth; sepals glabrous or at most scabrous on veins, with lobes tapering into prolonged bristle-tips*. **Leaves** at least lowest opposite, *glabrous*, coarsely toothed.
OCCURRENCE: Very rare, documented from only northern portion of Park.
NOTES: See *E. nemorosa*.
OTHER NAMES: *Euphrasia officinalis*

American Cow-wheat • *Melampyrum lineare*

Annual, parasitic herb of dry forests, 10–30 cm tall. **Flowers** *white with yellow tip*, strongly 2-lipped, 1–2 cm long, glabrous, *on short stalks in axils of upper leaves*. **Leaves** opposite, *entire, the lowest lance-shaped, uppermost with a few pointed teeth at base*; petioles short. **Fruit** a laterally flattened capsule.
OCCURRENCE: Occasional, widely distributed.

1 August Frost Pond Trail – MD

25 June Freezeout Trail – HC

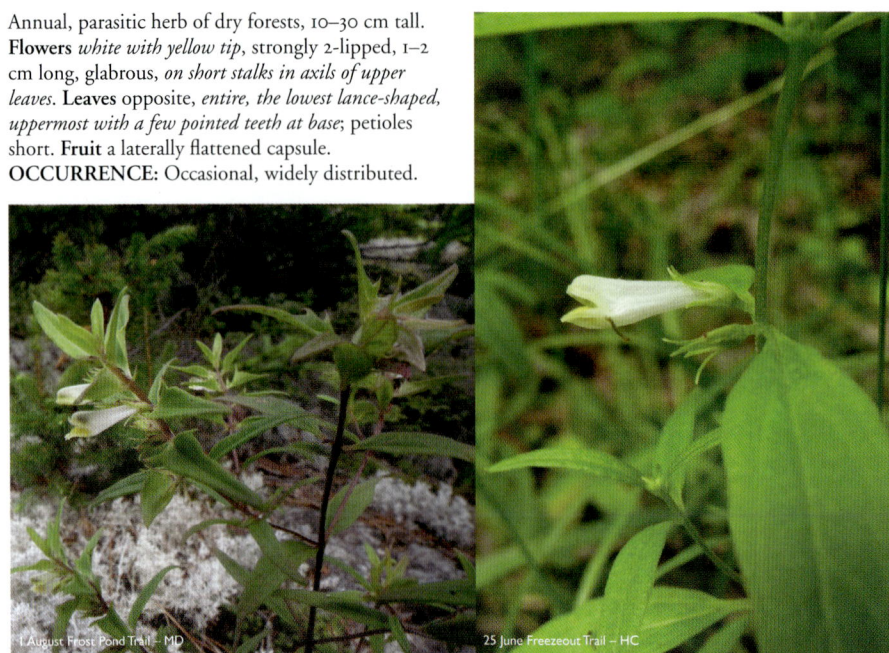

One-flowered Broom-rape • *Orobanche uniflora*

Parasitic, perennial herb of fields and forests, 1–3 cm tall. **Flowers** *white to violet, 5-parted, 10–25 mm long; sepals glandular-pubescent.* **Leaves** *absent.* **Fruit** a capsule, 6–12 mm long and 5–8 mm wide.
OCCURRENCE: Very rare, documented from only one township in Park.
OTHER NAMES: One-flowered Cancerroot

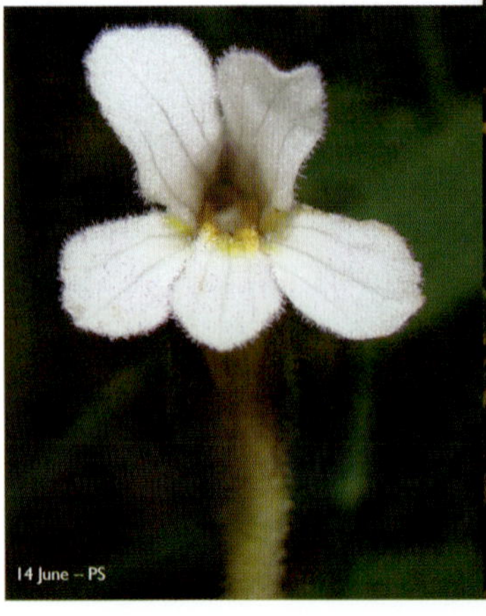
14 June – PS

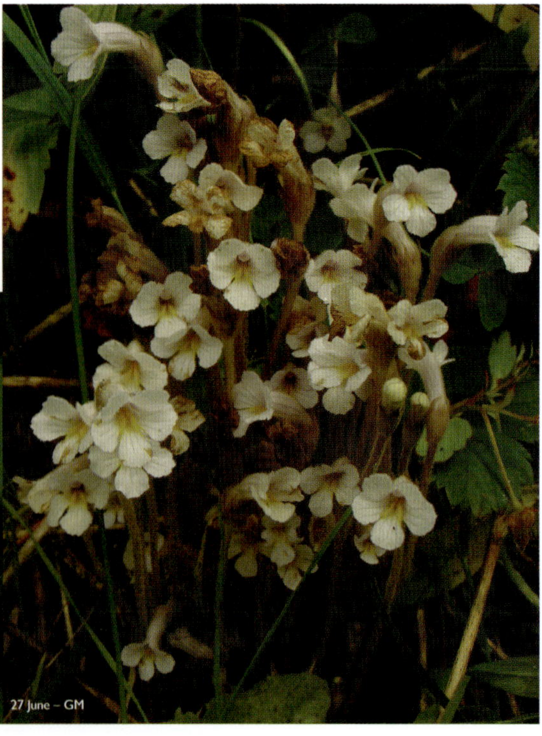
27 June – GM

OROBANCHACEAE • BROOM-RAPE FAMILY

Little Yellow-rattle • *Rhinanthus minor*

Parasitic, annual herb of dry, open sites, 10–60 cm tall. **Flowers** *yellow, 1–2 cm long; sepals bright green and partially inflated in flower.* **Leaves** *opposite, toothed, 2.1- to 3.7-times as long as wide.* **Fruit** a flat capsule with large seeds that rattle inside.
OCCURRENCE: Historical record, last documented in 1903, from southern portion of Park.
NOTES: The plant documented was the native alpine subspecies; the nonnative subspecies has not been documented in the Park.
OTHER NAMES: *Rhinanthus crista-galli*

OXALIDACEAE • WOOD SORREL FAMILY ▼

Northern Wood Sorrel • *Oxalis montana*

Creeping, perennial herb of wet, shady forests, 5–15 cm tall. **Flowers** *white with pink veins, ~2 cm wide, 1 per stalk.* **Leaves** *basal only,* divided, with 3 heart-shaped leaflets. **Fruit** a capsule, 2–4 mm long, splitting down sides.
OCCURRENCE: Common, widely distributed.

OXALIDACEAE • WOOD SORREL FAMILY

Common Yellow Wood Sorrel • *Oxalis stricta*

Annual or short-lived perennial herb of dry, open sites and roadsides, 7–60 cm tall. **Flowers** *yellow, 5-parted, 1–2 cm wide, up to 9 per stalk; petals 4–9 mm long.* **Leaves** alternate, divided; leaflets 3, 1–2 cm wide. **Stems** pubescent.
OCCURRENCE: Occasional, widely distributed.
OTHER NAMES: *Oxalis europaea*

PAPAVERACEAE • POPPY FAMILY ▼

Pink-corydalis • *Capnoides sempervirens*

Annual or biennial herb of dry, rocky outcrops, 30–80 cm tall. **Flowers** *pink with yellow tips, 10–17 mm long,* bilaterally symmetric, with a basal spur. **Leaves** alternate, *finely divided, light green or gray-green.* **Fruit** an erect, thin, cylindrical capsule, 3–5 cm long and up to 2 mm wide.
OCCURRENCE: Uncommon, widely distributed.
OTHER NAMES: Rock-harlequin, *Corydalis sempervirens*

PAPAVERACEAE • POPPY FAMILY

Greater Celandine • *Chelidonium majus*

Biennial or perennial herb of wet, disturbed areas, 30–90 cm tall. **Flowers** *yellow*, 20–25 mm wide, with 4 petals and 2 sepals, in small umbels. **Leaves** alternate, green with a *white cast*, pinnately divided; *leaflets irregularly lobed*. **Stems** *with bright yellow or orange latex*.
OCCURRENCE: Very rare, documented from only one township in Park.
NOTES: All parts of plant poisonous.

Dutchman's-breeches • *Dicentra cucullaria*

Perennial herb of floodplains, forests, talus, and rocky slopes, 10–30 cm tall. **Flowers** *white suffused with yellow at apex, 15–20 mm long, on pendulous pedicels, appearing like pants hung out to dry*. **Leaves** basal only, *divided into narrow segments*, 5–12 cm long. **Fruit** a capsule, 6–14 mm long and 2–3 mm wide.
OCCURRENCE: Very rare, documented from only one township in Park.

PAPAVERACEAE • POPPY FAMILY

Blood-root • *Sanguinaria canadensis*

Perennial herb of floodplains and hardwood forests, 10–40 cm tall. **Flowers** *white to pink, solitary, 2–5 cm wide; petals 8–12; sepals 2.* **Leaves** *basal only, solitary, 3- to 9-lobed,* up to 20 cm wide. **Fruit** a capsule, 2.5–6 cm long and 0.5–1 cm wide.
OCCURRENCE: Rare, with scattered distribution.
NOTES: Plant with red latex.

PHRYMACEAE • LOPSEED FAMILY ▼

Allegheny Monkey-flower • *Mimulus ringens*

Perennial herb of marshes, wetlands, and shorelines, 0.2–1.5 m tall. **Flowers** *violet or pinkish, irregular, 2–3.5 cm long; pedicels 2–4.5 cm long.* **Leaves** *opposite, sessile, toothed,* 5–15 cm long and up to 6 cm wide. **Stems** 4-angled. **Fruit** a capsule, 10–13 mm long.
OCCURRENCE: Uncommon, with scattered distribution.

PLANTAGINACEAE • SNAPDRAGON FAMILY ▼

Greater Water-starwort • *Callitriche heterophylla*

Aquatic herb of muddy ponds, 10–20 cm long. **Flowers** without petals or sepals, growing singly or in pairs in leaf axils. **Leaves** opposite, entire, crowded in floating rosettes. **Fruit** *a capsule, 0.6–1.3 mm long and 0.6–1.3 mm wide, usually as wide as long; apical margins without wings or sharp edges, without an evident groove dividing capsule.*
OCCURRENCE: Very rare, with scattered distribution.
NOTES: See *C. palustris*.

Vernal Water-starwort • *Callitriche palustris*

Aquatic herb of springs, small streams, and shorelines, 10–20 cm long. **Flowers** without petals or sepals, growing singly or in pairs in leaf axils. **Leaves** opposite, entire, crowded, in floating rosettes. **Fruit** *a capsule, 1–1.4 mm long and 0.8–1.2 mm wide, longer than wide; apical margins with wings, with an evident groove dividing capsule.*
OCCURRENCE: Uncommon, widely distributed.
NOTES: See *C. heterophylla*.

PLANTAGINACEAE • SNAPDRAGON FAMILY

White Turtlehead • *Chelone glabra*

Perennial herb of brooks and wet thickets, 0.4–2 m tall. **Flowers** *white*, sometimes with pink, purple, or green-yellow on top, in spikes 3–8 cm long; *lower lip with white or pale yellow hairs*. **Leaves** opposite, toothed. **Fruit** a capsule, 1–1.5 cm long.
OCCURRENCE: Uncommon, widely distributed.
OTHER NAMES: Balmony

Golden Hedge-hyssop • *Gratiola aurea*

Perennial herb of shorelines, 10–30 cm tall. **Flowers** *bright yellow, irregular, 12–16 mm long, solitary from axils*; upper lip 2-lobed and lower lip 3-lobed; sepals 5; pedicels 5–15 mm long. **Leaves** *opposite*, entire or barely toothed, 10–25 mm long. **Fruit** *a capsule, 2–3 mm long*.
OCCURRENCE: Rare, with scattered distribution.
OTHER NAMES: *Gratiola lutea*

PLANTAGINACEAE • SNAPDRAGON FAMILY

Clammy Hedge-hyssop • *Gratiola neglecta*

Annual herb of wetlands and shorelines, 3–40 cm tall. **Flowers** *yellowish to whitish, irregular, solitary from axils*; upper lip nearly 2-lobed and lower lip 3-lobed; sepals 5; *pedicels 1–4.5 cm long*. **Leaves** *opposite*, entire to toothed, 10–60 mm long. **Fruit** a capsule, 3–5 mm long.
OCCURRENCE: Rare, with scattered distribution.

Common Mare's-tail • *Hippuris vulgaris*

Perennial, aquatic herb of lakes, rivers, and wetlands, 0.2–1 m tall. **Flowers** tiny, sessile, in upper leaf axils. **Leaves** *in whorls of 6–12, entire, sessile*; emersed leaves thick, firm, 1–3 cm long and 1–3 mm wide, with blunt, hard tips. **Fruit** a capsule, 1.7–2.5 mm long.
OCCURRENCE: Rare, with scattered distribution.

WILDFLOWERS AND LOW SHRUBS 225

PLANTAGINACEAE • SNAPDRAGON FAMILY

Butter-and-eggs Toadflax • *Linaria vulgaris*

Perennial herb of roadsides and disturbed sites, 0.3–1 m tall. **Flowers** yellow with orange at base, 1.5–3 cm long, in a dense, terminal raceme; *spur straight; upper lip 2-lobed and lower lip 3-parted.* **Leaves** *alternate above*, whorled below, *blue-green*, 3–8 cm long and 2–4 mm wide. **Fruit** a capsule, 8–12 mm long.
OCCURRENCE: Rare, with scattered distribution.
OTHER NAMES: Wild Snapdragon

Common Plantain • *Plantago major*

Annual or perennial herb of roadsides and lawns, 1–50 cm tall. **Flowers** dirty-white, in dense spikes usually 5–30 cm long; *sepals blunt to rounded at tip, less than twice as long as wide.* **Leaves** basal only, somewhat fleshy, sometimes toothed, 4–18 cm long and 1.5–11 cm wide; *petioles long, usually green or pale pink and pubescent at base.* **Fruit** a capsule, 2.5–4 mm long, with 4–13 seeds, opening at or slightly below middle.
OCCURRENCE: Occasional, widely distributed.
NOTES: See *P. rugelii*.

PLANTAGINACEAE • SNAPDRAGON FAMILY

Rugel's Plantain • *Plantago rugelii*

Perennial herb of fields, shorelines, and disturbed sites, up to 50 cm tall. **Flowers** dirty-white, in dense spikes up to 50 cm long; *sepals sharply pointed at tip, about 2 or 3 times as long as wide*. **Leaves** basal only, somewhat fleshy, usually with a few marginal teeth, 5–33 cm long and 3–15 cm wide; *petioles usually deep red-purple and glabrous at base*. **Fruit** a capsule, 4–6 mm long, with 4–9 seeds, opening near base.
OCCURRENCE: Rare, with scattered distribution.
NOTES: See *P. major*.

American Speedwell • *Veronica americana*

Perennial herb of wetlands, up to 70 cm tall. **Flowers** blue or violet, somewhat irregular, 3–10 mm wide, *in racemes from leaf axils; pedicels 5–20 mm long*. **Leaves** opposite, toothed to subentire, *glabrous*, broadest near base, 1–10 cm long and 0.5–3.5 cm wide. **Stems** *glabrous, rooting at lower nodes*. **Fruit** a capsule, up to 3 mm long and 3 mm wide.
OCCURRENCE: Uncommon, with scattered distribution.

PLANTAGINACEAE • SNAPDRAGON FAMILY

Common Speedwell • *Veronica officinalis*

Mat-forming, perennial herb of dry fields and open forests, 5–25 cm tall. **Flowers** light blue to lavender, with darker lines, somewhat irregular, 3–8 mm wide, *in dense racemes from leaf axils; pedicels 1–2.5 mm long.* **Leaves** opposite, shallowly toothed, thick, *pubescent,* 1.5–5 cm long and 0.5–3.5 cm wide, *the uppermost on short, distinct petioles.* **Stems** *pubescent, trailing, rooting at nodes, with ascending tip.* **Fruit** a capsule, 3–4 mm long and 3.8–5 mm wide.
OCCURRENCE: Common, widely distributed.

Narrow-leaved Speedwell • *Veronica scutellata*

Perennial herb of sandy shorelines and swamps, 10–70 cm tall. **Flowers** bluish to lavender, somewhat irregular, 4–10 mm wide, *in racemes from leaf axils; pedicels 6–17 mm long.* **Leaves** opposite, *sessile, glabrous, long and narrow, entire or remotely toothed,* 1.5–9 cm long and 1–10 mm wide. **Fruit** *a flat capsule,* 2.5–3.5 mm long and 4–5 mm wide, *wider than long.*
OCCURRENCE: Occasional, widely distributed.
OTHER NAMES: Marsh Speedwell

PLANTAGINACEAE • SNAPDRAGON FAMILY

Thyme-leaved Speedwell • *Veronica serpyllifolia*

Perennial herb of forests, fields, and shorelines, 5–30 cm tall. **Flowers** *white or pale blue, with darker blue to purple lines,* somewhat irregular, 2–8 mm wide, *in loose, terminal racemes; pedicels pubescent, 2–8 mm long.* **Leaves** opposite, entire or obscurely toothed, *glabrous,* 0.5–2.5 cm long and 5–15 mm wide, *sessile or on short petioles.* **Stems** with fine, short hairs; *lower branches often creeping.* **Fruit** a notched capsule, 2.5–3.5 mm long and 3–5 mm wide.
OCCURRENCE: Uncommon, widely distributed.
NOTES: See *V. wormskjoldii.*

American Alpine Speedwell • *Veronica wormskjoldii*

Perennial herb of alpine and subalpine ravines, 5–30 cm tall. **Flowers** deep blue-violet, somewhat irregular, 6–10 mm wide, *in terminal racemes.* **Leaves** opposite below and alternate above, pubescent, 1–4 cm long and 0.5–2 cm wide. **Fruit** a capsule with glandular hairs, 4–7 mm long, notched at tip.
OCCURRENCE: Very rare, documented from only southern portion of Park.
NOTES: See *V. serpyllifolia.* Species is listed as Endangered in Maine.
OTHER NAMES: *Veronica alpina* var. *terrae-novae, Veronica alpina* var. *unalaschcensis*

WILDFLOWERS AND LOW SHRUBS 229

POLYGONACEAE • KNOTWEED FAMILY ▼

Alpine Bistort • *Bistorta vivipara*

Perennial herb of alpine areas, ridges, and ledges, 10–40 cm tall. **Flowers** *white or pale pink, in a dense inflorescence 20–90 mm long and 4–10 mm wide, producing flowers and fruits on upper half and red vegetative bulbils on lower half.* **Leaves** *alternate, basal ones on a long petiole.*
OCCURRENCE: Very rare, documented from only one township in Park.
NOTES: Species is listed as Endangered in Maine.
OTHER NAMES: *Persicaria vivipara, Polygonum viviparum*

Fringed Bindweed • *Fallopia cilinodis*

Perennial vine of dry, open sites, 1–5 m long. **Flowers** greenish white to white, 1.5–2 mm wide, in a spike-like raceme 4–10 cm long. **Leaves** alternate, sharp-pointed, 2–6 cm long and 2–5 cm wide, *with long, soft, straight hairs beneath.* **Stems** pubescent, often red; upper nodes *with a ring of reflexed bristles.* **Fruit** *a smooth, black, glossy achene, 3–4 mm long; outer 3 tepals without wings.*
OCCURRENCE: Uncommon, with scattered distribution.
NOTES: See *F. convolvulus* and *F. scandens*.
OTHER NAMES: Fringed Black-bindweed, *Polygonum cilinode*

POLYGONACEAE • KNOTWEED FAMILY

Black Bindweed • *Fallopia convolvulus*

Annual vine of fields, meadows, and waste places, 0.5–1 m long. **Flowers** greenish white, often with a pink or purple base, in a spike-like raceme 2–10 cm long. **Leaves** alternate, sharp-pointed, 2–6 cm long and 2–5 cm wide, *without long hairs beneath*. **Stems** pubescent, *upper nodes without a ring of reflexed bristles*. **Fruit** a granular, roughened, dull black achene, 3–4 mm long; outer 3 tepals merely keeled or with a very narrow wing less than 1 mm wide.
OCCURRENCE: Rare, documented from only northern portion of Park.
NOTES: See *F. cilinodis* and *F. scandens*.
OTHER NAMES: *Polygonum convolvulus*

Climbing Bindweed • *Fallopia scandens*

Annual or perennial vine of floodplains, forests, and shorelines, 1–5 m long. **Flowers** green to white or pinkish, in a raceme 1–28 cm long. **Leaves** alternate, sharp-pointed, 2–14 cm long and 2–7 cm wide, without long hairs beneath. **Stems** glabrous or scabrous; *upper nodes without a ring of reflexed bristles*. **Fruit** a smooth, glossy, black achene, 2–6 mm long; outer 3 tepals conspicuously winged in fruit with wings wider than 1.5 mm.
OCCURRENCE: Historical record, last documented in 1985, in the Park.
NOTES: See *F. cilinodis* and *F. convolvulus*.
OTHER NAMES: *Polygonum scandens*

POLYGONACEAE • KNOTWEED FAMILY

Water Smartweed • *Persicaria amphibia*

Perennial herb of swales, ditches, swamps, and shorelines, 0.2–3 m long. **Flowers** *deep pink to red, 2–3 mm long, in dense terminal racemes* 1–10 cm long and 0.8–2 cm wide. **Leaves** alternate, *pointed at tip*, 2–15 cm long and 1–6 cm wide, *glabrous in floating form, pubescent in terrestrial form*; stipules in terrestrial form pubescent, with an outward flange at apex.
OCCURRENCE: Occasional, with scattered distribution.
OTHER NAMES: *Polygonum amphibium*

Water-pepper Smartweed • **Persicaria hydropiper*

Annual herb of wet, open sites, 20–80 cm tall. **Flowers** greenish at base, white to pink distally, *4- or 5-parted, usually in a nodding inflorescence* 3–18 cm long and 5–9 mm wide; tepals dotted with numerous yellow to brown glandular dots, without an anchor-shaped branching vein. **Leaves** alternate, *with peppery taste*, up to 9 cm long; *tubular sheath (ocrea) with fringe of short bristles* 1–4 mm long around edge; petioles without prickles. **Stems** without prickles. **Fruit** a dull, dark brown to black achene, minutely granular, 1.9–3 mm long and 1.5–2 mm wide.
OCCURRENCE: Uncommon, with scattered distribution.
NOTES: See *P. lapathifolia* and *P. maculosa*.
OTHER NAMES: *Polygonum hydropiper*

POLYGONACEAE • KNOTWEED FAMILY

Dock-leaved Smartweed • *Persicaria lapathifolia*

Annual herb of swamps, shorelines, and swales, 0.1–1 m tall. **Flowers** white to pink, 4- or 5-parted, usually in a nodding inflorescence 3–8 cm long and 5–12 mm wide; *tepals without glandular dots, with an anchor-shaped branching vein.* **Leaves** alternate, without a peppery taste, up to 12 cm long; *tubular sheath (ocrea) without a fringe of short bristles or with bristles less than 1 mm long;* petioles without prickles. **Stems** without prickles; nodes conspicuously swollen. **Fruit** a shiny or dull achene, smooth, 1.5–3.2 mm long and 1.6–3 mm wide.
OCCURRENCE: Uncommon, with scattered distribution.
NOTES: See *P. hydropiper* and *P. maculosa*.
OTHER NAMES: Nodding Smartweed, *Polygonum lapathifolium*

Lady's-thumb Smartweed • **Persicaria maculosa*

Annual herb of meadows, fields, and shorelines, 0.1–1 m tall. **Flowers** pink or rose, 4- or 5-parted, 2–3.5 mm long, in a cylindric, dense, usually straight inflorescence 1–4 cm long and 7–12 mm wide. **Leaves** alternate, up to 10 cm long, *often with a dark triangular blotch on middle of upper surface;* petioles without prickles; *tubular sheath (ocrea) with fringe of short hairs 1–3.5 mm long around edge.* **Stems** *without prickles.* **Fruit** a shiny achene, smooth, 2–2.7 mm long and 1.8–2.2 mm wide.
OCCURRENCE: Very rare, documented from only northern portion of Park.
NOTES: See *P. hydropiper* and *P. lapathifolia*.
OTHER NAMES: *Persicaria persicaria, Persicaria vulgaris, Polygonum persicaria*

POLYGONACEAE • KNOTWEED FAMILY

Arrow-leaved Tearthumb • *Persicaria sagittata*

Annual herb of meadows, fens, swamps, and shorelines, 0.3–2 m long. **Flowers** white to pink, 5-parted, in an erect inflorescence 5–15 mm long and 4–10 mm wide. **Leaves** alternate, *arrowhead-shaped, with basal lobes directed back, 2–8.5 cm long and 1–3 cm wide; petioles with reflexed prickles.* **Stems** *4-angled, trailing, with reflexed prickles 1–1.5 mm long.* **Fruit** *a dull to shiny achene, 2.5–4 mm long and 1.8–2.5 mm wide.*
OCCURRENCE: Rare, with scattered distribution.
OTHER NAMES: *Polygonum sagittatum*

Sheep Dock • **Rumex acetosella*

Perennial herb of disturbed sites and rock outcrops, 10–40 cm tall. **Flowers** reddish, nodding, on short pedicels 1–3 mm long. **Leaves** alternate, 2–6 cm long and 3–20 mm wide, *with an acid taste; basal lobes spreading or forward pointing.*
OCCURRENCE: Occasional, widely distributed.
NOTES: A nonnative weed with edible leaves.
OTHER NAMES: Sheep Sorrel, Red Sorrel

POLYGONACEAE • KNOTWEED FAMILY

Greater Water Dock • *Rumex britannica*

Perennial herb of swales, swamps, and shorelines, 0.8–1.5 m tall. **Flowers** in whorls of 15–25; pedicels 1–4 mm long; *articulation point of pedicels indistinct, not swollen, in proximal third*. **Leaves** alternate, entire or with finely rounded teeth along margin, 20–55 cm long and 2–7 cm wide. **Fruit** with 3 wing-like tepals, each tepal longer than wide; *grain-like swellings (tubercles) well developed on midrib of all 3 inner tepals, equal in size or nearly so, distinctly above base of associated tepals.*
OCCURRENCE: Uncommon, with scattered distribution.
NOTES: See *R. crispus*, *R. longifolius*, and *R. obtusifolius*.
OTHER NAMES: *Rumex orbiculatus*

Curly Dock • **Rumex crispus*

Perennial or biennial herb of old fields and disturbed sites, 0.4–1 m tall. **Flowers** in whorls of 10–25; pedicels 4–8 mm long, *with a visibly swollen articulation point*. **Leaves** alternate, *with crumpled and crisped margins*, narrowing to base, the largest 15–30 cm long and 2–6 cm wide. **Fruit** with 3 wing-like tepals, each tepal longer than wide; *grain-like swellings (tubercles) usually well developed on midrib of all 3 inner tepals, one distinctly larger than others, the largest about half as long as associated tepals.*
OCCURRENCE: Uncommon, with scattered distribution.
NOTES: See *R. britanica*, *R. longifolius*, and *R. obtusifolius*.
OTHER NAMES: Yellow Dock

POLYGONACEAE • KNOTWEED FAMILY

Yard Dock • *Rumex longifolius*

Perennial herb of fields, roadsides, and disturbed sites, 0.5–1.2 m tall. **Flowers** in whorls of 10–20; pedicels 4–9 mm long, *with a visibly swollen articulation point*. **Leaves** alternate, widest near middle, the largest 25–50 cm long and 7–15 cm wide. **Fruit** with 3 wing-like tepals, *each tepal wider than long*; grain-like swellings (tubercles) minute or absent from midrib of inner tepals.
OCCURRENCE: Very rare, documented from only one township in Park.
NOTES: See *R. britanica*, *R. crispus*, and *R. obtusifolius*.
OTHER NAMES: *Rumex domesticus*

Bitter Dock • *Rumex obtusifolius*

Perennial herb of fields, roadsides, and disturbed sites, 0.6–1.2 m tall. **Flowers** in slightly remote whorls; pedicels 2.5–8.5 mm long, *with a visibly swollen articulation point*. **Leaves** alternate, thin, often with red veins, 20–40 cm long and 10–15 cm wide. **Fruit** with 3 wing-like *tepals, each tepal longer than wide, with distinctly toothed margins; teeth 0.8–1.5 mm long, visible both in flower and in fruit*; grain-like swellings (tubercles) usually well developed on midrib of only 1 of 3 inner tepals.
OCCURRENCE: Rare, with scattered distribution.
NOTES: See *R. britanica*, *R. crispus*, and *R. longifolius*.

PONTEDERIACEAE • WATER-HYACINTH FAMILY ▼

Pickerelweed • *Pontederia cordata*

Emergent, perennial or annual herb of ponds and coves of shallow lakes and rivers, 0.3–1 m tall. **Flowers** *purple-blue*, pubescent, 2-lipped with 3 lobes per lip, open for one day only, in a spike 7–10 cm long. **Leaves** heart-shaped at base, on a long petiole.
OCCURRENCE: Rare, with scattered distribution.

PORTULACACEAE • PURSLANE FAMILY ▼

Carolina Spring-beauty • *Claytonia caroliniana*

Perennial herbs of floodplains and hardwood forests, 6–35 cm tall. **Flowers** usually pale pink with darker veins, 8–12 mm wide; pedicels 1–3 cm long. **Leaves** opposite, 4–8 cm long and 0.5–2.5 cm wide, entire.
OCCURRENCE: Uncommon, with scattered distribution.

POTAMOGETONACEAE • PONDWEED FAMILY ▼

Big-leaved Pondweed • *Potamogeton amplifolius*

Perennial, aquatic herb of deeper lakes and ponds, 0.6–3 m long. **Flowers** in dense spikes of 9–16 whorls; spikes 3–7 cm long and 1–1.5 cm wide in fruit. **Floating leaves** *light green, with 27–49 veins, with blades 5–10 cm long and 2.5–5 cm wide*, and petioles 2–23 cm long. **Submerged leaves** light to dark green, *distinctly arched, with 19–49 veins, not clasping stem, entire, folded along midrib, tapering to a distinct petiole*; clear stripes of lacunae at midrib absent; *stipules persistent, conspicuous, strongly tinged with brown or green, free from blade, 1.5–12 cm long, not fibrous.*
OCCURRENCE: Uncommon, with scattered distribution.
NOTES: See *P. perfoliatus, P. praelongus*, and *P. richardsonii*.

Berchtold's Pondweed • *Potamogeton berchtoldii*

Perennial, aquatic herb of lakes, ponds, and rivers, up to 1.5 m long. **Flowers** in *rounded* spikes 1.5–10 mm long, *4 or more on each plant, spikes continuous without interruptions*. **Floating leaves** absent. **Submerged leaves** *green, 0.9–5.4 cm long and 0.2–2.5 mm wide, acutely to obtusely pointed at apex*; stipules distinct, delicate, *without united margins below middle; clear stripes of lacunae at midrib in 1–5 rows on each side of midrib*. **Stems** branched, with nodal glands on at least some nodes; glands up to 0.5 mm wide.
OCCURRENCE: Uncommon, with scattered distribution.
NOTES: See *P. gemiparus* and *P. obtusifolius*.

POTAMOGETONACEAE • PONDWEED FAMILY

Alga-like Pondweed • *Potamogeton confervoides*

Perennial, aquatic herb of peaty pools, 10–80 cm long. **Flowers** in rounded spikes 5–12 mm long; peduncle 5–25 cm long. **Floating leaves** absent. **Submerged leaves** *flaccid, sessile, uniform, in distinctive fan-shaped branches of extremely delicate leaves, 1.8–6.5 cm long and 0.1–0.5 mm wide*; stipules inconspicuous, free from blade, 0.5–1.2 cm long, not fibrous. **Stems** *thread-like, flaccid, forking repeatedly; nodal glands absent.*
OCCURRENCE: Rare, with scattered distribution.
OTHER NAMES: Tuckerman's Pondweed

Ribbon-leaved Pondweed • *Potamogeton epihydrus*

Perennial, aquatic herb of ponds and streams, up to 2 m long. **Flowers** numerous, in spikes 0.8–4 cm long. **Floating leaves** with blades 2–8 cm long and 4–20 mm wide, and petioles 2–12.5 cm long. **Submerged leaves** 1–10 mm wide, *usually 20 or more times as long as wide*, sessile, long and ribbon-like with essentially parallel margins for most of their length; *lacunae (prominent clear stripes at midrib) at least 1 mm wide and usually about a quarter or more of total leaf width*; stipules persistent, inconspicuous, free from blade, *1–3 cm long*, not fibrous. **Stems** *never branched above base; internodes usually at least 10 mm long.*
OCCURRENCE: Common, widely distributed.
NOTES: See *P. gramineus.*

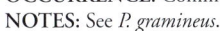

POTAMOGETONACEAE • PONDWEED FAMILY

Budding Pondweed • *Potamogeton gemmiparus*

Perennial, aquatic herb of lakes, ponds, and rivers, up to 1.5 m tall. **Flowers** in *cylindric* spikes 1.5–10 mm long, *1–3 on each plant; spikes continuous without interruptions.* **Floating leaves** absent. **Submerged leaves** 1.1–6 cm long and *0.2–0.7 mm wide, acutely pointed at apex;* stipules distinct, delicate, *without united margins below middle; clear stripes of lacunae at midrib absent or in 1 or 2 rows on each side of midrib.* **Stems** branched, with nodal glands on at least some nodes; glands up to 0.5 mm wide.
OCCURRENCE: Very rare, documented from only one township in Park.
NOTES: See *P. berchtoldii* and *P. obtusifolius*.

Grassy Pondweed • *Potamogeton gramineus*

Perennial, aquatic herb of deeper water, 0.3–1.5 m long. **Flowers** in spikes 1.5–3.5 cm long; peduncles 3.2–7.7 cm long. **Floating leaves** with blades 3.5–4 cm long and 16–20 mm wide; petioles 3–4.5 cm long. **Submerged leaves** 3–10 mm wide, *usually less than 15-times as long as wide,* sessile, entire; *clear stripes of lacunae at midrib very narrow, usually less than 1 mm wide;* stipules persistent, inconspicuous, free from blade, *1.3–1.6 cm long,* not fibrous. **Stems** *usually branched above base; internodes usually less than 6 mm long.*
OCCURRENCE: Occasional, widely distributed.
NOTES: See *P. epihydrus* and *P. nodosus*.

POTAMOGETONACEAE • PONDWEED FAMILY

Floating Pondweed • *Potamogeton natans*

Perennial, aquatic herb of lakes and slow-flowing streams, up to 1 m long. **Flowers** in spikes 2.5–5 cm long and 9–12 mm wide. **Floating leaves** *with blades 3.5–11 cm long and 1.5–6 cm wide, cordate at base, with 17–37 veins; petiole apex at junction with leaf blade distinctly pale in color, usually bent so that blade appears oriented in opposite direction of petiole.* **Submerged leaves** sessile, very narrow, *rigid, 0.7–2.5 mm wide, tapering to a blunt tip*; stipules persistent, conspicuous, large, free from blade, whitish. **Stems** *0.8–2.5 mm wide, often brown-spotted.*
OCCURRENCE: Occasional, with scattered distribution.
NOTES: See *P. nodosus* and *P. oakesianus*.

Long-leaved Pondweed • *Potamogeton nodosus*

Perennial, aquatic herb of lakes, ponds, and rivers, up to 2 m long. **Flowers** in spikes 2–7 cm long. **Floating leaves** with blades 3–11 cm long and 1.5–4.5 cm wide, wedge-shaped to rounded at base. **Submerged leaves** *lax, delicate, often deteriorating, on long petioles 2–13 cm long*; blades 9–20 cm long and 1–3.5 cm wide; lacunae (prominent clear stripes at midrib) in 2–5 rows on each side of midrib; stipules persistent, conspicuous, free from blade, 3–9 cm long. **Stems** branched, without spots.
OCCURRENCE: Very rare, documented from only one township in Park.
NOTES: See *P. gramineus*, *P. natans*, and *P. oakesianus*.

POTAMOGETONACEAE • PONDWEED FAMILY

Oakes' Pondweed • *Potamogeton oakesianus*

Perennial, aquatic herb of acidic, peaty ponds, up to 1 m long. **Flowers** in spikes 1–3.5 cm long and 7–9 mm wide. **Floating leaves** *with blades 2–4 cm long and 1–2 cm wide, rounded to tapering at base, not cordate, with 9–19 veins; petiole apex similar in color to rest of petiole.* **Submerged leaves** *sessile, lax, delicate, 0.3–1 mm wide, tapering to a pointed tip;* stipules persistent, conspicuous, small, free from blade, whitish. **Stems** *0.4–1 mm wide, branched, red-spotted.*
OCCURRENCE: Very rare, documented from only one township in Park.
NOTES: See *P. natans* and *P. nodosus*.

Blunt-leaved Pondweed • *Potamogeton obtusifolius*

Perennial, aquatic, herb of lakes, ponds, and rivers, up to 1 m long. **Flowers** in cylindric spikes 8–13 mm long. **Floating leaves** absent. **Submerged leaves** *usually suffused with red, 3–8.2 cm long and 1–3.5 mm wide, obtuse or rounded at apex;* stipules persistent, inconspicuous, free from blade, white, 0.6–1.8 cm long, fibrous; clear stripes of lacunae at midrib in 1–3 rows on each side of midrib. **Stems** branched, with nodal glands 0.2–1 mm wide.
OCCURRENCE: Very rare, documented from only one township in Park.
NOTES: See *P. berchtoldii* and *P. gemiparus*.

POTAMOGETONACEAE • PONDWEED FAMILY

Clasping-leaved Pondweed • *Potamogeton perfoliatus*

Perennial, aquatic herb of brackish waters, up to 2.5 m long. **Flowers** in spikes 1–4.5 cm long and 8 mm wide. **Floating leaves** absent. **Submerged leaves** olive-green, *0.9–7.6 cm long and 7–40 mm wide, entire, sessile, strongly cordate-clasping at base, delicate, not distinctly arched, with 3–25 veins, with wavy margins, flat apex, and obtuse tip; stipules deciduous, delicate, deteriorating, absent on proximal portion of stem, 3.5–6.5 cm long*.
OCCURRENCE: Rare, documented from only northern portion of Park.
NOTES: See *P. amplifolius*, *P. praelongus*, and *P. richardsonii*.

White-stemmed Pondweed • *Potamogeton praelongus*

Perennial, aquatic herb of lakes, ponds, and rivers, up to 3 m tall. **Flowers** in cylindric spikes 3–7.5 cm long. **Floating leaves** absent. **Submerged leaves** pale green, *weakly clasping at base, not distinctly arched, with 11–33 veins, 0.8–2.8 cm long and 1–4 cm wide, with hood-like leaf tips, splitting if flattened; margins wavy*; clear stripes of lacunae at midrib absent; *stipules persistent, conspicuous, white, free from blade, 3–8.1 cm long, fibrous, shredding at apex*. **Stems** *white, zigzagging*.
OCCURRENCE: Very rare, documented from only one township in Park.
NOTES: See *P. amplifolius*, *P. perfoliatus*, and *P. richardsonii*.

POTAMOGETONACEAE • PONDWEED FAMILY

Richardson's Pondweed • *Potamogeton richardsonii*

Perennial, aquatic herb of lakes, ponds, and rivers, up to 1 m long. **Flowers** in spikes 1.5–3 cm long. **Floating leaves** absent. **Submerged leaves** olive-green, *sessile, strongly clasping, not distinctly arched, with 3–35 veins, 1.6–13 cm long and 5–28 mm wide, with wavy margins, flat apex, and acuminate at tip; stipules disintegrating to persistent white fibers, even on proximal portion of stem, free from blade*; clear stripes of lacunae at midrib absent.
OCCURRENCE: Rare, with scattered distribution.
NOTES: See *P. amplifolius*, *P. perfoliatus*, and *P. praelongus*.

Robbins' Pondweed • *Potamogeton robbinsii*

Perennial, aquatic, herb of lakes, ponds, and rivers, up to 2 m long. **Flowers** in *branched* spikes 7–20 mm long. **Floating leaves** absent. **Submerged leaves** *stiff, dark green to reddish green, 2–7 cm long and 3–4 mm wide, in 2 vertical rows, with 20–60 fine veins, and minutely toothed margins, lobed at junction with stipule; stipules persistent, conspicuous, greenish brown to white, joined to base of blade for lower quarter of stipule*; clear stripes of lacunae at midrib absent.
OCCURRENCE: Uncommon, with scattered distribution.
NOTES: See *P. spirillus*.

244   THE PLANTS OF BAXTER STATE PARK

POTAMOGETONACEAE • PONDWEED FAMILY

Northern Snail-seed Pondweed • *Potamogeton spirillus*

Perennial, aquatic herb of quiet lakes and ponds, up to 1.5 m long. **Flowers** in either submerged or emersed spikes; submerged spikes 2–5 mm long; emersed spikes 4–13 mm long. **Floating leaves** rounded at tip, with blades 0.7–3.5 cm long and 2–13 mm wide. **Submerged leaves** sessile, *lax, often curved, not conspicuously in 2 vertical rows, tapering slightly to base, 0.8–8 cm long and 0.5–2 mm wide, with 5–15 veins*; stipules persistent, inconspicuous, *joined to base of blade for lower half of stipule; clear stripes of lacunae in a broad band on each side of midrib, often occupying most of leaf area.*
OCCURRENCE: Uncommon, with scattered distribution.
NOTES: See *P. robbinsii*.
OTHER NAMES: Spiral Pondweed

Flat-stem Pondweed • *Potamogeton zosteriformis*

Perennial, aquatic herb of lakes, ponds, and rivers, 0.6–1.2 m long. **Flowers** in cylindric spikes 1.5–3 cm long. **Floating leaves** absent. **Submerged leaves** sessile, rigid, light green, 10–20 cm long and *2–5 mm wide, with 15–35 veins*, rounded at base; stipules persistent, conspicuous, free from blade, white, 1.5–3.5 cm long, fibrous; clear stripes of lacunae at midrib absent. **Stems** *conspicuously flattened, usually winged*, branched, 1–3 mm wide.
OCCURRENCE: Very rare, documented from only one township in Park.

RANUNCULACEAE • BUTTERCUP FAMILY ▼

White Baneberry • *Actaea pachypoda*

Perennial herb of rich, hardwood forests, 40–80 cm tall. **Flowers** white, with 4–10 petals, narrowing abruptly near base, 3–5 mm long; *pedicels stout, 1–2.5 mm wide.* **Leaves** alternate, divided, *glabrous on underside.* **Fruit** *a white or rarely a red berry, 6.5–9 mm wide; pedicels of fruit bright red, stout, about as wide as axis of raceme.*
OCCURRENCE: Occasional, widely distributed.
OTHER NAMES: Doll's Eyes, *Actaea alba*

Red Baneberry • *Actaea rubra*

Perennial herb of acidic forests, 40–80 cm tall. **Flowers** white, with 4–10 petals, narrowing abruptly near base; *pedicels slender, 0.3–0.7 mm wide.* **Leaves** alternate, divided, glabrous or pubescent on underside. **Fruit** *a red or rarely a white berry, 5–11 mm wide; pedicels of fruit dull green or brown, slender, narrower than axis of raceme.*
OCCURRENCE: Common, widely distributed.
OTHER NAMES: Snakeberry

RANUNCULACEAE • BUTTERCUP FAMILY

Blunt-lobed Hepatica • *Anemone americana*

Perennial herb of forests, 5–18 cm tall. **Flowers** bluish to white or pink, 12–25 mm wide, solitary on scapes. **Leaves** basal only, thick, deeply 3-lobed, lobes rounded at tip; petioles villous.
OCCURRENCE: Historical record, last documented in 1984, from one township in Park.
OTHER NAMES: *Hepatica americana*

Red Columbine • *Aquilegia canadensis*

Perennial herb of rocky woods, outcrops, and forest edges, 15–90 cm tall. **Flowers** *bright red with yellow*, 3–5.3 cm long, nodding; *spurs red, essentially straight*, 1.3–2.5 cm long. **Leaves** alternate, divided; leaflets 1–4.5 cm long.
OCCURRENCE: Very rare, documented from only one township in Park.
NOTES: See *A. vulgaris*.
OTHER NAMES: Wild Columbine

WILDFLOWERS AND LOW SHRUBS

RANUNCULACEAE • BUTTERCUP FAMILY

European Columbine • *Aquilegia vulgaris*

Perennial herb of roadsides, fields, and disturbed sites, 30–80 cm tall. **Flowers** *blue, pink, or purple*, 2.5–5.5 cm long; *spurs blue or purple, strongly incurved*, 1.4–2.2 cm long. **Leaves** alternate, divided; leaflets 1–4.5 cm long.
OCCURRENCE: Rare, with scattered distribution.
NOTES: See *A. canadensis*.

Purple Virgin's-bower • *Clematis occidentalis*

Perennial vine of cliffs, rocky ledges, and talus slopes, up to 3.5 m long. **Flowers** *solitary in axils of leaves; petals absent; sepals 25–60 mm long, purple to reddish violet*. **Leaves** opposite, firm, divided into 3 leaflets; leaflets 3–10 cm long and 1.5–6 cm wide.
OCCURRENCE: Historical record, last documented in 1982, from one township in Park.
NOTES: Species is listed as Special Concern in Maine.

RANUNCULACEAE • BUTTERCUP FAMILY

Virginia Virgin's-bower • *Clematis virginiana*

Perennial vine of shorelines, wet roadsides, and disturbed sites, up to 7 m long. **Flowers** *in many-flowered cymes; petals absent; sepals 6–14 mm long, whitish.* **Leaves** opposite, divided into 3 leaflets; leaflets 3.5–9 cm long and 1.5–7.5 cm wide.
OCCURRENCE: Common, widely distributed.

Three-leaved Goldthread • *Coptis trifolia*

Perennial, evergreen herb of wet woods and swamps, up to 13 cm tall. **Flowers** *white*; petals 5- to 7-parted, *fleshy, 3–4 mm long*, with nectar in hollow tip. **Leaves** *basal only, shiny*, 1.5–3.5 cm long, *divided, with 3-toothed leaflets*. **Fruit** an achene, *in an umbel-like, stalked cluster*.
OCCURRENCE: Common, widely distributed.
OTHER NAMES: *Coptis groenlandica*

WILDFLOWERS AND LOW SHRUBS 249

RANUNCULACEAE • BUTTERCUP FAMILY

Kidney-leaved Crowfoot • *Ranunculus abortivus*

Annual or biennial herb of rich woods and meadows, 10–65 cm tall. **Flowers** pale yellow, 5-parted, solitary, *2–5 mm wide; petals 1.5–3.5 mm long and 1–2 mm wide.* **Leaves** *dimorphic; lowest simple, fleshy, with blades usually wider than long, 1.4–4.2 cm long and 2–5.2 cm wide; cauline leaves conspicuously lobed, sessile, gradually reduced toward top of stem.* **Fruit** an achene, in heads 2.5–5 mm long and 3–4 mm wide.
OCCURRENCE: Occasional, with scattered distribution.
OTHER NAMES: Small-flowered Crowfoot

Tall Crowfoot • **Ranunculus acris*

Perennial herb of fields, roadsides, and disturbed sites, 0.3–1 m tall. **Flowers** *yellow, shiny,* 5-parted, *up to 2 cm wide; petals 8–11 mm long and 7–13 mm wide;* sepals about half as long as petals. **Leaves** alternate, *the largest divided into 3–7 deeply cleft stalkless parts.* **Stems** *usually pubescent.* **Fruit** an achene, in heads 5–7 mm wide.
OCCURRENCE: Common, widely distributed.
NOTES: Fresh plants are harmful to livestock.
OTHER NAMES: Tall Buttercup

RANUNCULACEAE • BUTTERCUP FAMILY

Swamp Crowfoot • *Ranunculus caricetorum*

Perennial herb of wet areas, 50–90 cm tall. **Flowers** yellow; *styles elongate, with a terminal deciduous stigma*; sepals spreading or reflexed from base, 4–10 mm long and 2–5 mm wide; petals 8–16 mm long and 3–9 mm wide. **Leaves** alternate, divided, toothed, usually 3-parted, *without white blotches on upper surface*. **Stems** *not thickened at base*, decumbent, sometimes rooting at nodes. **Fruit** an achene, in heads 6–10 mm long and 7–10 mm wide; *beak on seed-like fruit mostly 1.5–2.6 mm long, straight or slightly curved.*
OCCURRENCE: Historical record, last documented in 1984, from northern portion of Park.
NOTES: See *R. repens*.

Greater Yellow Water Crowfoot • *Ranunculus flabellaris*

Perennial, aquatic herb of lakes, ponds, rivers, streams, and wetlands, up to 70 cm long. **Flowers** *yellow, 5- or 6-parted; petals 7–12 mm long and 7–8 mm wide*. **Leaves** alternate, 1- to 6-parted or -lobed, 1.2–7.3 cm long and 1.9–10.8 cm wide, the uppermost leaf segments flat.
OCCURRENCE: Historical record, last documented in 1985, in one township in Park.
NOTES: See *R. longirostris*.

RANUNCULACEAE • BUTTERCUP FAMILY

Creeping Crowfoot • *Ranunculus flammula*

Perennial herb of wet shorelines, 5–15 cm tall. **Flowers** yellow, 5- or 6-parted, solitary; sepals spreading or weakly reflexed, 1.5–4 mm long and 1–2 mm wide; petals 2.5–7 mm long and 1–4 mm wide. **Leaves** usually basal only, *unlobed, bluntly toothed; blades shorter than petioles.* **Stems** *rooting at nodes.*
OCCURRENCE: Occasional, widely distributed.
OTHER NAMES: Creeping Spearwort, *Ranunculus reptans*

Long-beaked Water Crowfoot • *Ranunculus longirostris*

Perennial, aquatic herb of lakes, ponds, rivers, and streams, up to 0.8 m long. **Flowers** *white with a yellowish base, 1–1.5 cm wide*, 5-parted; petals 4–7 mm long and 1–5 mm wide. **Leaves** alternate, 3-parted, 0.4–1.1 cm long and 0.7–2.3 cm wide, *finely divided into narrow fan-shaped segments; uppermost leaf segments capillary.*
OCCURRENCE: Uncommon, widely distributed.
NOTES: See *R. flabellaris*.
OTHER NAMES: *Ranunculus aquatilis*

RANUNCULACEAE • BUTTERCUP FAMILY

Bristly Crowfoot • *Ranunculus pensylvanicus*

Herb of marshes, swamps, shorelines, and disturbed sites, 30–70 cm tall. **Flowers** yellow; sepals 3–5 mm long and 1.5–2 mm wide; petals 2–4 mm long and 1–2.5 mm wide, *usually shorter than sepals*. **Leaves** alternate, *divided*, reduced toward top of stem. **Stems** pubescent with coarse, stiff hairs, not rooting at nodes. **Fruit** an achene, in heads *9–12 mm long* and 5–7 mm wide; *beak straight or nearly so*.
OCCURRENCE: Occasional, widely distributed.
NOTES: See *R. recurvatus*.

Hooked Crowfoot • *Ranunculus recurvatus*

Perennial herb of forests, wetlands, and shorelines, 10–50 cm tall. **Flowers** pale yellow; sepals 3–6 mm long and 1.5–2.5 mm wide; *petals 3–5 mm long and 1–2 mm wide*. **Leaves** alternate, *lobed but not divided*, reduced toward top of stem. **Stems** pubescent with coarse, stiff hairs, not rooting at nodes. **Fruit** an achene, in heads *5–6 mm long* and 5–6 mm wide; *beak curved, hooked*.
OCCURRENCE: Uncommon, with scattered distribution.
NOTES: See *R. pensylvanicus*.

Spot-leaved Crowfoot • *Ranunculus repens*

Perennial herb of wet areas, up to 80 cm tall. **Flowers** yellow; *styles short, recurved or hooked; stigma persistent and spread out over upper side of curved style;* sepals spreading or reflexed from base, 4–7 mm long and 1.5–3 mm wide; petals 6–18 mm long and 5–12 mm wide. **Leaves** alternate, dark green, divided, usually 3-parted, *usually with white blotches on upper surface.* **Stems** *not thickened at base,* decumbent, sometimes rooting at nodes. **Fruit** an achene, in heads 5–10 mm long and 5–8 mm wide; *beak on seed-like fruit mostly 0.8–1.2 mm long, recurved or hooked.*
OCCURRENCE: Uncommon, with scattered distribution.
NOTES: See *R. caricetorum*. Species has potential to spread aggressively in some natural habitats.

Early Meadow-rue • *Thalictrum dioicum*

Perennial herb of floodplains, forests, and shorelines, 30–80 cm tall, *flowering in April or May, before leaves fully expanded.* **Flowers** *greenish to purple;* sepals 1.8–4 mm long; *filaments yellow to greenish yellow,* 3.5–5.5 mm long; anthers 2–4 mm long. **Leaves** alternate; leaflets primarily with 4–6 apical lobes, lobe margins with rounded teeth; upper cauline leaves on long petioles, appearing to have 1 petiole at each node.
OCCURRENCE: Historical record, last documented in 1985, from northern portion of Park.
NOTES: See *T. pubescens*.

RANUNCULACEAE • BUTTERCUP FAMILY

Tall Meadow-rue • *Thalictrum pubescens*

Perennial herb of meadows, shorelines, and rich woods, up to 3 m tall, *flowering in June or July, after leaves fully expanded*. **Flowers** *white to purplish*; sepals 2–3.5 mm long; *filaments white to purplish*, 1.5–7 mm long; *anthers 0.5–1.5 mm long*. **Leaves** alternate; *leaflets primarily with 2–3(–5) apical lobes; lobe margins entire*; upper cauline leaves sessile or nearly so, with 3 main divisions on stalks, appearing to have 3 petioles at each node.
OCCURRENCE: Common, widely distributed.
NOTES: See *T. dioicum*.
OTHER NAMES: King-of-the-meadow, *Thalictrum polygamum*

RHAMNACEAE • BUCKTHORN FAMILY ▼

Alder-leaved Buckthorn • *Rhamnus alnifolia*

Deciduous shrub of fens and swamps, 0.2–1 m tall. **Flowers** *yellow-green, 5-parted, 1.5–3 mm wide, with staminate and carpellate flowers on separate plants*. **Leaves** alternate, simple, dark green above, with blades 5–12 cm long and 2–6 cm wide, pointed at tip. **Stems** without thorns; branchlets minutely pubescent. **Fruit** a black drupe, 6–9 mm wide.
OCCURRENCE: Rare, with scattered distribution.

WILDFLOWERS AND LOW SHRUBS

ROSACEAE • ROSE FAMILY ▼

Common Agrimony • *Agrimonia gryposepala*

Perennial herb of forests and disturbed sites, 0.5–1.5 m tall. **Flowers** yellow, 5-parted; *axis of raceme with abundant stipitate glands intermixed with long, spreading hairs; outer bristles at base of flower wide spreading.* **Leaves** alternate, divided, pubescent beneath with coarse, stiff, curving hairs; leaflets with small glandular dots beneath. **Fruit** an achene, *in clusters, with spreading bristles.*
OCCURRENCE: Rare, with scattered distribution.
NOTES: See *A. striata.*

Roadside Agrimony • *Agrimonia striata*

Perennial herb of forests, fields, and disturbed sites, up to 1 m tall. **Flowers** yellow, 5-parted; *axis of raceme densely pubescent with ascending hairs, without stipitate glands or glands sparse and hidden by pubescence; outer bristles at base of flower ascending to erect.* **Leaves** alternate, divided, sparsely pubescent beneath, especially on veins; leaflets with glandular dots beneath. **Fruit** an achene, *in clusters, with erect bristles.*
OCCURRENCE: Uncommon, with scattered distribution.
NOTES: See *A. gryposepala.*

ROSACEAE • ROSE FAMILY

Purple Chokeberry • *Aronia floribunda*

Deciduous shrub of mountaintops, exposed areas, and peatlands, 0.3–2 m tall. **Flowers** white, 5-parted; petals 4–6 mm long; *pedicels sparsely to densely pubescent.* **Leaves** alternate, simple, finely and minutely toothed, *sparsely to densely pubescent beneath*; upper surface of midrib with a row of dark glands. **Stems** *with sparsely to densely pubescent branchlets.* **Fruit** *a red or purple pome, 8–10 mm wide.*
OCCURRENCE: Occasional, widely distributed.
NOTES: See *A. melanocarpa, Prunus virginiana* (Rosaceae), and *Ilex verticillata* (Aquifoliaceae).
OTHER NAMES: *Photinia floribunda, Pyrus floribunda*

Black Chokeberry • *Aronia melanocarpa*

Deciduous shrub of mountaintops, exposed areas, and peatlands, 0.1–1 m tall. **Flowers** white, 5-parted; petals 4–6 mm long; *pedicels glabrous or with few hairs.* **Leaves** alternate, simple, finely and minutely toothed, *without or with few hairs beneath*; upper surface of midrib with a row of dark glands. **Stems** *with glabrous branchlets.* **Fruit** *a black pome, 6–8 mm wide.*
OCCURRENCE: Occasional, widely distributed.
NOTES: See *A. floribunda, Prunus virginiana* (Rosaceae), and *Ilex verticillata* (Aquifoliaceae).
OTHER NAMES: *Photinia melanocarpa, Pyrus melanocarpa*

ROSACEAE • ROSE FAMILY

Marsh-cinquefoil • *Comarum palustre*

Perennial herb of bogs, swamps, swales, and shallow water, 20–60 cm tall. **Flowers** *purple to red*, 2–3 cm wide, generally more than one per stalk; *petals shorter than purple sepals*. **Leaves** *pinnately divided*, with 5–7 sharply toothed, blunt-tipped leaflets.
OCCURRENCE: Uncommon, with scattered distribution.
OTHER NAMES: Marsh-five-finger, *Potentilla palustris*

Shrubby-cinquefoil • *Dasiphora floribunda*

Deciduous shrub of rocky outcrops, rocky shorelines and fens, 0.2–1 m tall. **Flowers** yellow, 5-parted, 1.8–3 cm wide. **Leaves** alternate, divided; *leaflets 3–7, sessile, narrow lance-shaped, covered with silky hairs on both sides, giving them a grayish hue.*
OCCURRENCE: Rare, with scattered distribution.
OTHER NAMES: Buckbrush, *Potentilla fruticosa*, *Pentaphylloides floribunda*, *Dasiphora fruticosa*

ROSACEAE • ROSE FAMILY

Tall Wood-beauty • *Drymocallis arguta*

Perennial herb of rocky areas and shorelines, 0.3–1 m tall. **Flowers** white, 5-parted, 10–18 mm wide. **Leaves** alternate, divided, the lowest with 7–11 toothed leaflets; leaflets downy beneath. **Stems** *with long, sticky hairs.*
OCCURRENCE: Very rare, documented from only northern portion of Park.
OTHER NAMES: *Potentilla arguta*

Strawberry • *Fragaria virginiana*

Perennial herb of fields, meadows, and roadsides, up to 30 cm tall. **Flowers** *white, 5-parted, 1–2.5 cm wide*, in clusters usually shorter than leaves; stamens numerous. **Leaves** basal only, *divided, with 3 leaflets, toothed with terminal tooth usually shorter than adjacent teeth.* **Fruit** an aggregation of achenes embedded in surface of a round, juicy, red, receptacle.
OCCURRENCE: Common, widely distributed.
OTHER NAMES: Wild Strawberry

WILDFLOWERS AND LOW SHRUBS 259

ROSACEAE • ROSE FAMILY

Yellow Avens • *Geum aleppicum*

Perennial herb of forests, wet meadows, and shorelines, 0.3–1.5 m tall. **Flowers** *bright yellow to orange-yellow*; petals 5–9 mm long and 5–8 mm wide, equaling or slightly longer than sepals; *styles without glandular hairs at base; pedicels with both dense minute hairs and abundant longer stout hairs.* **Leaves** alternate, divided, variable in shape; basal leaves with terminal leaflet usually wedge-shaped at base; terminal leaflet usually longer than wide. **Stems** pubescent.
OCCURRENCE: Uncommon, with scattered distribution.
NOTES: See *G. macrophyllum* and *G. rivale*.

White Avens • *Geum canadense*

Slender, perennial herb of dry woods, meadows, and roadsides, 0.2–1 m tall. **Flowers** *white*; petals 4–7 mm long, longer than or equaling green sepals; pedicels with only tiny short hairs or occasionally with scattered longer hairs. **Leaves** alternate, divided; *lowest usually with 3 principal leaflets*; terminal leaflet much larger than lateral ones. **Stems** *glabrous or with a few scattered hairs.*
OCCURRENCE: Historical record, last documented in 1985, from northern portion of Park.
NOTES: See *G. laciniatum*.

ROSACEAE • ROSE FAMILY

Floodplain Avens • *Geum laciniatum*

Perennial herb of floodplains, forests, fields, and disturbed sites, 0.4–1 m tall. **Flowers** *white to pale yellow; petals 2.5–4 mm long, distinctly shorter than sepals; pedicels usually 1 mm or more wide, with both dense, stiff, coarse, spreading hairs and tiny short hairs.* **Leaves** *alternate, divided; lowest usually with 3–7 principal leaflets; terminal leaflet much larger than lateral ones.* **Stems** *densely pubescent with spreading hairs over 2 mm long.*
OCCURRENCE: Very rare, documented from only one township in Park.
NOTES: See *G. canadense*.

Large-leaved Avens • *Geum macrophyllum*

Perennial herb of forests, meadows, and wetlands, up to 1 m tall. **Flowers** *bright yellow to orange-yellow; petals 4–7 mm long; style with scattered, minute, glandular hairs at base; pedicels with dense minute hairs and sometimes with scattered, longer, stout hairs.* **Leaves** *alternate, divided, variable in shape; basal leaves with terminal leaflet rounded to cordate at base; terminal leaflet round or wider than long, much larger than lateral segments.* **Stems** *pubescent.*
OCCURRENCE: Very rare, documented from only one township in Park.
NOTES: See *G. aleppicum* and *G. rivale*.

WILDFLOWERS AND LOW SHRUBS

ROSACEAE • ROSE FAMILY

Water Avens • *Geum rivale*

Perennial herb of wet areas, 0.3–1 m tall.
Flowers *nodding; sepals red or purplish, in a bell-shaped cluster; petals yellowish with purple veins, usually shorter than sepals.*
Leaves alternate, divided, progressively reduced toward top of stem, the lowest with 3–5 principal leaflets; terminal leaflet much larger than others and sometimes lobed or cleft.
OCCURRENCE: Uncommon, with scattered distribution.
NOTES: See *G. aleppicum* and *G. macrophyllum*.
OTHER NAMES: Purple Avens

Silver-leaved Cinquefoil • **Potentilla argentea*

Perennial herb of dry fields and roadsides, 10–50 cm tall. **Flowers** yellow, 7–10 mm wide, few to many in a terminal cyme; petals 2.5–4 mm long, not much longer than sepals. **Leaves** alternate, *densely pubescent and silver beneath; leaflets 5–7, deeply toothed, with 2–4 tooth-like lobes near tip.* **Stems** woolly.
OCCURRENCE: Rare, with scattered distribution.

ROSACEAE • ROSE FAMILY

Norwegian Cinquefoil • *Potentilla norvegica*

Annual to short-lived perennial herb of forests, fields, and roadsides, 0.1–1.1 m tall. **Flowers** yellow, 5-parted, *0.6–1.5 cm wide, in terminal branched clusters; petals 2–5 mm long, slightly shorter than or equaling sepals.* **Leaves** alternate, green, *with long hairs beneath, the lowest palmately divided with 3 leaflets, uppermost up to 8 cm long.* **Stems** *branched, stout, pubescent.*
OCCURRENCE: Occasional, widely distributed.
NOTES: See *P. recta.*
OTHER NAMES: Rough Cinquefoil

Sulphur Cinquefoil • **Potentilla recta*

Perennial herb of dry fields and roadsides, 15–80 cm tall. **Flowers** pale yellow, *1.2–3 cm wide, few to many in a terminal cyme; petals 8–15 mm long, slightly to evidently longer than sepals.* **Leaves** alternate, green, *the lowest palmately divided with 5–9 coarsely toothed leaflets.*
OCCURRENCE: Rare, with scattered distribution.
NOTES: See *P. norvegica.*
OTHER NAMES: Rough-fruited Cinquefoil

ROSACEAE • ROSE FAMILY

Old-field Cinquefoil • *Potentilla simplex*

Perennial herb of forests, fields, and thickets, 20–70 cm tall. **Flowers** yellow, 1–1.5 cm wide; petals 4–7 mm long, *borne solitary on leafless stalks from leaf axils*. **Leaves** alternate, divided; *leaflets 3–5*, toothed from tip to well below middle, pubescent on underside, the largest 5–7.5 cm long. **Stems** *trailing, slender, often red.*
OCCURRENCE: Uncommon, widely distributed.
OTHER NAMES: Five-finger

Smooth Rose • *Rosa blanda*

Deciduous shrub of shorelines, meadows, and rocky slopes, up to 2 m tall. **Flowers** pink to red, *usually 1–3 per inflorescence; pedicels glabrous, without stipitate glands*; sepals 20–30 mm long and 2.5–3.5 mm wide. **Leaves** alternate, 8.5–11 cm long, divided; leaflets singly toothed; terminal leaflet with 10–26 teeth on each side. **Stems** *with distal branches unarmed, occasionally armed on older growth.*
OCCURRENCE: Rare, with scattered distribution.
NOTES: See *R. virginiana*.

ROSACEAE • ROSE FAMILY

Shining Rose • *Rosa nitida*

Deciduous shrub of *wet soils*, up to 1 m tall. **Flowers** pink, usually solitary; *pedicels with stipitate glands*. **Leaves** alternate, divided; *leaflets finely toothed, teeth near middle of leaflet 0.4–0.7 mm long; terminal leaflet with 14–20 teeth on each side*. **Stems** *with slender, small-based, straight prickles at nodes; internodal prickles numerous, similar in size to nodal prickles*.
OCCURRENCE: Very rare, documented from only one township in Park.
NOTES: See *R. palustris*.
OTHER NAMES: Bog Rose

Swamp Rose • *Rosa palustris*

Deciduous shrub of *wet soils*, up to 2.5 m tall. **Flowers** pink, very fragrant, *usually 2 or more per inflorescence; pedicels with stipitate glands; sepals often pinnately lobed*. **Leaves** alternate, divided; *leaflets finely toothed, teeth near middle of leaflet 0.4–0.7 mm long; terminal leaflet with 21–25 teeth on each side*. **Stems** *with a pair of stout, broad-based, downward-curving prickles at most nodes; internodal prickles few, more slender than nodal prickles*.
OCCURRENCE: Rare, with scattered distribution.
NOTES: See *R. nitida*.

WILDFLOWERS AND LOW SHRUBS

ROSACEAE • ROSE FAMILY

Virginia Rose • *Rosa virginiana*

Deciduous shrub of moist to dry thickets and marshes, up to 2 m tall. **Flowers** pink to red, *usually 3 or more per inflorescence; pedicels usually with stipitate glands*; sepals 20–40 mm long and 2.5–4 mm wide. **Leaves** alternate, 5–8 cm long, divided; leaflets singly or doubly toothed; terminal leaflet with 10–18 teeth on each side. **Stems** *with distal branches usually armed with a pair of stout, broad-based, curving prickles at most nodes; internodal prickles usually absent.*
OCCURRENCE: Very rare, documented from only one township in Park.
NOTES: See *R. blanda.*

Common Blackberry • *Rubus allegheniensis*

Deciduous shrub of openings and thickets, 0.5–3 m tall. **Flowers** white, 2.5–4.8 cm wide; *inflorescence 8–22 cm long with 9–22 flowers; rachis with stipitate glands.* **Leaves** alternate, divided, *soft-pubescent beneath*, with 3 leaflets on green first-year stems and with 5 leaflets on red to brown second-year stems. **Stems** woody at base, clearly differentiated into green first-year stems and red to brown second-year stems; green first-year stems erect or arching, not rooting at tip, *armed with stout, broad-based prickles wider than 1 mm at base, with glandular hairs.* **Fruit** a cluster of drupes, black, not easily separating from receptacle.
OCCURRENCE: Uncommon, with scattered distribution.
NOTES: See *R. canadensis, R. elegantulus,* and *R. pensilvanicus.*

ROSACEAE • ROSE FAMILY

Smooth Blackberry • *Rubus canadensis*

Deciduous shrub of openings and thickets, 0.5–3 m tall. **Flowers** white, 2.5–4 cm wide, *inflorescence 3–15 cm long with 3–25 flowers; rachis without stipitate glands.* **Leaves** alternate, divided, *glabrous beneath or with hairs only along primary veins*, with 3–7 leaflets. **Stems** woody at base, clearly differentiated into green first-year stems and red to brown second-year stems; green first-year stems erect or arching, not rooting at tip, *usually unarmed, occasionally armed with stout, broad-based prickles wider than 1 mm at base and numbering 0–10 per 10 cm of stem.* **Fruit** a cluster of drupes, black, not easily separating from receptacle.
OCCURRENCE: Occasional, widely distributed.
NOTES: See *R. allegheniensis, R. elegantulus,* and *R. pensilvanicus.*
OTHER NAMES: Canada Blackberry

Dewdrop • *Rubus dalibarda*

Perennial herb of moist forests, up to 10 cm tall. **Flowers** white, solitary; petals 4–8 mm long, sometimes absent. **Leaves** basal only, *simple, unlobed, evergreen,* 1.5–3 cm long and 1.8–3 cm wide, *with margins shallowly scalloped.* **Stems** *without bristles or prickles.*
OCCURRENCE: Common, widely distributed.
NOTES: See *Mitella nuda* (Saxifragaceae).
OTHER NAMES: *Dalibarda repens*

WILDFLOWERS AND LOW SHRUBS

ROSACEAE • ROSE FAMILY

Showy Blackberry • *Rubus elegantulus*

Deciduous shrub of open areas and thickets, up to 1 m tall. **Flowers** white; *inflorescence 3–11 cm long, with 3–15 flowers; rachis without stipitate glands.* **Leaves** alternate, divided, *glabrous beneath or with hairs only along primary veins*, with 3–5 leaflets; *central petiolules of leaves of green first-year stems 25–40 mm long.* **Stems** woody at base, clearly differentiated into green first-year stems and red to brown second-year stems; green first-year stems erect or arching, not rooting at tip, *armed with moderately abundant, stout, broad-based prickles wider than 1 mm at base and numbering 10–60 per 10 cm of stem.* **Fruit** a cluster of drupes, black, not easily separating from receptacle.
OCCURRENCE: Historical record, last documented in 1938, from one township in Park.
NOTES: See *R. allegheniensis, R. canadensis,* and *R. pensilvanicus.*

Bristly Blackberry • *Rubus hispidus*

Deciduous, *evergreen* shrub of open forests, up to 2.5 m long. **Flowers** white; *petals 5–12 mm long; inflorescence with few flowers; rachis rarely sparsely glandular; pedicels strongly ascending.* **Leaves** alternate, divided, *firm, glossy, with 3 leaflets on both first-year and second-year stems; leaflets 3.5–5 cm long, terminal one short-stalked, widest above middle.* **Stems** woody at base, clearly differentiated into green first-year stems and pale red to brown second-year stems; *green first-year stems prostrate or trailing, armed with hairs, bristles, or slender, small-based prickles less than 1 mm wide at base.* **Fruit** a cluster of drupes, black, not easily separating from receptacle.
OCCURRENCE: Uncommon, with scattered distribution.
NOTES: See *R. recurvicaulis* and *R. vermontanus.* A hybrid with *R. vermontanus* has been documented in Baxter State Park.

ROSACEAE • ROSE FAMILY

Red Raspberry • *Rubus idaeus*

Deciduous shrub of thickets and disturbed sites, up to 2 m tall. **Flowers** white or greenish white; *inflorescence with 2–5 flowers; pedicels with straight bristles or minute prickles, usually with stipitate glands.* **Leaves** alternate, divided, *densely pubescent and appearing white beneath, with 3–9 pinnately arranged leaflets.* **Stems** woody at base, clearly differentiated into green first-year stems and red to brown second-year stems; *green first-year stems arched, not rooting at tip, not strongly white-glacous.* **Fruit** a cluster of drupes, *red, easily separating from dry receptacle.*
OCCURRENCE: Common, widely distributed.
NOTES: See *R. occidentalis*.

Black Raspberry • *Rubus occidentalis*

Deciduous shrub of fields and forest edges, 1–3 m tall. **Flowers** white; *inflorescence with 5–15 flowers; pedicels with stout, curved prickles, without stipitate glands.* **Leaves** alternate, divided, densely pubescent beneath, *with 3–5 leaflets.* **Stems** woody at base, clearly differentiated into green first-year stems and red to brown second-year stems; *green first-year stems arched, occasionally rooting at tip, strongly white-glaucous.* **Fruit** a cluster of drupes, *black, easily separating from dry receptacle; individual drupes separated by bands of woolly hairs.*
OCCURRENCE: Historical record, last documented in 1985, from northern portion of Park.
NOTES: See *R. idaeus*.

WILDFLOWERS AND LOW SHRUBS

ROSACEAE • ROSE FAMILY

Pennsylvania Blackberry • *Rubus pensilvanicus*

Deciduous shrub of openings and thickets, 0.5–3 m tall. **Flowers** white; *inflorescence loosely corymbose, 3–11 cm long with 7–12 flowers; rachis without stipitate glands.* **Leaves** alternate, divided, *evidently pubescent beneath*, with 3–5 leaflets. **Stems** woody at base, clearly differentiated into green first-year stems and red to brown second-year stems; green first-year stems erect or arching, not rooting at tip, *armed with stout, broad-based prickles wider than 1 mm at base and numbering up to 20 per 10 cm of stem.* **Fruit** a cluster of drupes, black, not easily separating from receptacle.
OCCURRENCE: Very rare, documented from only one township in Park.
NOTES: See *R. allegheniensis, R. canadensis,* and *R. elegantulus.*

Dwarf Raspberry • *Rubus pubescens*

Annual or biennial herb of wet forests and thickets, *10–40 cm tall*. **Flowers** white or pale pink; *inflorescence with 2–4 flowers; petals 4–8 mm long and 1.5–3 mm wide; sepals reflexed.* **Leaves** alternate, divided, with central leaflet tapering to pointed base and tip. **Stems** *without bristles or prickles, not woody, not differentiated into green first-year and red to brown second-year stems, with erect, leafy branches.* **Fruit** a cluster of drupes, *dark red*, 1–2 cm wide, juicy, *difficult to separate from receptacle.*
OCCURRENCE: Common, widely distributed.
NOTES: The bright red fruit has the flavor of a wild raspberry, but the few drupelets do not readily separate from the receptacle.
OTHER NAMES: Swamp Red Raspberry

ROSACEAE • ROSE FAMILY

Arching Blackberry • *Rubus recurvicaulis*

Deciduous shrub of dry, open sites, up to 5 m long. **Flowers** white; *petals 10–25 mm long*; inflorescence with 2–8 flowers; pedicels pubescent, 5–40 mm long, ascending to spreading. **Leaves** alternate, divided, with 3–5 leaflets, glabrous on lower surface; *leaflets 6–15 cm long*. **Stems** woody at base, clearly differentiated into green first-year stems and red to brown second-year stems; *green first-year stems prostrate or trailing, armed with stout, broad-based prickles wider than 1 mm at base*. **Fruit** a cluster of drupes, black, not easily separating from receptacle.
OCCURRENCE: Very rare, documented from only one township in Park.
NOTES: See *R. hispidus* and *R. vermontanus*.

8 September – ML

Vermont Blackberry • *Rubus vermontanus*

Deciduous shrub of fields and forest edges, up to 1.1 m tall. **Flowers** white, 2–3 cm wide; inflorescence with 5–10 flowers; *rachis usually with stipitate glands*. **Leaves** alternate, divided, *with 5 leaflets on red to brown second-year stems; central petiolules of leaves of green first-year stems 14–20 mm long*. **Stems** woody at base, clearly differentiated into green first-year stems and red to brown second-year stems; *green first-year stems erect or arching, sometimes trailing near tip but not rooting, armed with hairs, bristles, or slender, small-based prickles less than 1 mm wide at base and numbering 10–500 per 10 cm of stem*. **Fruit** a cluster of drupes, black, not easily separating from receptacle.
OCCURRENCE: Uncommon, with scattered distribution.
NOTES: See *R. hispidus* and *R. recurvicaulis*. A hybrid with *R. hispidus* has been documented in Baxter State Park.

22 June – AH

21 September – AH

WILDFLOWERS AND LOW SHRUBS 271

ROSACEAE • ROSE FAMILY

Three-toothed-cinquefoil • *Sibbaldia tridentata*

Perennial herb of exposed headlands, mountains, and rock outcrops, 2–27 cm tall. **Flowers** white, 5-parted, 0.9–1.4 cm wide, in branched clusters. **Leaves** alternate, divided, light to dark green; *leaflets 3, with a 3-toothed tip, evergreen.*
OCCURRENCE: Occasional, widely distributed.
OTHER NAMES: *Potentilla tridentata, Sibbaldiopsis tridentata*

White Meadowsweet • *Spiraea alba*

Deciduous shrub of wetlands and thickets, 0.3–1.2 m tall. **Flowers** *white or pink-tinged, 4–7 mm wide,* in a wide terminal panicle. **Leaves** alternate, coarsely toothed, 3–7.5 cm long and 0.8–1.8 cm wide, *green, glabrous beneath.* **Twigs** *reddish to purple-brown, glabrous.*
OCCURRENCE: Common, widely distributed.
NOTES: See *S. tomentosa.*
OTHER NAMES: *Spiraea latifolia, Spiraea septentrionalis*

ROSACEAE • ROSE FAMILY

Rosy Meadowsweet • *Spiraea tomentosa*

Deciduous shrub of fields and wetlands, 0.3–1.2 m tall. **Flowers** *pink, 3–4 mm wide*, in an elongate, slim, terminal cluster. **Leaves** alternate, *rugose, white-woolly beneath,* 3–5.5 cm long and 1–2.2 cm wide. **Twigs** *white-woolly.*
OCCURRENCE: Occasional, widely distributed.
NOTES: See *S. alba*.
OTHER NAMES: Steeple-bush

RUBIACEAE • MADDER FAMILY ▼

Scratch Bedstraw • *Galium aparine*

Annual herb of wet woods and thickets, up to 1 m long. **Flowers** white; *ovary pubescent or bristly with hooked hairs*. **Leaves** whorled, *with an abrupt, projecting tip; lowest leaves in whorls of 6–10, 1–8 cm long and 2–6 mm wide, with bristle-tipped hairs on margins bent toward base*. **Stems** weak, reclining, with tiny, downward-pointing, stiff hairs. **Fruit** *pubescent or bristly with hooked hairs*.
OCCURRENCE: Uncommon, with scattered distribution.
OTHER NAMES: Spring Cleavers

RUBIACEAE • MADDER FAMILY

Rough Bedstraw • *Galium asprellum*

Perennial herb of wet woods and thickets, 0.5–2 m long. **Flowers** white; *ovary smooth, glabrous, or with bristles.* **Leaves** whorled, *with an abrupt, projecting tip; lowest leaves in whorls of 6, 0.5–1.7 cm long and 2–4 mm wide*, prickly, with stiff hairs on margins bent toward base. **Stems** weak, usually reclining, *with downward-pointing prickles.* **Fruit** *smooth, glabrous, or with bristles.*
OCCURRENCE: Occasional, widely distributed.
NOTES: See *G. triflorum*.

Northern Bedstraw • *Galium boreale*

Perennial herb of fens, fields, and forests, 0.2–0.9 m tall. **Flowers** white; *ovary with or without spreading bristles; pedicels clearly visible, longer than 1 mm.* **Leaves** whorled or opposite, *blunt to pointed at tip, without an abrupt, projecting tip; lowest leaves in whorls of 4, 1.5–4.5 cm long, 4- to 14-times as long as wide, with 3–5 prominent parallel veins visible at base.* **Stems** *erect*, with downward-pointing prickles. **Fruit** a schizocarp, *with or without spreading bristles*, the bristles, if present, not hooked.
OCCURRENCE: Historical record, last documented in 1984, from northern portion of Park.

RUBIACEAE • MADDER FAMILY

Forest Licorice Bedstraw • *Galium circaezans*

Perennial herb of cliffs, ledges, and forests, 0.1–0.6 m tall. **Flowers** greenish; *ovary with spreading bristles; pedicels less than 1 mm long.* **Leaves** whorled, *blunt to pointed at tip, without an abrupt, projecting tip; lowest leaves in whorls of 4, 1.5–5 cm long and 7–25 mm wide.* **Stems** erect or ascending. **Fruit** a schizocarp, *with spreading, hooked bristles.*
OCCURRENCE: Very rare, documented from only one township in Park.
NOTES: See *G. kamtschaticum*.

2 September – DC

Boreal Bedstraw • *Galium kamtschaticum*

Perennial herb of forests, shorelines, and swamps, 10–20 cm tall. **Flowers** yellow-white; *ovary with spreading bristles; pedicels clearly visible, longer than 1 mm.* **Leaves** whorled or opposite, *blunt to pointed at tip, without an abrupt, projecting tip; lowest leaves in whorls of 4, 1–3 cm long, less than 3-times as long as wide.* **Stems** glabrous. **Fruit** a schizocarp, *with spreading, hooked bristles.*
OCCURRENCE: Historical record, last documented in 1901, from southern portion of Park.
NOTES: See *G. circaezans*. Species is listed as Threatened in Maine.

11 September – DC

21 July – DC   11 September – DC

WILDFLOWERS AND LOW SHRUBS

RUBIACEAE • MADDER FAMILY

Whorled Bedstraw • *Galium mollugo*

Perennial herb of fields and disturbed sites, 0.3–1.2 m tall. **Flowers** white; *ovary smooth, glabrous.* **Leaves** whorled, *with an abrupt, projecting tip; lowest leaves in whorls of 8, 0.5–1.5 cm long and 1–3 mm wide*, with short, upward-pointing, marginal hairs. **Stems** *firm, erect, becoming tangled, smooth to pubescent but not prickly.* **Fruit** a schizocarp, *smooth, glabrous.*
OCCURRENCE: Rare, with scattered distribution.
NOTES: See *G. sylvaticum.*
OTHER NAMES: Smooth Bedstraw

Marsh Bedstraw • *Galium palustre*

Perennial herb of swales, wet thickets, and swamps, 25–60 cm tall. **Flowers** white; *ovary smooth, without spreading bristles; inflorescence with 10–25 flowers beyond distal, reduced, leaf-like bract.* **Leaves** whorled, *blunt to pointed at tip, without an abrupt, projecting tip; lowest leaves in whorls of 4, 1.2–1.5 cm long, with only 1 prominent vein.* **Stems** *weak, reclining.* **Fruit** a schizocarp, *smooth, without spreading bristles.*
OCCURRENCE: Occasional, widely distributed.
NOTES: See *G. tinctorium* and *G. trifidum.*

RUBIACEAE • MADDER FAMILY

Wood Bedstraw • *Galium sylvaticum*

Perennial herb of forest edges, meadows, and fields, 0.4–0.8 m tall. **Flowers** white; *ovary smooth, glabrous*. **Leaves** whorled, *with an abrupt, projecting tip; lowest leaves in whorls of 6–8, longer than 2.5 cm, broadest near or below middle*. **Stems** *firm, erect, smooth to pubescent but not prickly*. **Fruit** a schizocarp, *smooth, glabrous*.
OCCURRENCE: Very rare, documented from only one township in Park.
NOTES: See *G. mollugo*.

Stiff Three-petaled Bedstraw • *Galium tinctorium*

Perennial herb of swales, wet thickets, and swamps, up to 0.6 m long. **Flowers** white; *ovary smooth, without spreading bristles; inflorescence with 1–6 flowers beyond distal, reduced, leaf-like bract*. **Leaves** whorled, *blunt to pointed at tip, without an abrupt, projecting tip; lowest leaves in whorls of 4–6, 0.5–1.6 cm long and 1.5–4 mm wide, with only 1 prominent vein*. **Stems** *weak, reclining*. **Fruit** a schizocarp, *smooth, without spreading bristles; mature pedicels usually straight, less than 8 mm long*.
OCCURRENCE: Uncommon, widely distributed.
NOTES: See *G. palustre* and *G. trifidum*.
OTHER NAMES: Small Bedstraw

WILDFLOWERS AND LOW SHRUBS 277

RUBIACEAE • MADDER FAMILY

Three-petaled Bedstraw • *Galium trifidum*

Perennial herb of swamps and bogs. **Flowers** greenish white; *ovary smooth, without spreading bristles; inflorescence with 1–6 flowers beyond distal, reduced, leaf-like bract.* **Leaves** whorled, *blunt to pointed at tip, without an abrupt, projecting tip; lowest leaves in whorls of 4, 0.5–2 cm long, with only 1 prominent vein.* **Stems** *weak, reclining,* with matted basal offshoots. **Fruit** a schizocarp, *smooth, without spreading bristles; mature pedicels usually curved, 7–18 mm long, with dense but minute downward-pointing hairs.*
OCCURRENCE: Occasional, widely distributed.
NOTES: See *G. palustre* and *G. tinctorium*.

Fragrant Bedstraw • *Galium triflorum*

Perennial herb of wet forests and thickets, 0.2–1 m long. **Flowers** greenish white, sweet-scented; *ovary pubescent or bristly.* **Leaves** whorled, *with an abrupt, projecting tip; lowest leaves in whorls of 6–8, 1.5–6 cm long and 5–12 mm wide, with bristle-tipped hairs on margins bent toward tip.* **Stems** weak, often with stiff, hooked hairs on angles. **Fruit** a schizocarp, *bristly.*
OCCURRENCE: Occasional, widely distributed.
NOTES: See *G. asprellum*.

RUBIACEAE • MADDER FAMILY

Little Bluet • *Houstonia caerulea*

Perennial herb of fields and forest openings, 5–20 cm tall. **Flowers** *pale blue with yellow center,* on slender, erect peduncles 2–6 cm long. **Leaves** opposite, entire, *mostly basal,* 5–15 mm long. **Stems** *delicate.* **Fruit** *a flat capsule,* 2.5–4.5 mm wide.
OCCURRENCE: Uncommon, with scattered distribution.
OTHER NAMES: Quaker Ladies

Partridge-berry • *Mitchella repens*

Perennial herb of forests, 10–30 cm long. **Flowers** *white, in terminal pairs, fragrant,* 10–14 mm wide; *petals with long, soft, shaggy hairs.* **Leaves** opposite, entire, 1–2 cm long, *with white midvein.* **Stems** *creeping, forming mats, rooting at nodes.* **Fruit** a solitary, fleshy, red drupe, 4–8 mm wide.
OCCURRENCE: Common, widely distributed.

WILDFLOWERS AND LOW SHRUBS 279

RUSCACEAE • BUTCHER'S BROOM FAMILY ▼

Canada-mayflower • *Maianthemum canadense*

Perennial herb of woods, 5–20 cm tall. **Flowers** small, white, *with 4 tepals, in a simple raceme 1–3.5 cm long.* **Leaves** alternate, light green, *heart-shaped at base, 2 or 3 per plant,* 2.5–8 cm long and 1.5–5 cm wide. **Stems** erect, with several sheathing bracts near base. **Fruit** *a red, speckled berry,* 4–7 mm wide.
OCCURRENCE: Common, widely distributed.
OTHER NAMES: Wild Lily-of-the-valley

280   THE PLANTS OF BAXTER STATE PARK

RUSCACEAE • BUTCHER'S BROOM FAMILY

Feathery False Solomon's-seal • *Maianthemum racemosum*

Perennial herb of rich woods and clearings, 0.3–1 m tall.
Flowers white, *with 6 tepals, in a terminal panicle 4–17 cm long; tepals 0.5–1 mm long.* **Leaves** alternate, 7–18 cm long and 2.5–8 cm wide, *narrowed to base.* **Stems** *zigzagging, often arching.* **Fruit** *a green berry, becoming light red with maturity,* 3–6 mm wide.
OCCURRENCE: Occasional, widely distributed.
OTHER NAMES: False Spikenard, *Smilacina racemosa*

Star-like False Solomon's-seal • *Maianthemum stellatum*

Perennial herb of forests, fields, and fens, 30–65 cm tall.
Flowers white, *with 6 tepals, in a simple raceme, 2–4.5 cm long; tepals 4–5 mm long.* **Leaves** alternate, *4 or more per stem,* 4–15 cm long and 1–5 cm wide, *narrowed to base.* **Stems** erect to arching. **Fruit** a green to red berry *striped with red to dark purple,* 6–10 mm wide.
OCCURRENCE: Historical record, last documented in 1983, from southern portion of Park.
OTHER NAMES: *Smilacina stellata*

WILDFLOWERS AND LOW SHRUBS

RUSCACEAE • BUTCHER'S BROOM FAMILY

Three-leaved False Solomon's-seal • *Maianthemum trifolium*

Perennial herb of *bogs and forested wetlands*, 5–20 cm tall. **Flowers** small, white, *with 6 tepals, in slender racemes 2–6 cm long; tepals 2.5–6 mm long*. **Leaves** alternate, 2–4 per stem, clasping at base, ascending, 6–12 cm long and 1–4 cm wide. **Fruit** a green berry, turning dark red.
OCCURRENCE: Uncommon, widely distributed.
OTHER NAMES: *Smilacina trifolia*

Hairy Solomon's-seal • *Polygonatum pubescens*

Perennial herb of moist woods and thickets, 0.3–1 m tall. **Flowers** *yellowish green, bell-shaped, 7–15 mm long, solitary or paired, drooping from stem*. **Leaves** alternate, 4–15 cm long and 1–6 cm wide, *with 3–9 pubescent veins on underside*. **Stems** unbranched, arching, slender. **Fruit** *a dark blue to black berry, 6–9 mm wide*.
OCCURRENCE: Occasional, widely distributed.
NOTES: See *Streptopus lanceolatus* (Liliaceae).

SALICACEAE • WILLOW FAMILY ▼

Northern Willow • *Salix arctophila*

Deciduous, colonial, dwarf shrub of *alpine areas, 3–15 cm tall.* **Carpellate flowers** expanding with or after leaves; ovaries pubescent; floral bracts brown or black, at least at apex; *inflorescence 3–8 cm long.* **Leaves** alternate, dark green, shiny above, gray or white beneath from dense hairs or bloom; margins entire, undulate, or with irregular, blunt teeth; blades 2–4 cm long, *1.2- to 2-times as long as wide*; petioles 5–12 mm long. **Fruit** *a pubescent capsule, 5–9 mm long.*
OCCURRENCE: Very rare, documented from only southern portion of Park.
NOTES: See *S. argyrocarpa, S. herbacea,* and *S. uva-ursi.* Species is listed as Endangered in Maine.
OTHER NAMES: Arctic Willow

Labrador Willow • *Salix argyrocarpa*

Deciduous shrub of *alpine areas, up to 0.7 m tall.* **Carpellate flowers** expanding with or after leaves; ovaries pubescent; floral bracts brown or black, at least at apex; *inflorescence 1.1–2 cm long.* **Leaves** alternate, gray or white beneath from dense hairs or bloom, pubescent on one or both surfaces; margins entire, undulate, or with irregular, blunt teeth, rolled under; blades 2–5 cm long, *3- to 17-times as long as wide*; stipules absent; new leaves with silky, long, neatly appressed hairs, red-brown hairs absent. **Fruit** *a pubescent capsule, 2–4 mm long.*
OCCURRENCE: Historical record, last documented in 1934, from southern portion of Park.
NOTES: See *S. arctophila, S. herbacea, S. planifolia,* and *S. uva-ursi.* Species is listed as historical in Maine.

SALICACEAE • WILLOW FAMILY

Snow-bed Willow • *Salix herbacea*

Deciduous dwarf shrub of *alpine areas, 1–15 cm tall.* **Carpellate flowers** with a glabrous ovary; *floral bracts yellow to black, at least at apex, 0.5–1 mm long; inflorescence 0.3–1.3 cm long, with fewer than 12 flowers.* **Leaves** alternate, *green beneath, without dense hairs or bloom; blades 1–2.5 cm long, rounded at apex.* **Fruit** *a glabrous capsule, 2.2–7.5 mm long.*
OCCURRENCE: Very rare, documented from only one township in Park.
NOTES: See *S. arctophila*, *S. argyrocarpa*, and *S. uva-ursi*. A hybrid between *S. herbacea* and *S. uva-ursi* has been documented in Baxter State Park. Species is listed as Threatened in Maine.

Bearberry Willow • *Salix uva-ursi*

Deciduous, dwarf shrub *of alpine areas, usually less than 15 cm tall.* **Carpellate flowers** with a glabrous ovary; floral bracts brown or black, at least at apex, *1.1–1.8 mm long; inflorescence 1–5 cm long.* **Leaves** alternate, gray or white beneath from dense hairs or bloom, occasionally green beneath without hairs or bloom; *margins entire, undulate, or with irregular, blunt teeth; blades 0.5–2 cm long, pointed to obtuse at apex, 1.7- to 3.6-times as long as wide; petioles 2–4 mm long.* **Stems** *prostrate, matted.* **Fruit** *a glabrous capsule, 3–5 mm long.*
OCCURRENCE: Very rare, documented from only one township in Park.
NOTES: See *S. arctophila*, *S. argyrocarpa*, and *S. herbacea*. Species is listed as Threatened in Maine.

Purple Pitcherplant • *Sarracenia purpurea*

Perennial herb of bogs, up to 75 cm tall. **Flowers** *deep red, ~5 cm wide, nodding.* **Leaves** basal only, *funnel-shaped*, with purple veins and downward-pointing hairs; *margins connate, forming a water-holding structure.*
OCCURRENCE: Common, widely distributed.
NOTES: This insectivorous plant lures insects and other small organisms into leafy pitchers, where some eventually drown and decompose. The plant absorbs nutrients from this liquid.

17 June – PS

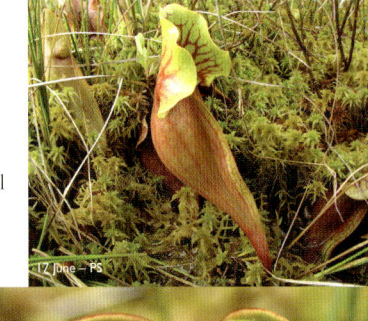

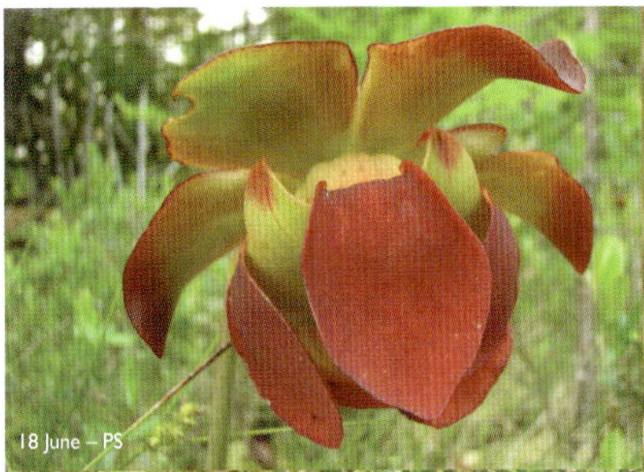
18 June – PS

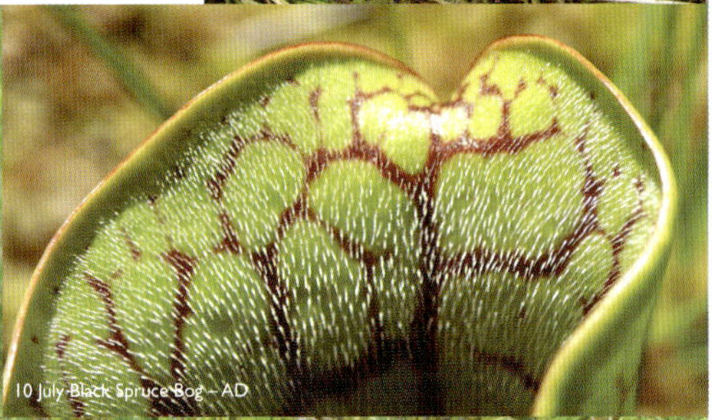
10 July Black Spruce Bog – AD

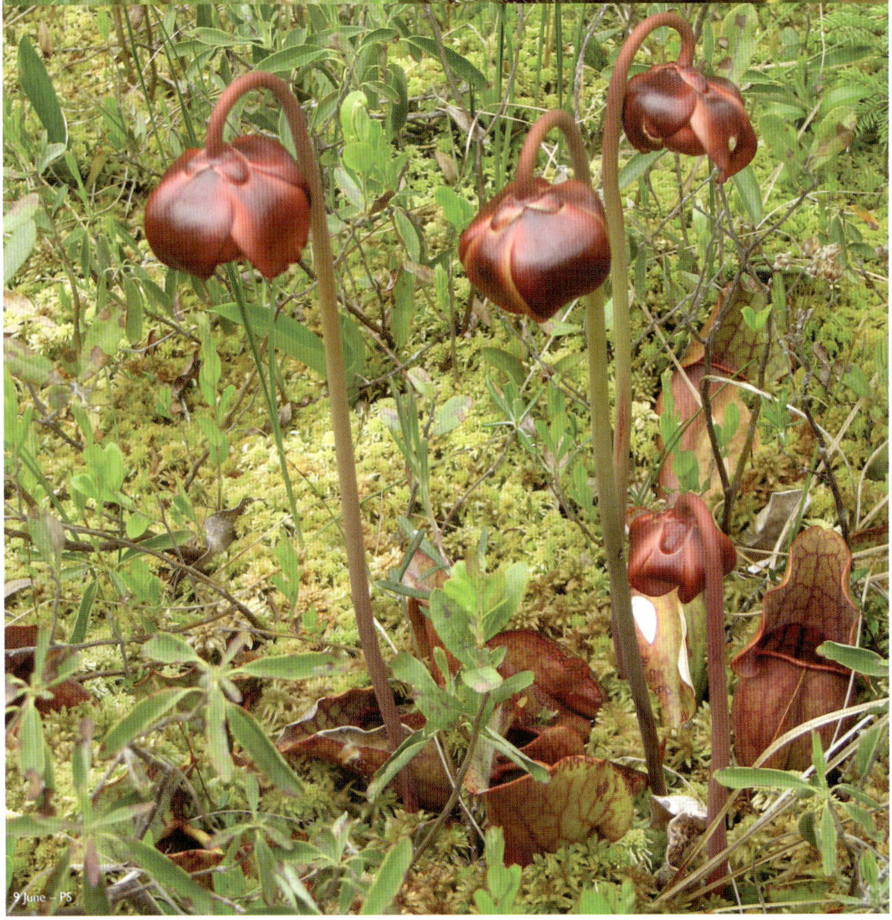
9 June – PS

Golden-saxifrage • *Chrysosplenium americanum*

Perennial, mat-forming, creeping herb of stream beds and wet sites, 4–30 cm tall. **Flowers** *yellow, sometimes greenish, 4-parted,* borne at tip of branches; anthers purple, red, or orange. **Leaves** *opposite, obscurely toothed, 2.5–15 mm long and 2–10 mm wide.*
OCCURRENCE: Occasional, widely distributed.
OTHER NAMES: Water Carpet

20 May – DC

13 July Doubletop Mountain – AD

20 May – DC

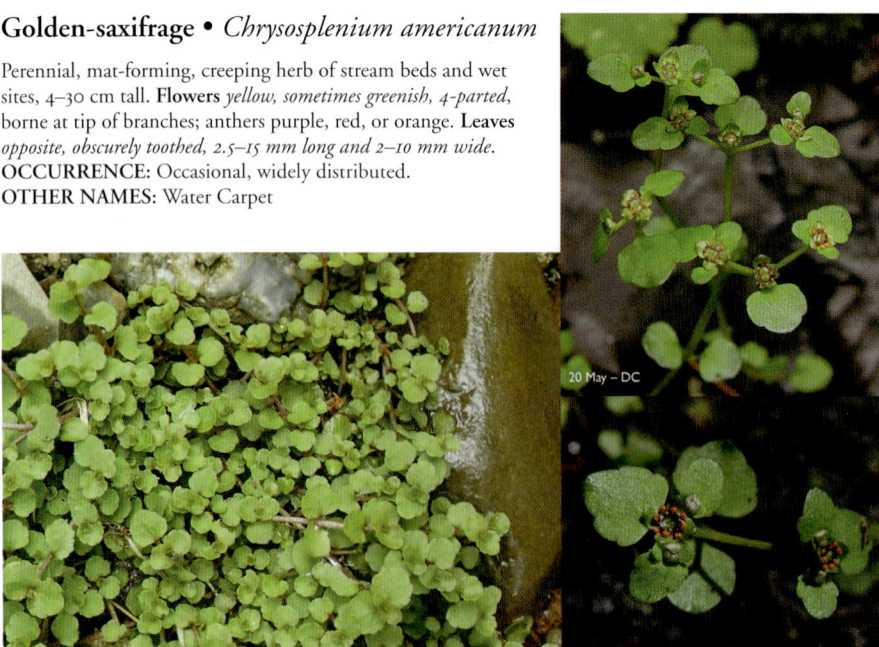

Naked-bulbil Small-flowered-saxifrage • *Micranthes foliolosa*

Perennial herb of alpine ravines, up to 15 cm tall. **Flowers** *white, each petal with 1 or 2 basal yellow spots*; sepals reflexed; *inflorescence narrow, branched, producing flowers and fruits on upper half and vegetative bulbils on lower half.* **Leaves** *clustered at base, fleshy, toothed only near apex*, glabrous.
OCCURRENCE: Very rare, documented from only one township in Park.
NOTES: See *M. virginiensis.* Species is listed as Endangered in Maine.
OTHER NAMES: Leafy Saxifrage, *Saxifraga foliolosa, Saxifraga stellaris*

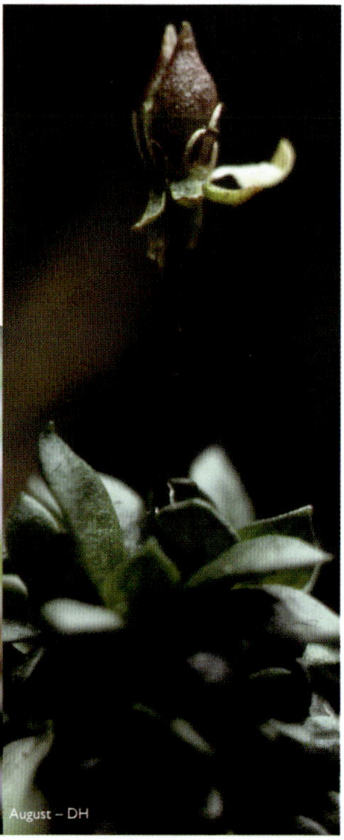
August – DH

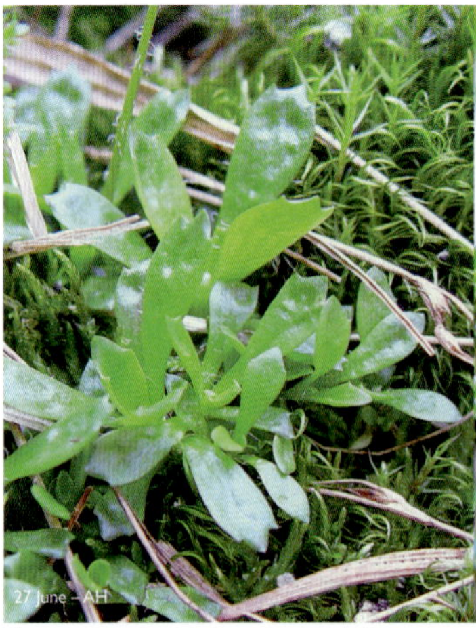
27 June – AH

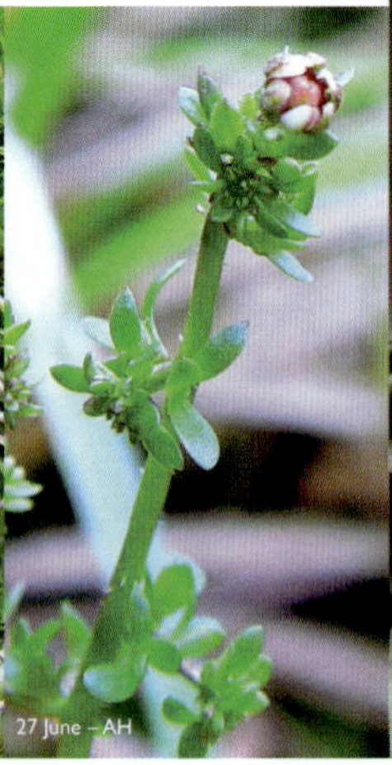
27 June – AH

SAXIFRAGACEAE • SAXIFRAGE FAMILY

Early Small-flowered-saxifrage • *Micranthes virginiensis*

Perennial herb of wet, rocky ledges and outcrops, 5–40 cm tall, *flowering in early spring*. **Flowers** *white, 5-parted, in branched clusters*. **Leaves** basal only, *irregularly toothed throughout*, narrowing at base, often with a reddish hue, somewhat fleshy, *1–5 cm long*.
OCCURRENCE: Very rare, documented from only one township in Park.
NOTES: See *M. foliolosa*.
OTHER NAMES: *Saxifraga virginiensis*

Naked Bishop's-cap • *Mitella nuda*

Perennial herb of cool, wet forests and swamps, up to 50 cm tall. **Flowers** *greenish, 5-parted, few per plant, with fringed petals*. **Leaves** basal only, light green, heart-shaped, with blunt teeth. **Fruit** a capsule, opening flat with two clusters of shiny, black seeds.
OCCURRENCE: Occasional, widely distributed.
NOTES: See *Rubus dalibarda* (Rosaceae).
OTHER NAMES: Naked Mitterwort, Bugs-on-a-plate

SAXIFRAGACEAE • SAXIFRAGE FAMILY

White Mountain Saxifrage • *Saxifraga paniculata*

Perennial, cushion-forming herb of *alpine areas*, up to 40 cm tall. **Flowers** white to cream to pink, sometimes orange- or purple-spotted; sepals erect, often reddish; inflorescence with 2–20 flowers, narrow, branched distally. **Leaves** alternate, sessile, leathery, 8–35 mm long; margins with fine, white teeth.
OCCURRENCE: Very rare, documented from only one township in Park.
NOTES: Species is listed as Special Concern in Maine.
OTHER NAMES: *Saxifraga aizoon*

Foam-flower • *Tiarella cordifolia*

Perennial herb of forests, swamps, and wetlands, up to 50 cm tall. **Flowers** *white, numerous, up to 6 mm wide, in a raceme 3–15 cm long*; flowering scape without bracts. **Leaves** *basal only, 3- to 7-lobed, 4–14 cm long*.
OCCURRENCE: Very rare, documented from only northern portion of Park.
OTHER NAMES: False Miterwort

SCHEUCHZERIACEAE • POD-GRASS FAMILY ▼

Pod-grass • *Scheuchzeria palustris*

Perennial herb of bogs and peaty shorelines, 20–40 cm tall. **Flowers** *yellow-green*, in a terminal raceme 3–10 cm long, with tepals 2–3 mm long. **Leaves** *alternate, erect, with a pore at tip, 2–21 cm long and 1–3 mm wide*, the lowest clustered at base. **Stems** zigzagging, with hairs at nodes.
OCCURRENCE: Occasional, widely distributed.

SCROPHULARIACEAE • FIGWORT FAMILY ▼

Common Mullein • **Verbascum thapsus*

Biennial herb of fields, roadsides, and disturbed sites, up to 2 m tall. **Flowers** *yellow, 5-parted, sessile, 1–2.8 cm wide, in a dense, solitary spike.* **Leaves** alternate, *woolly on both sides with branched hairs*, up to 30 cm long. **Stems** *woolly with branched hairs*.
OCCURRENCE: Rare, with scattered distribution.
OTHER NAMES: Flannel-plant

THYMELAEACEAE • MEZEREUM FAMILY ▼

Eastern Leatherwood • *Dirca palustris*

Deciduous shrub of hardwood forests, up to 2 m tall, *flowering in April.* **Flowers** *pale yellow, 4-parted, in clusters of 2–4.* **Leaves** alternate, *blunt at apex,* 4–8 cm long and 2.5–5 cm wide. **Branchlets** *with conspicuously swollen nodes, very pliable.* **Bark** gray to brown, smooth.
OCCURRENCE: Very rare, documented from only one township in Park.

TYPHACEAE • CAT-TAIL FAMILY ▼

American Bur-reed • *Sparganium americanum*

Perennial, aquatic herb of muddy or peaty shorelines, up to 1 m tall. **Flowers** in a branched, emergent inflorescence with 3–10 staminate heads and 1–6 carpellate heads on each branch; *carpellate heads sessile in leaf axils or bracts*; stigmas solitary on all carpellate flowers. **Leaves** alternate, emergent, erect, not especially stiff, 6–12 mm wide. **Fruit** an achene, tan to dark greenish brown, in spikes 15–25 mm wide; *beak usually slightly curved, 3–5 mm long.*
OCCURRENCE: Uncommon, with scattered distribution.
NOTES: See *S. emersum* and *S. natans*.
OTHER NAMES: Lesser Bur-reed

TYPHACEAE • CAT-TAIL FAMILY

Narrow-leaved Bur-reed • *Sparganium angustifolium*

Perennial, aquatic herb of deep or shallow water and wet shorelines, up to 2 m long. **Flowers** in an unbranched, floating inflorescence, usually partially erect at water surface, with *2–4 staminate heads* and *2–5 carpellate heads*; *staminate heads contiguous, appearing as one elongate head; some carpellate heads borne a short distance above associated leaf axils or bracts*; stigmas solitary on all carpellate flowers. **Leaves** alternate, *floating, limp, not emergent*, 2–3(–5) mm wide. **Fruit** an achene, *reddish to brownish*, in spikes 10–30 mm wide; *beak straight, 1.5–2 mm long*.
OCCURRENCE: Occasional, widely distributed.
NOTES: See *S. emersum* and *S. fluctuans*.

Simple-stemmed Bur-reed • *Sparganium emersum*

Perennial, aquatic herb of peaty, shallow water, up to 2 m long. **Flowers** in an unbranched, emergent inflorescence with *3–7 staminate heads* and *1–6 carpellate heads*; staminate portion of flowering stalk 4–10 cm long, *at least some heads not contiguous; some carpellate heads borne a short distance above associated leaf axils or bracts*; stigmas solitary on all carpellate flowers. **Leaves** alternate, usually emergent, erect, stiff, occasionally some or all floating and limp; erect leaves rather stiff, flat, 4–10 mm wide; floating leaves, if present, limp, 4–18 mm wide, *triangular at least at base*. **Fruit** an achene, *green* to red-brown, in spikes 16–35 mm wide; *beak straight to slightly curved, 2–4.5 mm long*.
OCCURRENCE: Occasional, widely distributed.
NOTES: See *S. americanum*, *S. angustifolium*, and *S. fluctuans*.
OTHER NAMES: *Sparganium erectum, Sparganium chlorocarpum*

TYPHACEAE • CAT-TAIL FAMILY

Great Bur-reed • *Sparganium eurycarpum*

Perennial, aquatic herb of lakes, ponds, and streams, up to 2.5 m long. **Flowers** in a branched, emergent inflorescence with 10–40 staminate heads and 2–6 carpellate heads on each branch; carpellate heads sessile in leaf axils or bracts; *stigmas 2 on all or many carpellate flowers (stigmas can break off on mature fruit)*. **Leaves** alternate, emergent, erect, 6–20 mm wide. **Fruit** an achene, *straw-colored*, in spikes 15–50 mm wide; *beak straight, 2–4 mm long*.
OCCURRENCE: Historical record, last documented in 1958, from one township in Park.

Floating Bur-reed • *Sparganium fluctuans*

Perennial, aquatic herb of cold lakes and ponds, up to 1 m long. **Flowers** in a branched or unbranched floating inflorescence with distal portion partially erect at water surface and with 1–6 staminate heads and 1–2 carpellate heads on each branch; carpellate heads either sessile in leaf axils or bracts or borne a short distance above; *stigmas solitary on all carpellate flowers*. **Leaves** alternate, *floating, limp, unkeeled, flat*, 4–10 mm wide. **Fruit** an achene, *dark reddish brown, in spikes 15–23 mm wide; beak flat, strongly curving, 2–3.5 mm long*.
OCCURRENCE: Uncommon, with scattered distribution.
NOTES: See *S. angustifolium*, *S. emersum*, and *S. natans*.

292   THE PLANTS OF BAXTER STATE PARK

TYPHACEAE • CAT-TAIL FAMILY

Arctic Bur-reed • *Sparganium natans*

Perennial, aquatic herb of shallow water in lakes, ponds, streams, and fens, up to 60 cm long. **Flowers** *in an unbranched, floating inflorescence with 1 staminate head and 1–3 carpellate heads*; stigmas solitary on all carpellate flowers. **Leaves** alternate, *floating, limp, not emergent, 2–6 mm wide*. **Fruit** an achene, dark greenish or brownish, *in spikes 5–12 mm wide; beak curved slightly, 0.5–1.5 mm long*.
OCCURRENCE: Uncommon, with scattered distribution.
NOTES: See *S. americanum* and *S. fluctuans*.
OTHER NAMES: *Sparganium minimum*

Broad-leaved Cat-tail • *Typha latifolia*

Perennial, aquatic herb of marshes and shallow water, up to 3 m tall. **Flowers** *in a spike with staminate flowers above and carpellate flowers at base, without space between them*. **Leaves** alternate, long, lance-shaped, usually glacous, *10–23 mm wide*.
OCCURRENCE: Occasional, widely distributed.

WILDFLOWERS AND LOW SHRUBS   293

URTICACEAE • NETTLE FAMILY ▼

Small-spiked False Nettle • *Boehmeria cylindrica*

Perennial herb of floodplains, swamps, and wetlands, *usually less than 1 m tall*. **Flowers** green, *in stiff, dense, unbranched, axillary spikes*; spikes with a few staminate and several carpellate flowers. **Leaves** *opposite or nearly so*, coarsely toothed, *without stinging hairs, glabrous to pubescent beneath*, 5–18 cm long and 2–10 cm wide. **Stems** sparsely to densely pubescent, *without stinging hairs*.
OCCURRENCE: Historical record, last documented in 1984, from one township in Park.
NOTES: See *Pilea pumila*.
OTHER NAMES: *Urtica cylindrica*

Canada Wood-nettle • *Laportea canadensis*

Annual or perennial herb of rich, hardwood forests and shorelines, 0.3–1.5 m tall. **Flowers** white to greenish, in lower staminate panicles and upper carpellate panicles. **Leaves** *alternate, prominently and regularly toothed*, 6–30 cm long and 3–18 cm wide. **Stems** *covered with both conspicuous, bulbous-based stinging hairs and nonstinging hairs*.
OCCURRENCE: Rare, with scattered distribution.
NOTES: See *Urtica gracilis*.
OTHER NAMES: *Urtica canadensis*

URTICACEAE • NETTLE FAMILY

Canada Clearweed • *Pilea pumila*

Annual or perennial herb of wet woods and shorelines, *up to 70 cm tall*. **Flowers** white to greenish, *in loose, spreading, branched clusters*. **Leaves** *opposite, wedge-shaped at base, toothed, with 3–11 rounded teeth on each side*, 2–13 cm long and 1–9 cm wide. **Stems** *without stinging hairs*.
OCCURRENCE: Historical record, last documented in 1982, from one township in Park.
NOTES: See *Boehmeria cylindrica*.
OTHER NAMES: *Urtica pumila*

Slender Stinging Nettle • *Urtica gracilis*

Perennial herb of woods, fields, shorelines, and roadsides, *usually over 1 m tall*. **Flowers** greenish white, very small; petals absent; staminate and carpellate flowers borne separately in slender, spreading or drooping clusters from leaf axils. **Leaves** *opposite, sharply and coarsely toothed, usually with stinging hairs on both upper and lower surfaces*, 6–20 cm long and 2–13 cm wide. **Stems** usually covered with conspicuous, bulbous-based, stinging hairs less than 2 mm long.
OCCURRENCE: Rare, with scattered distribution.
NOTES: See *Laportea canadensis*.

VERBENACEAE • VERBENA FAMILY ▼

Blue Vervain • *Verbena hastata*

Perennial herb of marshes, fields, and shorelines, 0.5–1.5 m tall. **Flowers** *blue to purple, small, on branched, erect, narrow spikes 3–20 cm long and 5–9 mm wide.* **Leaves** *opposite, coarsely toothed, 4–20 cm long, the lowest sometimes lobed, on a short stalk.*
OCCURRENCE: Very rare, documented from only one township in Park.
OTHER NAMES: Simpler's-joy

VIOLACEAE • VIOLET FAMILY ▼

Hook-spurred Violet • *Viola adunca*

Perennial herb of forests, fields, and roadsides, 2–15 cm tall. **Flowers** pale blue to violet; lateral petals pubescent with hairs tapering to tip; spur 5–7 mm long, usually 2- to 3-times as long as wide. **Leaves** *alternate, dark green, thick, borne on aerial stems, usually with sparse to dense pubescence beneath, truncate at base; petioles distinctly winged.* **Rhizomes** usually 2–3 mm wide. **Stolons** absent. **Fruit** a capsule, 4–5 mm long.
OCCURRENCE: Historical record, last documented in 1901, from one township in Park.
NOTES: See *V. labradorica*.

VIOLACEAE • VIOLET FAMILY

Sweet White Violet • *Viola blanda*

Perennial herb of rich woods, wet slopes, and shaded ravines, 3–11 cm tall. **Flowers** white, 7–12 mm long, on peduncles 3–11 cm long, *at same height or shorter than leaves*; spur short, less than 2 mm long, less than 2-times as long as wide. **Leaves** basal only, *sharply toothed, thick, with hairs on upper surface, usually rounded or longer than wide, 2–4 cm wide*. **Rhizomes** usually 1–3 mm wide. **Stolons** usually produced by mid-summer, these naked or with tiny leaves, with less conspicuous self-pollinated (cleistogamous) flowers from nodes. **Fruit** a capsule, *purple to purple-brown*, 4–6 mm long.
OCCURRENCE: Common, widely distributed.
NOTES: See *V. pallens* and *V. renifolia*.
OTHER NAMES: Large-leaved White Violet

Blue Marsh Violet • *Viola cucullata*

Perennial herb of bogs, swamps, and wet meadows, 4–20 cm tall. **Flowers** light violet with white center, 13–25 mm long; *lateral petals pubescent with hairs less than 1 mm long and widened at tip; spur short, 1–2.5 mm long, less than 2-times as long as wide; sepals without cilia.* **Leaves** basal only, *glabrous or rarely with a few hairs near basal lobes on upper surface, pointed at apex; blades up to 10 cm wide, longer than wide; teeth at base not larger than those along middle*; petioles ascending, held above ground. **Rhizomes** usually 4–6 mm wide. **Stolons** absent. **Fruit** a capsule, 10–15 mm long, green.
OCCURRENCE: Occasional, widely distributed.
NOTES: See *V. selkirkii* and *V. sororia*.
OTHER NAMES: Marsh Blue Violet

VIOLACEAE • VIOLET FAMILY

American Dog Violet • *Viola labradorica*

Perennial herb of forests, fields, roadsides, swamps, and subalpine areas, up to 20 cm tall. **Flowers** violet to purple; lateral petals pubescent with hairs tapering to tip; spur 2.8–4.7 mm long, usually more than 2-times as long as wide. **Leaves** *alternate, light green or yellowish green, thin, on aerial stems, glabrous beneath, heart-shaped at base; petioles without wings*. **Rhizomes** usually 2–4 mm wide. **Stolons** absent. **Fruit** a capsule, 4–5.5 mm long.
NOTES: See *V. adunca*.
OCCURRENCE: Very rare, documented from only one township in Park.

Lance-leaved Violet • *Viola lanceolata*

Perennial herb of wet, sandy to peaty soil of open areas, 2–17 cm tall. **Flowers** white, 7–12 mm long; lateral petals glabrous; spur short, less than 2 mm long, less than 2-times as long as wide. **Leaves** *basal only; blades greater than 3-times as long as wide, 0.7–2.5 cm wide*. **Rhizomes** 1–2 mm wide. **Stolons** usually produced by mid-summer, naked or with tiny leaves, with less conspicuous self-pollinated (cleistogamous) flowers from nodes. **Fruit** a capsule, green, sometimes with orange spots, 5–8 mm long.
OCCURRENCE: Very rare, documented from only one township in Park.

VIOLACEAE • VIOLET FAMILY

Smooth White Violet • *Viola pallens*

Perennial herb of wet springy meadows, thickets, woods, and shallow water, 2–9 cm tall. **Flowers** white, 7–12 mm long, on peduncles 1.5–8.5 cm long, *overtopping leaves*; spur short, up to 2 mm long, less than 2-times as long as wide. **Leaves** basal only, *thin, strictly glabrous*; blades less than 1.5-times as long as wide, *1–3 cm wide; margins with rounded teeth*. **Rhizomes** 1–2 mm wide. **Stolons** usually produced by mid-summer, naked or with tiny leaves, with less conspicuous self-pollinated (cleistogamous) flowers from nodes. **Fruit** a capsule, *green, often with orange dots*, 4–6 mm long.
OCCURRENCE: Common, widely distributed.
NOTES: See *V. blanda* and *V. renifolia*.
OTHER NAMES: Small White Violet, *Viola macloskeyi*

Yellow Forest Violet • *Viola pubescens*

Perennial herb of rich woods and roadsides, 10–45 cm tall. **Flowers** *yellow; lateral petals pubescent with hairs widened at tip; spur less than 2 mm long, less than 2-times as long as wide*. **Leaves** alternate, *on aerial stems, with toothed margins*. **Rhizomes** usually 3–5 mm wide. **Stolons** absent. **Fruit** a capsule, green, 10–12 mm long.
OCCURRENCE: Occasional, with scattered distribution.

VIOLACEAE • VIOLET FAMILY

Kidney-leaved Violet • *Viola renifolia*

Perennial herb of forests, swamps, and roadsides, up to 10 cm tall. **Flowers** white, 7–12 mm long, *at same height or shorter than leaves*; peduncles 3–10 cm long; spur short, up to 2 mm long, less than 2-times as long as wide. **Leaves** basal only, *thick, with hairs on lower surface; blades 1.5–8 cm wide, wider than long; margins with sharp teeth*. **Rhizomes** usually less than 3 mm wide. **Stolons** absent. **Fruit** a capsule, *with purple spotting or coloration*, 4–5 mm long.
OCCURRENCE: Occasional, widely distributed.
NOTES: See *V. blanda* and *V. pallens*.

Round-leaved Violet • *Viola rotundifolia*

Perennial herb of forests, up to 7 cm tall. **Flowers** yellow; *lateral petals pubescent with hairs widened at tip*. **Leaves** basal only; petioles spreading, nearly flat on ground; stipules entire. **Rhizomes** usually 4–6 mm wide. **Stolons** absent. **Fruit** a capsule, often spotted with purple, 5–8 mm long.
OCCURRENCE: Rare, with scattered distribution.

VIOLACEAE • VIOLET FAMILY

Arrowhead Violet • *Viola sagittata*

Perennial herb of forests, fields, roadsides, and swamps, 3–20 cm tall. **Flowers** violet-purple with a darker center, 13–22 mm long; *lateral petals pubescent with hairs longer than 1 mm and tapering to tip; spur short, 1.7–2.5 mm long, less than 2 times as long as wide*. **Leaves** basal only; blades greater than 1.5-times as long as wide, *coarsely toothed or lobed and wedge-shaped to subcordate at base*. **Rhizomes** 4–6 mm wide. **Stolons** absent. **Fruit** a capsule, green, 6–10 mm long.
OCCURRENCE: Very rare, documented from only one township in Park.

Great-spurred Violet • *Viola selkirkii*

Perennial herb of moist forests, up to 10 cm tall. **Flowers** violet with a white center; *lateral petals glabrous; sepals without cilia; spur elongate, 4–6 mm long, 2 or more times as long as wide*. **Leaves** basal only, *minutely pubescent above, glabrous beneath; blades 1.5–3 cm wide, with rounded teeth; basal lobes converging or overlapping*; petioles spreading. **Rhizomes** usually 2–3 mm wide. **Stolons** absent. **Fruit** a capsule, often spotted with purple, 4–6 mm long.
OCCURRENCE: Historical record, last documented in 1901, from one township in Park.
NOTES: See *V. cucullata* and *V. sororia*.

WILDFLOWERS AND LOW SHRUBS 301

VIOLACEAE • VIOLET FAMILY

Woolly Blue Violet • *Viola sororia*

Perennial herb of woods, clearings, meadows, and slopes, 3–25 cm tall. **Flowers** light purple, 13–22 mm long; *lateral petals pubescent with hairs longer than 1 mm and tapering to tip; spur short, 1.5–3.2 mm long, less than 2-times as long as wide; sepals usually with cilia.* **Leaves** basal only, *pubescent; blades as wide as or wider than long, blunt at apex,* unlobed, *with teeth at base not larger than those along middle, cordate at base*; petioles ascending, held above ground. **Rhizomes** usually 4–6 mm wide. **Stolons** absent. **Fruit** a capsule, green, with purple spots, 6–10 mm long.
OCCURRENCE: Occasional, widely distributed.
NOTES: See *V. cucullata* and *V. selkirkii*.
OTHER NAMES: *Viola septentrionalis*

VISCACEAE • CHRISTMAS-MISTLETOE FAMILY ▼

Dwarf Mistletoe • *Arceuthobium pusillum*

Shrublet, parasitic on white and black spruce branches, up to 2 cm tall. **Flowers** solitary in axils; staminate flowers usually 3-parted; carpellate flowers 2-parted. **Leaves** scale-like on rectangular branches. **Stems** olive-green, rust, or purple.
OCCURRENCE: Very rare, documented from only one township in Park.

VITACEAE • GRAPE FAMILY ▼

Virginia-creeper • *Parthenocissus quinquefolia*

Perennial liana of forests, shorelines, and rocky slopes, up to 30 m long. **Flowers** *yellowish green, very small, in terminal panicles longer than wide from upper axils.* **Leaves** alternate, *palmately divided, with 5 coarsely toothed leaflets; tip of tendrils with a disc-like foot that attaches to walls and tree trunks, allowing plant to climb.* **Stems** trailing or climbing. **Fruit** a dark blue to black berry, 5–7 mm wide.
OCCURRENCE: Very rare, documented from only one township in Park.

XYRIDACEAE • YELLOW-EYED-GRASS FAMILY ▼

Bog Yellow-eyed-grass • *Xyris difformis*

Perennial herb of peaty and sandy shorelines, 20–70 cm tall. **Flowers** yellow, 3-parted, *in spikes 10–20 mm long; largest floral bracts 5–7 mm long, with an entire margin,* usually with a well-defined, green, central patch 2–3 mm long; *sepal tips covered by subtending bract; scape winged at apex, 0.5–3 mm wide.* **Leaves** basal only, linear, deep green, erect, the largest 10–30 cm long and 2–7 mm wide; sheaths smooth.
OCCURRENCE: Very rare, documented from only one township in Park.
NOTES: See *X. montana*.

XYRIDACEAE • YELLOW-EYED-GRASS FAMILY

Northern Yellow-eyed-grass • *Xyris montana*

Perennial herb of peaty and sandy shorelines, 5–30 cm tall. **Flowers** yellow, 3-parted, *in spikes 4–8 mm long; largest floral bracts 3–4 mm long, with a ragged or fibrous margin,* usually lacking a well-defined green, central patch or patch less than 2 mm long; *sepal tips visible above apex of adjacent bract; scape not winged at apex, 0.5–0.8 mm wide.* **Leaves** basal only, linear, deep green, erect, 4–15 cm long and 0.8–2 mm wide; sheaths covered with short, rounded projections.
OCCURRENCE: Rare, with scattered distribution.
NOTES: See *X. difformis*.

ADOXACEAE • ELDERBERRY FAMILY

Black Elderberry • *Sambucus nigra*

Deciduous shrub of *wet areas*, 1–4 m tall, *flowering July and August*. **Flowers** white, 5-parted, 3–5 mm wide, fragrant, *in clusters wider than tall with 4–6 primary rays*. **Leaves** opposite, pinnately divided; leaflets 5–11, usually 7. **Twigs** *with white pith in second-year branches*. **Fruit** a *purple-black*, berry-like drupe, *ripening late summer*.
OCCURRENCE: Rare, with scattered distribution.
NOTES: See *S. racemosa*.
OTHER NAMES: Common Elderberry, *Sambucus canadensis*

Red Elderberry • *Sambucus racemosa*

Deciduous shrub of *dry areas*, up to 8 m tall, *flowering May and June*. **Flowers** creamy white, 5-parted, 3–4 mm wide, ill-scented, *in clusters taller than wide with a main stalk extending up through flowers to near apex*. **Leaves** opposite, pinnately divided, with 5–7 leaflets. **Twigs** *with orange-brown pith in second-year branches*. **Fruit** a *bright red*, berry-like drupe, *ripening early summer*.
OCCURRENCE: Occasional, widely distributed.
NOTES: See *S. nigra*.
OTHER NAMES: Red-berried Elder, Stinking Elder, *Sambucus pubens*

ADOXACEAE • ELDERBERRY FAMILY

Maple-leaved Viburnum • *Viburnum acerifolium*

Deciduous shrub of hardwood forests, 1–2 m tall. **Flowers** white, 5-parted; outer flowers in inflorescence same size as inner flowers. **Leaves** opposite, the lowest deeply 3-lobed, with *outer lobes curving outward; petioles without raised glands near junction with leaf blade.* **Twigs** *of new branchlets minutely pubescent.* **Fruit** a dark blue to dark purple berry-like drupe.
OCCURRENCE: Very rare, documented from only one township in Park.
NOTES: See *V. edule* and *V. opulus.*
OTHER NAMES: Dockmackie

Smooth Arrowwood • *Viburnum dentatum*

Deciduous shrub of wet to dry soils, 1–3 m tall. **Flowers** white, 5-parted; outer flowers in inflorescence same size as inner flowers. **Leaves** opposite, *coarsely toothed, the lowest unlobed.* **Twigs** *new branchlets densely pubescent.* **Fruit** a dark blue to black berry-like drupe.
OCCURRENCE: Rare, documented from only southern portion of Park.
OTHER NAMES: Southern Arrow-wood, *Viburnum recognitum*

ADOXACEAE • ELDERBERRY FAMILY

Squashberry • *Viburnum edule*

Deciduous shrub of shorelines, wetland edges, and forests in boreal and subalpine regions, up to 2 m tall. **Flowers** white, 5-parted; outer flowers in inflorescence same size as inner flowers. **Leaves** opposite, round in outline, *roughly as long as wide*, the lowest 3-lobed, with shallow sinuses, *with outer lobes curving forward*; petioles occasionally with glands near junction with leaf blade. **Twigs** *with new branchlets glabrous*. **Fruit** a red berry-like drupe.
OCCURRENCE: Rare, with scattered distribution.
NOTES: See *V. acerifolium* and *V. opulus*.

Hobblebush • *Viburnum lantanoides*

Deciduous shrub of shady, wet woods, up to 3 m tall. **Flowers** white, 5-parted; outer flowers in inflorescence sterile, larger than inner flowers. **Leaves** opposite, *rounded in outline, roughly as long as wide, finely and sharply toothed*, the lowest unlobed. **Twigs** with new branchlets densely pubescent or with small scales. **Fruit** a red to black berry-like drupe.
OCCURRENCE: Common, widely distributed.
OTHER NAMES: Mooseberry, Witch-hobble, Tangle-legs, *Viburnum alnifolium*

ADOXACEAE • ELDERBERRY FAMILY

Nannyberry • *Viburnum lentago*

Deciduous shrub of field margins, swamp edges, and riparian forests, up to 10 m tall. **Flowers** white, 5-parted; outer flowers in inflorescence same size as inner flowers. **Leaves** opposite, *finely and sharply toothed*, the lowest unlobed; *petiole irregularly winged*. **Twigs** with new branchlets glabrous. **Fruit** a dark blue to black berry-like drupe.
OCCURRENCE: Historical record, last documented in 1984, from northern portion of Park.
NOTES: See *V. nudum*.

Withe-rod • *Viburnum nudum*

Deciduous shrub of woods and swamps, 1–4 m tall. **Flowers** white, 5-parted; *array of flowers on a stalk 5–50 mm long*; outer flowers in inflorescence same size as inner flowers. **Leaves** opposite, *bluntly or obscurely toothed*, the lowest unlobed. **Twigs** with new branchlets covered with small scales. **Fruit** a dark blue berry-like drupe.
OCCURRENCE: Common, widely distributed.
NOTES: See *V. lentago*.
OTHER NAMES: Wild-raisin, *Viburnum cassinoides*

ADOXACEAE • ELDERBERRY FAMILY

Highbush-cranberry • *Viburnum opulus*

Deciduous shrub of wet areas, up to 4 m tall. **Flowers** white, 5-parted; outer flowers in inflorescence sterile, larger than inner flowers. **Leaves** opposite, *longer than wide*, with deep sinuses, the lowest 3-lobed, *with outer lobes curving outward; petioles with several raised glands near junction with leaf blade*. **Twigs** *with new branchlets glabrous*. **Fruit** a shiny, red, berry-like drupe.
OCCURRENCE: Uncommon, with scattered distribution.
NOTES: See *V. acerifolium* and *V. edule*.
OTHER NAMES: Pimbina, *Viburnum trilobum*

ANACARDIACEAE • CASHEW FAMILY ▼

Smooth Sumac • *Rhus glabra*

Deciduous shrub of roadsides, forest edges, and dry fields, up to 6 m tall. **Flowers** greenish white, 5-parted, unisexual. **Leaves** pinnately divided; *petioles glabrous*. **Twigs** *glabrous*. **Fruit** a red drupe *with club-shaped hairs 0.1–0.3 mm long*.
OCCURRENCE: Historical record, last documented in 1982, from northern portion of Park.
NOTES: See *R. typhina*.
OTHER NAMES: Red Sumac

TREES AND TALL SHRUBS

ANACARDIACEAE • CASHEW FAMILY

Staghorn Sumac • *Rhus typhina*

Deciduous shrub of roadsides, forest edges, and dry fields, up to 11 m tall. **Flowers** greenish white, 5-parted, unisexual. **Leaves** pinnately divided; *petioles densely pubescent*. **Twigs** *densely velvety-pubescent*. **Fruit** a red drupe *with slender hairs 0.5–3 mm long*.
OCCURRENCE: Occasional, widely distributed.
NOTES: See *R. glabra*.
OTHER NAMES: Velvet Sumac, Vinegar-tree, *Rhus hirta*

AQUIFOLIACEAE • HOLLY FAMILY ▼

Mountain Holly • *Ilex mucronata*

Deciduous shrub of alpine areas, bogs, shorelines, and disturbed sites, 0.3–3 m tall. **Flowers** yellowish or greenish, 4- or 5-parted, unisexual. **Leaves** alternate, 3–5.5 cm long and 1.5–2.5 cm wide, widest at middle or above, entire throughout or toothed near apex; petioles glabrous, purple. **Fruit** a deep red, berry-like drupe, ~7 mm wide; pedicels 1–3 cm long.
OCCURRENCE: Common, widely distributed.
NOTES: See *I. verticillata*.
OTHER NAMES: Catberry, *Nemopanthus mucronatus*

AQUIFOLIACEAE • HOLLY FAMILY

Common Winterberry • *Ilex verticillata*

Deciduous shrub of wet areas, 1–4 m tall. **Flowers** whitish, 5- to 8-parted, bisexual or unisexual. **Leaves** alternate, *5–10 cm long and 2–4.5 cm wide, widest at middle or below, finely toothed throughout; petioles short-pubescent, green.* **Fruit** a shiny, bright red, berry-like drupe, ~7 mm wide, conspicuous and persistent into winter; *pedicels less than 1 cm long.*
OCCURRENCE: Common, widely distributed.
NOTES: See *I. mucronata, Aronia floribunda* (Rosaceae), and *Aronia melanocarpa* (Rosaceae). Drupes are poisonous to humans.
OTHER NAMES: Black Alder, Winterberry

BETULACEAE • BIRCH FAMILY ▼

Green Alder • *Alnus alnobetula*

Deciduous shrub of bogs, rocky shorelines, and mountains, up to 3 m tall. **Staminate catkins** shedding pollen in late May to early June with expansion of leaves. **Carpellate catkins** *all on stalks 8–30 mm long.* **Leaves** toothed, *without shallow lobes; cross-veins (those between parallel secondary veins) inconspicuous on lower surface, not extending all the way between parallel secondary veins.*
OCCURRENCE: Occasional, widely distributed.
NOTES: See *A. incana.*
OTHER NAMES: Mountain Alder, *Alnus crispa, Alnus viridis*

Speckled Alder • *Alnus incana*

Deciduous shrub of wet areas, up to 3 m tall. **Staminate catkins** shedding pollen in April before leaves appear. **Carpellate catkins** *sessile except for terminal one on a 4–8-mm-long stalk.* **Leaves** toothed, *usually shallowly lobed to 6 mm deep; cross-veins (those between parallel secondary veins) conspicuous on lower surface.*
OCCURRENCE: Common, widely distributed.
NOTES: See *A. alnobetula.*
OTHER NAMES: *Alnus rugosa*

BETULACEAE • BIRCH FAMILY

Yellow Birch • *Betula alleghaniensis*

Deciduous tree of hardwood forests and swamps, up to 30 m tall. **Carpellate catkins** erect when mature, usually less than 3 cm long; stalks 1–3 mm long. **Leaves** with 8–12 pairs of distinct, lateral veins, coarsely and irregularly toothed. **Twigs** *with a strong wintergreen odor when crushed.* **Bark** *yellowish brown, exfoliating in thin strips.*
OCCURRENCE: Common, widely distributed.
OTHER NAMES: *Betula lutea*

Heart-leaved Paper Birch • *Betula cordifolia*

Deciduous tree of mixed woods, subalpine forests, alpine ravines, and talus slopes, up to 20 m tall. **Carpellate catkins** pendulous when mature, in clusters of 2–4; stalks 10–23 mm long. **Leaves** *cordate at base*, with 9–12 pairs of lateral veins, pointed at apex but not long-tapering. **Twigs** nonaromatic. **Bark** *orange-tinged, exfoliating in large sheets.*
OCCURRENCE: Common, widely distributed.
NOTES: A hybrid (*Betula ×caerulea*), similar to this species and also *B. populifolia*, has been documented at Baxter State Park. It can be recognized by having leaf blades with (6–)7–9 pairs of lateral veins (versus (8–)9–12 in *B. cordifolia* and 5–7 in *B. populifolia*) and carpellate catkins 25–50 mm long at maturity (versus 40–60 mm long in *B. cordifolia* and 10–25(–30) mm long in *B. populifolia*).
OTHER NAMES: Mountain Paper Birch, *Betula papyrifera* var. *cordifolia*

BETULACEAE • BIRCH FAMILY

Glandular Birch • *Betula glandulosa*

Deciduous shrub of alpine plateaus and ravines, up to 3 m tall. **Carpellate catkins** erect when mature, 1–2.5 cm long. **Leaves** firm, with resin glands and blunt teeth; blades 0.5–3 cm long, *rounded at apex.* **Twigs** *with large and abundant resin glands.*
OCCURRENCE: Very rare, documented from only one township in Park.
NOTES: Species is listed as Endangered in Maine.

Dwarf Birch • *Betula minor*

Deciduous shrub of alpine plateaus and ravines, up to 5 m tall. **Carpellate catkins** erect when mature, 1–3 cm long. **Leaves** firm, often with resin glands, with blunt to sharp teeth; blades 1.5–5.5 cm long, *acute to acuminate at apex.* **Twigs** often with resin glands.
OCCURRENCE: Very rare, documented from only one township in Park.
NOTES: Species is listed as Endangered in Maine.
OTHER NAMES: *Betula pubescens* ssp. *minor*

BETULACEAE • BIRCH FAMILY

Paper Birch • *Betula papyrifera*

Deciduous tree of mixed woods, up to 25 m tall. **Carpellate catkins** pendulous when mature, in clusters of 2–4; stalks 7–17 mm long. **Leaves** *rounded to truncate at base*, with 7–9 pairs of lateral veins, pointed at apex but not long-tapering. **Twigs** nonaromatic. **Bark** *white, exfoliating in large sheets.*
OCCURRENCE: Common, widely distributed.
OTHER NAMES: Canoe Birch, White Birch

Gray Birch • *Betula populifolia*

Deciduous tree of forests and wetlands, up to 10 m tall. **Carpellate catkins** erect to pendulous when mature, solitary or infrequently paired. **Leaves** *triangular, truncate at base*, doubly toothed except at base, mostly with 5–7 pairs of lateral veins, glabrous on underside, *with a long-acuminate tip.* **Twigs** with warty resin glands. **Bark** chalky, not exfoliating.
OCCURRENCE: Uncommon, with scattered distribution.
OTHER NAMES: Fire Birch

BETULACEAE • BIRCH FAMILY

Beaked Hazelnut • *Corylus cornuta*

Deciduous shrub of upland forests, up to 3 m tall. **Carpellate catkins** resembling leaf buds, but with red stigmas. **Staminate catkins** sessile, stout, slightly curved. **Leaves** *short-acuminate, coarsely and doubly toothed, with a pale green underside.* **Fruit** *a nut inside a husk-like covering (involucre) 4–7 cm long, with a pronounced, tubular beak* covered with tiny, stiff, bristly protrusions.
OCCURRENCE: Common, widely distributed.

Hop-hornbeam • *Ostrya virginiana*

Deciduous, shade-tolerant tree of rich forests and rocky slopes, up to 10 m tall. **Carpellate catkins** borne singly at tip of first-year branchlets. **Leaves** alternate, *finely toothed, teeth sharply and often doubly toothed; leaf bases symmetric.* **Bark** furrowed in flat, narrow ridges, *exfoliating in vertical strips.* **Fruit** a small nutlet *enclosed in an inflated sac*, in clusters borne at branch tips.
OCCURRENCE: Uncommon, with scattered distribution.
NOTES: See *Ulmus americana* in the Ulmaceae family
OTHER NAMES: Ironwood, Leverwood

CORNACEAE • DOGWOOD FAMILY ▼

Alternate-leaved Dogwood • *Swida alternifolia*

Deciduous shrub or small tree, up to 8 m tall. **Flowers** creamy white, 4-parted, in dense, terminal cymes 3–6 cm wide. **Leaves** *alternate*, often crowded, with 5 or 6 veins on each side of leaf; *petioles usually longer than 2.5 cm.* **Twigs** green; *new branchlets glabrous.* **Fruit** *a blue to black drupe.*
OCCURRENCE: Common, widely distributed.
OTHER NAMES: Pagoda Dogwood, Green Osier, *Cornus alternifolia*

Round-leaved Dogwood • *Swida rugosa*

Deciduous shrub, 1–4 m tall. **Flowers** white, 4-parted, in dense, terminal cymes 2–7 cm wide. **Leaves** *opposite, with 7–9 veins on each side of leaf; petioles less than 2 cm long.* **Twigs** yellow-green; *new branchlets usually pubescent and with dark purple streaks.* **Fruit** *a pale blue drupe.*
OCCURRENCE: Occasional, widely distributed.
OTHER NAMES: *Cornus rugosa*

TREES AND TALL SHRUBS 317

CORNACEAE • DOGWOOD FAMILY

Red-osier Dogwood • *Swida sericea*

Deciduous shrub of shorelines, wet fields, and swamps, 1–3 m tall. **Flowers** white, 4-parted, in terminal cymes 2–5.5 cm wide. **Leaves** *opposite, with 5–7 veins on each side of leaf; petioles less than 2.5 cm long.* **Twigs** green-red to red, with white pith; *new branchlets usually pubescent and without dark purple streaks.* **Fruit** a white drupe.
OCCURRENCE: Occasional, widely distributed.
OTHER NAMES: *Cornus sericea, Cornus stolonifera, Swida stolonifera*

CUPRESSACEAE • CYPRESS FAMILY ▼

Common Juniper • *Juniperus communis*

Low, evergreen shrub of dry fields and shorelines, usually less than 0.5 m tall. **Leaves** whorled, often in groups of 3, needle-like, *blue-green with white or yellowish central stripe, tapering to a sharp point.* **Seed cones** fleshy, resembling a berry, *borne in axils of branches.*
OCCURRENCE: Uncommon, with scattered distribution.
OTHER NAMES: Ground Juniper

CUPRESSACEAE • CYPRESS FAMILY

Northern White-cedar • *Thuja occidentalis*

Evergreen tree of wet slopes and woods, up to 15 m tall. **Leaves** scale-like, *yellowish green, keeled and overlapping*. **Bark** shredding, light reddish brown. **Seed cones** *leathery-woody, erect*, with winged seeds inside larger, middle scales.
OCCURRENCE: Common, widely distributed.
OTHER NAMES: Arborvitae

ERICACEAE • HEATH FAMILY ▼

Highbush Blueberry • *Vaccinium corymbosum*

Deciduous, crown-forming shrub of wet areas, 1–4 m tall. **Flowers** white or pink, urn-shaped, 5–10 mm long; anthers without projecting horns. **Leaves** alternate, thin, not shiny, *pale green, glabrous or sparsely pubescent beneath with white to gray-white hairs*; mature blades 3–8 cm long and 2–3 cm wide. **Branchlets** with rounded bumps on surface, glabrous or pubescent in two lines down branchlet with whitish hairs. **Bark** peeling and shredding from base of stem. **Fruit** a *dark blue to blue-black, glaucous* berry, *sweet*.
OCCURRENCE: Occasional, widely distributed.
NOTES: See *V. fuscatum*.
OTHER NAMES: Swamp Blueberry

TREES AND TALL SHRUBS   319

ERICACEAE • HEATH FAMILY

Black Highbush Blueberry • *Vaccinium fuscatum*

Deciduous, crown-forming shrub of wet areas, 1–4 m tall. **Flowers** white or pink, urn-shaped, 5–8 mm long; anthers without projecting horns. **Leaves** alternate, thin, not shiny, *dark green, moderately to densely pubescent beneath with brown to gray hairs*; mature blades 3–8 cm long and 1.5–4 cm wide. **Branchlets** with rounded bumps on surface, pubescent throughout with brown to gray hairs. **Bark** peeling and shredding from base of stem. **Fruit** a *shiny, black* berry, *semi-sweet*.
OCCURRENCE: Very rare, documented from only one township in Park.
NOTES: See *V. corymbosum*.
OTHER NAMES: *Vaccinium atrococcum*

FABACEAE • LEGUME FAMILY ▼

Bristly Locust • **Robinia hispida*

Stoloniferous shrub of woods, thickets, and slopes, up to 3.5 m tall. **Flowers** magenta-pink, 5-parted, ~2 cm long, crowded in somewhat erect racemes; *peduncles sticky, glandular*. **Leaves** alternate, pinnately divided; leaflets 9–17, *bristle-tipped*, glandular-bristly. **Twigs** *sticky with warty glands*. **Fruit** a flat, *glandular legume covered with few short, stiff hairs*, several-seeded.
OCCURRENCE: Very rare, documented from only one township in Park.
NOTES: Planted by Park in one location in the 1980s.
OTHER NAMES: Rose-acacia, Mossy Locust

320   THE PLANTS OF BAXTER STATE PARK

FAGACEAE • BEECH FAMILY ▼

American Beech • *Fagus grandifolia*

Deciduous tree of rich uplands, up to 30 m tall. **Carpellate flowers** in pairs above involucral bracts. **Leaves** *alternate, sharply and coarsely pointed with hooked teeth*. **Buds** 8–25 mm long, slender, *sharply pointed*. **Bark** *light gray*, often pockmarked. **Fruit** a hardened involucre with 2 triangular nuts.
OCCURRENCE: Common, widely distributed.

Northern Red Oak • *Quercus rubra*

Deciduous tree of uplands, up to 30 m tall. **Staminate flowers** in long, slender, conspicuous catkins. **Leaves** alternate, *with 2–4 pairs of primary lobes with bristle-tips on each lobe*. **Bark** hard and furrowed in older trees, sometimes with reddish fissures. **Fruit** *an acorn, with a scaly dome-shaped cup usually covering about one-quarter of nut*.
OCCURRENCE: Occasional, widely distributed.

TREES AND TALL SHRUBS 321

HAMAMELIDACEAE • WITCH-HAZEL FAMILY ▼

American Witch-hazel • *Hamamelis virginiana*

Deciduous tree or shrub of dry, rocky woods, up to 6 m tall. **Flowers** bright yellow, *flowering in small axillary clusters in autumn*; petals 1.5–2 cm long and 1 mm wide. **Leaves** alternate, 5–15 cm long, *with stellate hairs on veins beneath, with asymmetric bases and wavy margins*. **Fruit** a woody, pubescent capsule, 1–1.5 cm long.
OCCURRENCE: Uncommon, with scattered distribution.

MALVACEAE • MALLOW FAMILY ▼

American Linden • *Tilia americana*

Deciduous tree of floodplains and forests, up to 40 m tall. **Flowers** white to yellow, 5-parted, fragrant; *peduncle joined with a tongue-shaped bract with inflorescence beneath*. **Leaves** alternate, simple, palmately veined, sharply toothed with glandular teeth. **Bark** gray, furrowed.
OCCURRENCE: Very rare, with scattered distribution.
OTHER NAMES: American Basswood

OLEACEAE • OLIVE FAMILY ▼

White Ash • *Fraxinus americana*

Deciduous tree of rich sites, up to 40 m tall. **Flowers** without petals; *sepals persistent*. **Leaves** opposite, pinnately divided; leaflets 5–9, glabrous beneath; *lateral leaflets on a glabrous stalk 5–15 mm long, not noticeably winged*. **Twigs** smooth, glabrous; terminal buds low and rounded, wider than long. **Bark** of mature trunk *with distinctive, diamond-patterned ridges*. **Fruit** a samara 1–5 cm long, with a long terminal wing.
OCCURRENCE: Common, widely distributed.
NOTES: See *F. nigra* and *F. pennsylvanica*.

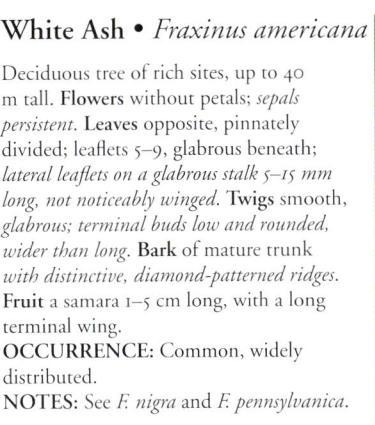

Black Ash • *Fraxinus nigra*

Deciduous tree of swamps, up to 25 m tall. **Flowers** without petals; *sepals deciduous*. **Leaves** opposite, pinnately divided; leaflets 7–13, 11–14 cm long; *lateral leaflets sessile, with a tangle of brown hairs at base*. **Twigs** round, smooth, glabrous; terminal buds dark brown, conical. **Bark** of mature trunks with soft, corky ridges. **Fruit** a samara, winged nearly to base, nearly flat.
OCCURRENCE: Uncommon, widely distributed.
NOTES: See *F. americana* and *F. pennsylvanica*.

TREES AND TALL SHRUBS 323

OLEACEAE • OLIVE FAMILY

Green Ash • *Fraxinus pennsylvanica*

Deciduous tree of hardwood and mixed forests, up to 25 m tall. **Flowers** without petals; *sepals deciduous*. **Leaves** opposite, pinnately divided; leaflets 5–9, 11–14 cm long, *often pubescent on underside; lateral leaflets on a pubescent, winged stalk 1–4 mm long*. **Twigs** round, *pubescent; terminal buds dark brown, conical*. **Fruit** a samara, winged almost halfway to base, nearly flat.
OCCURRENCE: Common, widely distributed.
NOTES: See *F. americana* and *F. nigra*.

Common Lilac • **Syringa vulgaris*

Deciduous shrub of fields, roadsides, and abandoned homesteads, up to 6 m tall. **Flowers** *lilac to purple, occasionally white, up to 1 cm wide, fragrant, 4-parted*. **Leaves** opposite, simple, *entire*, glabrous beneath, 5–10 cm long. **Twigs** usually less than 2 mm wide.
OCCURRENCE: Rare, with scattered distribution.

PINACEAE • PINE FAMILY ▼

Balsam Fir • *Abies balsamea*

Evergreen tree of uplands and wetlands, up to 23 m tall. **Leaves** flat, needle-like, with whitened lower surface and prominent midrib, blunt at apex, sessile, *leaving circular scar*, 1.2–2.5 cm long and 1.5–2 mm wide. **Bark** thin, *with resin-filled blisters*. **Seed cones** erect, on upper sides of branches, 3–8 cm long; scales deciduous.
OCCURRENCE: Common, widely distributed.
NOTES: See *Tsuga canadensis* and *Taxus canadensis*.

American Larch • *Larix laricina*

Deciduous tree or shrub of wetlands, up to 20 m tall. **Leaves** soft, needle-like, turning vibrant yellow in autumn, in circular clusters on short, woody spurs on older branches or scattered spirally on younger branches, 1–2 cm long and 0.5–0.8 mm wide. **Twigs** with many spurs. **Bark** thin and scaly. **Seed cones** erect, 1–2 cm long.
OCCURRENCE: Occasional, widely distributed.
NOTES: This is our only deciduous conifer. The wood is resistant to decay and is often used in construction of docks and wharves.
OTHER NAMES: Hackmatack, Larch, Tamarack

TREES AND TALL SHRUBS 325

PINACEAE • PINE FAMILY

White Spruce • *Picea glauca*

Evergreen tree of bogs, streambanks, and wooded slopes, up to 30 m tall. **Leaves** needle-like, *blue-green*, 4-angled in cross section, *sharp-pointed at tip*, with an odor of cat urine when crushed, *1.5–2 cm long*. **Youngest branchlets** *glaucous, glabrous*. **Bark** thin and scaly; inner bark whitish. **Seed cones** *2.5–6 cm long; scales flexuous, with margin at apex usually entire*.
OCCURRENCE: Occasional, widely distributed.
NOTES: See *P. mariana*, *P. rubens*, and *Tsuga canadensis*.
OTHER NAMES: Cat Spruce

Black Spruce • *Picea mariana*

Evergreen tree or shrub of wetlands, and subalpine and alpine areas, up to 25 m tall. **Leaves** needle-like, *pale blue-green*, 4-angled, *blunt at tip*, *0.6–1.5 cm long*. **Youngest branchlets** *pubescent, with glandular hairs*. **Bark** thin, scaly; inner bark yellowish. **Seed cones** *1.5–2.5 cm long; scales rigid, with margin at apex irregularly toothed*.
OCCURRENCE: Occasional, widely distributed.
NOTES: See *P. glauca*, *P. rubens*, and *Tsuga canadensis*.
OTHER NAMES: Bog Spruce

PINACEAE • PINE FAMILY

Red Spruce • *Picea rubens*

Evergreen tree of forests, up to 40 m tall. **Leaves** needle-like, *yellowish green to dark green*, 4-angled, *sharp-pointed at tip, 0.8–2.5 cm long*. **Youngest branchlets** *densely pubescent, without glandular hairs*. **Bark** thin, scaly; inner bark yellowish. **Seed cones** 2.3–4.5 cm long; scales rigid, with margin at apex entire to irregularly toothed.
OCCURRENCE: Common, widely distributed.
NOTES: See *P. glauca*, *P. mariana*, and *Tsuga canadensis*.

Jack Pine • *Pinus banksiana*

Evergreen tree of forests, dry hills, and lakeshores, up to 27 m tall. **Leaves** needle-like, twisted, thick, *stiff, in fascicles of 2, 2–5 cm long* and 1–1.5 mm wide; *pairs widely forking*. **Bark** orange- to red-brown, scaly. **Seed cones** *asymmetric*, usually remaining closed, 3–5.5 cm long.
OCCURRENCE: Uncommon, with scattered distribution.
NOTES: See *P. resinosa* and *P. rigida*.
OTHER NAMES: Scrub Pine

TREES AND TALL SHRUBS 327

PINACEAE • PINE FAMILY

Red Pine • *Pinus resinosa*

Evergreen tree of deep, sandy soils, up to 37 m tall. **Leaves** needle-like, straight or slightly twisted, *brittle, in fascicles of 2, 10–20 cm long* and 1 mm wide. **Bark** light red-brown, furrowed into scaly plates. **Seed cones** *symmetric*, nearly stalkless, opening and shedding from tree soon after maturity, 3.5–6 cm long.
OCCURRENCE: Occasional, widely distributed.
NOTES: See *P. banksiana* and *P. rigida*.

Pitch Pine • *Pinus rigida*

Evergreen tree of granite outcrops, up to 31 m tall. **Leaves** needle-like, usually twisted, *in fascicles of 3, 5–10 cm long* and 1–1.5 mm wide. **Bark** red-brown, rough, furrowed. **Seed cones** *symmetric*, often clustered, persistent, *3–9 cm long*.
OCCURRENCE: Very rare, documented from only one township in Park.
NOTES: See *P. banksiana* and *P. resinosa*.

328 THE PLANTS OF BAXTER STATE PARK

PINACEAE • PINE FAMILY

Eastern White Pine • *Pinus strobus*

Evergreen tree of forests and fields, up to 60 m tall. **Leaves** needle-like, straight or slightly twisted, *flexible, in fascicles of 5, 6–10 cm long* and 0.7–1 mm wide. **Bark** gray-brown, deeply furrowed on older trees. **Seed cones** *symmetric*, much longer than wide, 8–20 cm long.
OCCURRENCE: Occasional, widely distributed.

Eastern Hemlock • *Tsuga canadensis*

Evergreen tree of cool, moist sites, up to 30 m tall. **Leaves** needle-like, shiny green, *with 2 white lines beneath, flattened, 15–20 mm long; petioles present.* **Bark** brownish, fissured; inner bark rose-colored. **Seed cones** *pendent from ends of twigs, 1.5–2.5 cm long.*
OCCURRENCE: Occasional, widely distributed.
NOTES: See *Abies balsamea* and *Picea* spp.

ROSACEAE • ROSE FAMILY ▼

Downy Shadbush • *Amelanchier arborea*

Deciduous tree or tall shrub of forests, fields, and roadsides, 2–20 m tall. **Flowers** white, *borne in drooping racemes of 4–11 flowers with an axis 4–12 cm long; petals 10–18 mm long and 2–5 mm wide*; ovary apex rounded and glabrous. **Leaves** alternate, green, *4–10 cm long and 2.2–5 cm wide, less than 50% expanded at flowering time, densely white-woolly on underside when young*, retaining some pubescence at maturity; petioles at maturity 10–25 mm long. **Fruit** a purple-black pome, *6–10 mm wide, mildly sweet; fruiting pedicels pubescent, 8–17 mm long*.
OCCURRENCE: Rare, with scattered distribution.
NOTES: See *A. laevis*.
OTHER NAMES: Downy Serviceberry

Mountain Shadbush • *Amelanchier bartramiana*

Deciduous shrub of mountains, bogs, and shorelines, up to 2.5 m tall. **Flowers** white, *borne in bundles of 1–4*; petals 6–12 mm long and 3–5 mm wide; *ovary apex conical and densely pubescent*. **Leaves** alternate, brownish or greenish, *wedge-shaped at base, glabrous at flowering time*, 2–6 cm long and 1.5–3.5 cm wide; *petioles at maturity 2–10 mm long*. **Fruit** a *semi-sweet*, dark purple pome, 1–1.5 cm wide.
OCCURRENCE: Common, widely distributed.
NOTES: See *A. spicata*. Hybrids with *A. laevis* have been documented in Baxter State Park.
OTHER NAMES: Mountain Serviceberry

ROSACEAE • ROSE FAMILY

Intermediate Shadbush • *Amelanchier intermedia*

Deciduous shrub or small tree of swamps, bogs, and thickets, up to 7 m tall. **Flowers** white, *borne in straight, erect or ascending racemes of 4–9 flowers with an axis 2.5–6 cm long; petals 9–12 mm long and 3–6 mm wide; ovary apex rounded and glabrous.* **Leaves** alternate, *often reddish*, rounded to subcordate at base, *sparsely pubescent beneath when young*; petioles at maturity 10–15 mm long. **Fruit** a sweet, dark purple pome, 7–12 mm wide.
OCCURRENCE: Uncommon, with scattered distribution.
OTHER NAMES: *Amelanchier ×intermedia*

Smooth Shadbush • *Amelanchier laevis*

Deciduous tree or shrub of forests, fields, and roadsides, up to 17 m tall. **Flowers** white, *borne in drooping racemes of 4–11 flowers with an axis 4–12 cm long; petals 10–20 mm long and 3–7 mm wide*; ovary apex rounded and glabrous. **Leaves** alternate, *strongly tinged with red-purple to purple, 4–6 cm long and 2.5–4 cm wide, 50–75% expanded at flowering time, nearly glabrous on underside when young*; petioles at maturity 12–25 mm long. **Fruit** a sweet and juicy pome, 10–15 mm wide; fruiting pedicels glabrous, 15–30 mm long.
OCCURRENCE: Occasional, widely distributed.
NOTES: See *A. arborea*. Hybrids with *A. sanguinea* have been documented in Baxter State Park.

ROSACEAE • ROSE FAMILY

Nantucket Shadbush • *Amelanchier nantucketensis*

Deciduous, colonial shrub of open and rocky sites, often forming dense clumps, up to 2.5 m tall. **Flowers** white, *borne in racemes of 7–10 flowers with an axis 2.1–2.9 cm long; petals 2.6–5.8 mm long and 1–3 mm wide, sometimes bearing pollen*; ovary apex rounded, downy or glabrous. **Leaves** alternate, green, rounded to wedge-shaped at base, folded and white-woolly at flowering, *1–1.8 cm long* and 1.1–3.5 cm wide; petioles at maturity 10–15 mm long. **Fruit** a sweet, dark purple pome, 7.5–10 mm wide.
OCCURRENCE: Very rare, documented from only one township in Park.
NOTES: See *A. spicata*. Species is listed as Threatened in Maine.

Dwarf Shadbush • *Amelanchier spicata*

Deciduous, colonial shrub of open and rocky sites, often forming dense thickets, up to 1.2 m tall. **Flowers** white, *borne in erect racemes of 4–10 flowers with an axis 1.5–4 cm long; petals 6–10 mm long and 2.5–4 mm wide*; ovary apex rounded and densely pubescent. **Leaves** alternate, pale green, *rounded to subcordate at base, white-woolly at flowering time, 1.5–6.5 cm long* and 1–4 cm wide; *petioles at maturity 8–18 mm long*. **Fruit** a *sweet* pome, 7–12 mm wide.
OCCURRENCE: Occasional, widely distributed.
NOTES: See *A. bartramiana* and *A. nantucketensis*.
OTHER NAMES: Running Shadbush, Thicket Shadbush, *Amelanchier stolonifera*

ROSACEAE • ROSE FAMILY

Keep's Hawthorn • *Crataegus keepii*

Deciduous, small tree or shrub of thickets adjacent to rivers, up to 7 m tall. **Flowers** white, 13–19 mm wide; pedicels with long hairs; sepals entire or with scattered glandular teeth; anthers white when fresh, turning brown with age. **Leaves** alternate, pubescent above, 48–78 mm long, with a less than 90-degree angle at base. **Stems** *with thorns.* **Fruit** a bright red pome, ~9 mm wide.
OCCURRENCE: Rare, with scattered distribution.

Cultivated Apple • **Malus pumila*

Deciduous tree of old home sites and fields, up to 15 m tall. **Flowers** *white or light pink, fragrant, 3 cm wide*; sepals pubescent; pedicels pubescent. **Leaves** alternate, finely toothed, densely pubescent when young, retaining hairs beneath on older leaves. **Fruit** *6–12 cm wide, juicy, edible.*
OCCURRENCE: Uncommon, with scattered distribution.
OTHER NAMES: *Pyrus malus, Malus sylvestris*

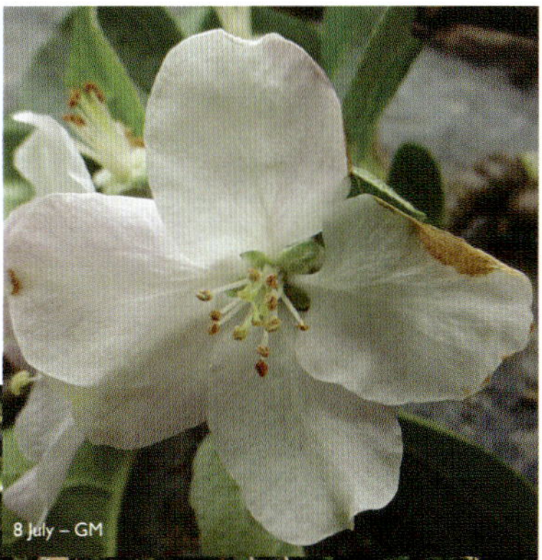

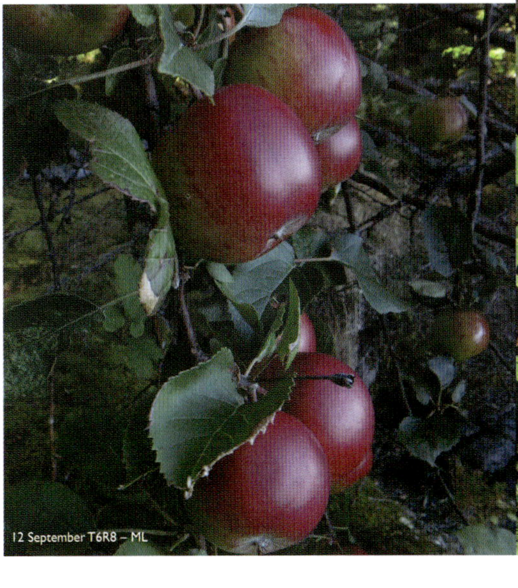

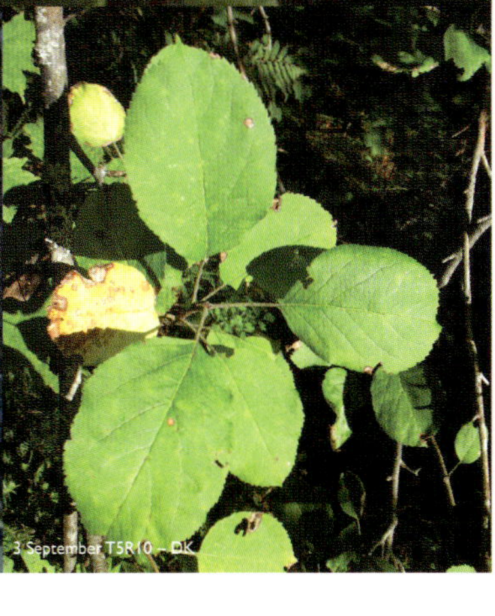

ROSACEAE • ROSE FAMILY

Pin Cherry • *Prunus pensylvanica*

Deciduous shrub or small tree of disturbed sites, often becoming established after fire, up to 10 m tall. **Flowers** white, 5-parted, *in short umbel-like clusters of 3–6 flowers*; petals 3.5–6 mm long; *sepals without glands*. **Leaves** alternate, *widest near base*, 5–9 cm long and 2–4 cm wide, with or without hairs beneath; petioles with 1–3 sessile red glands near apex. **Fruit** a fleshy drupe, *bright red at maturity, 5–8 mm wide*.
OCCURRENCE: Common, widely distributed.
NOTES: See *P. serotina* and *P. virginiana*.
OTHER NAMES: Bird Cherry, Fire Cherry

Black Cherry • *Prunus serotina*

Deciduous tree of hardwood and mixed forests, up to 30 m tall. **Flowers** white, 5-parted, *in an elongated raceme of 20–60 flowers*; petals 2.5–3.5 mm long; *sepals glandless or with at most 2–4 marginal glands*. **Leaves** alternate, *widest at middle*, 6–12 cm long and 2.5–5 cm wide, *with more than 15 pairs of lateral veins, usually with red to brown hairs on lower portion of midrib beneath, finely toothed, the teeth blunt with a thickened, firm tip*; petioles with 1–3 sessile glands near apex. **Fruit** a fleshy drupe, *black at maturity, 7–11 mm wide; sepals persistent on fruit*.
OCCURRENCE: Uncommon, with scattered distribution.
NOTES: See *P. pensylvanica* and *P. virginiana*.
OTHER NAMES: Rum Cherry

ROSACEAE • ROSE FAMILY

Choke Cherry • *Prunus virginiana*

Deciduous shrub or small tree of forest edges, up to 8 m tall. **Flowers** white, 5-parted, *in a dense cylindrical raceme of 20–60 flowers*; petals 2.5–4.5 mm long; *sepals with 10 or more marginal glands*. **Leaves** alternate, *widest above middle*, 5.5–10 cm long and 3–6 cm wide, with 8–11 pairs of lateral veins, *finely and sharply toothed, glabrous beneath or with scattered whitish to yellowish hairs in axils of lateral veins*; petioles with 1–3 glands near apex. **Fruit** a fleshy drupe, *dark red to purple, 8–11 mm wide; sepals not persistent on fruit*.
OCCURRENCE: Occasional, widely distributed.
NOTES: See *P. pensylvanica* and *P. serotina*.

American Mountain-ash • *Sorbus americana*

Deciduous shrub or small tree of forests and mountaintops, up to 12 m tall. **Flowers** white, 5-parted, *5–8 mm wide; inflorescence glabrous or with at most a few long, soft, shaggy hairs*. **Leaves** alternate, divided; *leaflets 4–8 mm long, 3- to 5-times as long as wide, with upper surface glabrous or sparsely pubescent, tapering to a long, slender point at tip; bud scales glabrous or sparsely ciliate, sticky*. **Branchlets** *glabrous or with sparse long hairs*. **Fruit** a berry-like pome, red, 4–7 mm wide, often remaining on tree late in winter.
OCCURRENCE: Common, widely distributed.
NOTES: See *S. aucuparia* and *S. decora*.
OTHER NAMES: *Pyrus americana*

ROSACEAE • ROSE FAMILY

European Mountain-ash • *Sorbus aucuparia*

Deciduous shrub or small tree of forests and fields, up to 10 m tall. **Flowers** white, 5-parted, 7.5–12 mm wide; *inflorescence with dense, long, soft, shaggy hairs.* **Leaves** alternate, divided; *leaflets 2.5–4 cm long, the upper surface with dense, long, soft, shaggy hairs, tapering to a short, abrupt point at tip; bud scales not sticky, with long, soft, shaggy hairs.* **Branchlets** *with dense, long, soft, shaggy hairs.* **Fruit** a berry-like pome, orange, 7–11 mm wide.
OCCURRENCE: Very rare, documented from only one township in Park.
NOTES: See *S. americana* and *S. decora*.
OTHER NAMES: *Pyrus aucuparia*, Rowan

Showy Mountain-ash • *Sorbus decora*

Deciduous shrub or small tree of forests and mountaintops, up to 8 m tall. **Flowers** white, 5-parted, *7–12 mm wide; inflorescence glabrous or with at most a few long, soft, shaggy hairs.* **Leaves** alternate, divided; *leaflets 4–8 mm long, 2- to 3.5-times as long as wide, with upper surface glabrous or sparsely pubescent, tapering to a long, slender point at tip; bud scales ciliate, sticky.* **Branchlets** *glabrous or with sparse long hairs.* **Fruit** a berry-like pome, red to red-orange, 8–12 mm wide, often remaining on tree late in winter.
OCCURRENCE: Common, widely distributed.
NOTES: See *S. americana* and *S. aucuparia*.
OTHER NAMES: *Pyrus decora*

SALICACEAE • WILLOW FAMILY ▼

Balsam Poplar • *Populus balsamifera*

Deciduous tree of rich, wet sites, up to 25 m tall. **Flowers** in drooping catkins appearing before leaves. **Leaves** alternate, finely and regularly toothed, *shiny above, streaked with orange resin beneath, longer than wide; petioles round or channeled in cross section*; bud scales glabrous. **Youngest branchlets** usually pubescent. **Bark** furrowed, the upper trunk smooth.
OCCURRENCE: Occasional, widely distributed.
NOTES: See *P. tremuloides*.

Big-toothed Poplar • *Populus grandidentata*

Deciduous tree of rocky soils, up to 30 m tall. **Flowers** in drooping catkins appearing before leaves. **Leaves** alternate, with fine, matted, soft, woolly hairs beneath when young, *not shiny above, coarsely toothed with 4–12 teeth per side*, nearly round in shape, pale green beneath; *petioles distinctly flattened in cross section; bud scales pubescent.* **Youngest branchlets** pubescent. **Bark** greenish gray, smooth, darkening and furrowing with age.
OCCURRENCE: Common, widely distributed.

SALICACEAE • WILLOW FAMILY

Quaking Poplar • *Populus tremuloides*

Deciduous tree of dry, rocky soils, up to 30 m tall. **Flowers** in drooping catkins appearing before leaves. **Leaves** alternate, glabrous, *nearly round in shape, finely toothed with 20–50 teeth per side, shiny above, pale green beneath; petioles distinctly flattened in cross section;* bud scales glabrous. **Youngest branchlets** glabrous. **Bark** grayish green, smooth, darkening and furrowing with age.
OCCURRENCE: Common, widely distributed.
NOTES: See *P. balsamifera*.
OTHER NAMES: Trembling Aspen

Long-beaked Willow • *Salix bebbiana*

Deciduous shrub of disturbed fields, shorelines, and swamps, up to 7 m tall. **Carpellate flowers** *with a pubescent ovary, the ovary visible through hairs; floral bracts pale yellow to light brown, sometimes suffused with red at apex.* **Leaves** alternate, without red-brown hairs, gray or white beneath from dense hairs or bloom, not leathery, *with impressed veins above; margins flat (not rolled under), with 1–3 irregular, blunt teeth per cm; petioles 3–17 mm long; blades 1.7- to 3.9-times as long as wide.* **Branchlets** *persistently pubescent even on second-year branches, red to red-brown.* **Fruit** *a pubescent capsule, 5–9 mm long.*
OCCURRENCE: Uncommon, with scattered distribution.
NOTES: See *S. discolor* and *S. humilis*. A hybrid with *S. discolor* has been documented in Baxter State Park.

SALICACEAE • WILLOW FAMILY

Sage Willow • *Salix candida*

Deciduous shrub of *fens and swamps, up to 5 m tall.* **Carpellate flowers** expanding with or after leaves; ovaries pubescent, *often nearly or fully concealed by bent, woolly hairs; floral bracts pale yellow to black at apex.* **Leaves** alternate, gray or white beneath from dense hairs; margins entire, undulate, or with irregular, blunt teeth, rolled under; *blades 4–17 cm long, 3- to 40-times as long as wide; stipules present; new leaves pubescent with tangled hairs, not red-brown.* **Branchlets** *pubescent with white, yellow-white, or gray-white hairs.* **Fruit** *a pubescent capsule, 4–6 mm long.*
OCCURRENCE: Rare, with scattered distribution.
NOTES: See *S. pellita.* Species is listed as Endangered in Maine.

Pussy Willow • *Salix discolor*

Deciduous shrub of wet thickets, swamps, and swales, 2–7 m tall. **Carpellate flowers** *expanding before leaves; ovaries pubescent, tapering to a distinct beak; floral bracts brown or black, at least at apex, 1.4–4 mm long; stipe 0.6–2.7 mm long; inflorescence 4–11 cm long.* **Leaves** alternate, gray or white beneath from dense hairs or bloom, not leathery; veins not notable; *margins flat (not rolled under), with 1–3 irregular, blunt teeth per cm; petioles 3–17 mm long.* **Branchlets** *commonly glabrate with small patches of hairs near nodes*, not becoming glaucous later in season. **Fruit** *a pubescent capsule, 5–12 mm long.*
OCCURRENCE: Uncommon, widely distributed.
NOTES: See *S. bebbiana, S. humilis,* and *S. pyrifolia.*

SALICACEAE • WILLOW FAMILY

Heart-leaved Willow • *Salix eriocephala*

Deciduous shrub or tree of wetlands and shorelines, *0.2–4 m tall*. **Carpellate flowers** with a glabrous ovary; *floral bracts brown or black, at least at apex, 0.8–1.6 mm long*. **Leaves** alternate, gray or white beneath from thin bloom, *thick, leathery, rounded to cordate at base, with short, curving hairs; margins flat (not rolled under), with regular, sharp teeth; stipules large, 4–10 mm long; blades 3.5–17 cm long, widest below middle; petiole without dark glands*. **Branchlets** *yellow to brown*. **Fruit** *a glabrous capsule, 3.5–7 mm long*.
OCCURRENCE: Uncommon, with scattered distribution.
NOTES: See *S. lucida*, *S. nigra*, *S. petiolaris*, and *S. sericea*.

Prairie Willow • *Salix humilis*

Deciduous shrub of *dry areas, sandy flats, and lower slopes*, 0.3–3 m tall. **Carpellate flowers** expanding before leaves; *ovaries pubescent, tapering to a distinct beak; floral bracts brown or black, at least at apex, 0.8–2 mm long; stipe 0.6–2.7 mm long; inflorescence 1–5 cm long*. **Leaves** alternate, gray or white beneath from dense hairs or bloom, pubescent on one or both surfaces; margins entire, undulate, or with irregular, blunt teeth, rolled under; *blades 2–17 cm long, 2.3- to 7.5-times as long as wide*; stipules absent or present; *new leaves with tangled hairs; red-brown hairs usually present*. **Branchlets** *pubescent with long gray or brown hairs*. **Fruit** *a pubescent capsule, 5–12 mm long*.
OCCURRENCE: Occasional, widely distributed.
NOTES: See *S. bebbiana* and *S. discolor*.

SALICACEAE • WILLOW FAMILY

Shining Willow • *Salix lucida*

Deciduous shrub or tree of swales, swamps, and shorelines, *up to 7 m tall*. **Carpellate flowers** with a glabrous ovary; floral bracts pale yellow to light brown, sometimes suffused with red at apex; *inflorescence densely flowered, 1.7–8 cm long, with more than 11 flowers*. **Leaves** alternate, *green beneath, glossy deep green above, without dense hairs or bloom, sharply and uniformly toothed*; blades 3.5–17 cm long and 1.1–4.3 cm wide, widest below middle; margins flat (not rolled under); petioles with one or more glands; stipules 1–6 mm long. **Fruit** *a glabrous capsule, 5–7 mm long*.
OCCURRENCE: Occasional, widely distributed.
NOTES: See *S. eriocephala*, *S. nigra*, *S. petiolaris*, and *S. sericea*.

Black Willow • *Salix nigra*

Deciduous tree of floodplains, shorelines, and swamps, *up to 20 m tall*. **Carpellate flowers** with a glabrous ovary; floral bracts pale yellow to light brown, sometimes suffused with red at apex; *inflorescence loosely flowered, 1.7–8 cm long, with more than 11 flowers*. **Leaves** alternate, *green beneath, without dense hairs or bloom, closely toothed*; margins flat (not rolled under); mature blades 0.7–1.7 cm wide, usually 4- to 10-times as long as wide; stipules up to 12 mm long. **Fruit** *a glabrous capsule, 3–5 mm long*.
OCCURRENCE: Rare, with scattered distribution.
NOTES: See *S. eriocephala*, *S. lucida*, *S. petiolaris*, and *S. sericea*.

SALICACEAE • WILLOW FAMILY

Bog Willow • *Salix pedicellaris*

Deciduous shrub of fens, shorelines, and swamps, *up to 1.5 m tall*. **Carpellate flowers** *with a glabrous ovary; floral bracts pale yellow to light brown, sometimes suffused with red at apex, 0.8–1.6 mm long; inflorescence densely flowered, 1.7–8 cm long, with more than 11 flowers*. **Leaves** alternate, *gray or white beneath from bloom, glabrous; margins usually rolled under, entire; petioles 2–7 mm long; blades 3.5–17 cm long, usually widest above middle; stipules absent*. **Branchlets** not becoming glaucous later in season; winter branchlets glabrous, not red. **Fruit** *a glabrous capsule, 4–8 mm long*.
OCCURRENCE: Rare, with scattered distribution.

Satiny Willow • *Salix pellita*

Deciduous shrub or small tree *of shorelines and wetlands, 3–5 m tall*. **Carpellate flowers** expanding before leaves; *ovaries pubescent; floral bracts brown or black, at least at apex; stipe 0.1–0.5 mm long*. **Leaves** alternate, gray or white beneath from dense hairs; margins undulate or with irregular, blunt teeth, rolled under; blades 4–17 cm long, 3- to 40-times as long as wide; stipules usually absent; new leaves pubescent with straight, neatly aligned hairs. **Fruit** *a pubescent capsule, 3.5–6.5 mm long*.
OCCURRENCE: Rare, with scattered distribution.
NOTES: See *S. candida*.

SALICACEAE • WILLOW FAMILY

Meadow Willow • *Salix petiolaris*

Deciduous shrub of fens, fields, and shorelines, up to 4 m tall. **Carpellate flowers** expanding with or after leaves; *ovaries pubescent with straight, silky hairs; floral bracts brown or black, at least at apex; stipe 1.5–4 mm long.* **Leaves** alternate, *gray or white beneath from dense, neatly aligned hairs, thick, leathery, wedge-shaped at base; margins flat (not rolled under), with regular, sharp teeth, sometimes entire at base*; stipules usually absent or up to 4 mm long; blades 2.5–7 cm long and 0.3–2 cm wide. **Branchlets** *yellow to brown.* **Fruit** *a pubescent capsule, 5–9 mm long.*
OCCURRRENCE: Very rare, documented from only one township in Park.
NOTES: See *S. eriocephala, S. lucida, S. nigra,* and *S. sericea*.

Tea-leaved Willow • *Salix planifolia*

Deciduous shrub of *alpine areas*, 1–3 m tall. **Carpellate flowers** expanding before leaves; *ovaries pubescent; floral bracts brown or black, at least at apex; stipe 0.1–0.8 mm long.* **Leaves** alternate, *gray or white beneath from dense hairs or bloom; margins entire, undulate, or with irregular, blunt teeth, flat (not rolled under).* **Branchlets** *becoming glaucous later in season.* **Fruit** *a pubescent capsule, 5.5–6 mm long.*
OCCURRENCE: Very rare, documented from only one township in Park.
NOTES: See *S. argyrocarpa*. Species is listed as Threatened in Maine.

SALICACEAE • WILLOW FAMILY

Balsam Willow • *Salix pyrifolia*

Deciduous shrub or tree of thickets and forest edges, up to 4 m tall. **Carpellate flowers** with a glabrous ovary; floral bracts pale yellow to light brown, sometimes suffused with red at apex, *1.5–4 mm long; inflorescence densely flowered, 1.7–8 cm long, with more than 11 flowers.* **Leaves** alternate, *gray or white beneath from dense hairs or bloom, thin, somewhat translucent; margins entire, undulate, or with irregular teeth, flat (not rolled under).* **Branchlets** *not becoming glaucous later in season; winter branchlets red, glabrous.* **Fruit** *a glabrous capsule, 7–8 mm long.*
OCCURRENCE: Uncommon, with scattered distribution.
NOTES: See *S. discolor*.

25 July – AH   25 April – DC

Silky Willow • *Salix sericea*

Deciduous shrub or tree of thickets and wetland edges, up to 4 m tall. **Carpellate flowers** expanding before or after leaves; *ovaries pubescent with straight, silky hairs, short-beaked or blunt at apex; floral bracts brown or black, at least at apex; stipe 0.6–1.5 mm long.* **Leaves** alternate, *gray or white beneath from dense, neatly aligned hairs, thick, leathery, wedge-shaped or rounded at base; margins flat (not rolled under), with regular, sharp teeth, toothed to base; stipules up to 4 mm long; blades 4–15 cm long and 1–4 cm wide.* **Branchlets** *yellow to brown.* **Fruit** *a pubescent capsule, 2.5–4 mm long.*
OCCURRENCE: Uncommon, with scattered distribution.
NOTES: See *S. eriocephala, S. lucida, S. nigra,* and *S. petiolaris*.

10 June – AH   10 June – AH

SAPINDACEAE • SOAPBERRY FAMILY ▼

Striped Maple • *Acer pensylvanicum*

Deciduous tree in understory of mixed forests, up to 9 m tall. **Flowers** *bright yellow*, appearing after leaf emergence, *in slender, drooping racemes*. **Leaves** opposite, pale green, pubescent beneath, *3-lobed, 12–18 cm long; sinuses between lobes V-shaped; margins of sinuses toothed at least part way to base*. **Bark** *green or red-brown, with slender, white stripes*. **Fruit** a samara-like schizicarp, 2–3.3 cm long.
OCCURRENCE: Common, widely distributed.
NOTES: See *A. spicatum*.
OTHER NAMES: Moosewood, Goose-foot Maple

TREES AND TALL SHRUBS   345

SAPINDACEAE • SOAPBERRY FAMILY

Red Maple • *Acer rubrum*

Deciduous tree of upland and wetland forests, up to 29 m tall. **Flowers** *red or yellow-red, 5-parted, with staminate and carpellate flowers on separate trees, appearing before leaf emergence, in rounded clusters on short pedicels elongating at maturity.* **Leaves** opposite, 3- or 5-lobed; sinuses between lobes V-shaped; margins of sinuses toothed at least part way to base; apical lobe broadest at base. **Fruit** a samara-like schizicarp, *1.7–2.4 cm long.*
OCCURRENCE: Common, widely distributed.
NOTES: See *A. spicatum*, *A. saccharinum*, and *A. saccharum*.
OTHER NAMES: Swamp Maple

Silver Maple • *Acer saccharinum*

Deciduous tree of floodplains, forests, and shorelines, up to 32 m tall. **Flowers** *reddish, with staminate and carpellate flowers on separate trees, appearing before leaf emergence; pedicels drooping.* **Leaves** opposite, minutely pubescent beneath, 5-lobed; sinuses between lobes U- or V-shaped; margins of sinuses toothed at least part way to base; apical lobe widest above base; petioles glabrous or pubescent. **Fruit** a samara-like schizicarp, *3.5–5.5 cm long.*
OCCURRENCE: Rare, with scattered distribution.
NOTES: See *A. rubrum*.

SAPINDACEAE • SOAPBERRY FAMILY

Sugar Maple • *Acer saccharum*

Deciduous tree of deep, rich soils, up to 25 m tall. **Flowers** *greenish yellow, appearing during leaf emergence*; pedicels drooping. **Leaves** opposite, glabrous or nearly so beneath, *3- to 5-lobed; sinuses between lobes U-shaped; margins of sinuses entire*; petioles glabrous. **Fruit** a samara-like schizicarp, 2.5–4.5 cm long.
OCCURRENCE: Common, widely distributed.
NOTES: See *A. rubrum*.
OTHER NAMES: Rock Maple

27 June – CWG

6 June – DC   20 April – AH

Mountain Maple • *Acer spicatum*

Deciduous tree or shrub of cool, wet forests, up to 7 m tall. **Flowers** *yellow-green*, appearing after leaf emergence, *in slender, erect panicles*. **Leaves** opposite, finely pubescent beneath, 6–12.5 cm long, *3-lobed; sinuses between lobes V-shaped; margins of sinuses toothed at least part way to base*. **Bark** *gray or mottled*, smooth to rough. **Fruit** a samara-like schizicarp, 1.8–2.8 cm long.
OCCURRENCE: Common, widely distributed.
NOTES: See *A. pensylvanicum* and *A. rubrum*.

8 June – DC   29 August – GM

TAXACEAE • YEW FAMILY ▼

American Yew • *Taxus canadensis*

Evergreen shrub of forests, up to 2 m tall. **Leaves** flat, needle-like, *with pale green lower surface, sharply pointed at tip*, 1–2.5 cm long and 1–2.4 mm wide. **Stems** diffusely branched, straggling, spreading to prostrate. **Fruit** *a single brown seed surrounded by a fleshy, scarlet, cup-shaped seed coat*.
OCCURRENCE: Uncommon, widely distributed.
NOTES: See *Abies balsamea*.
OTHER NAMES: Canada Yew, Ground-hemlock

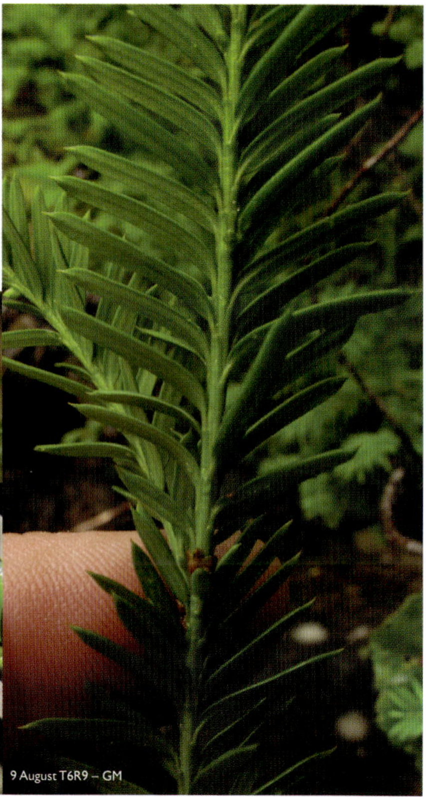

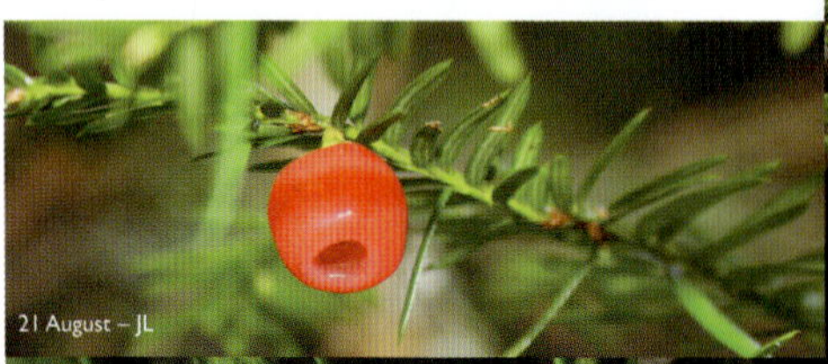

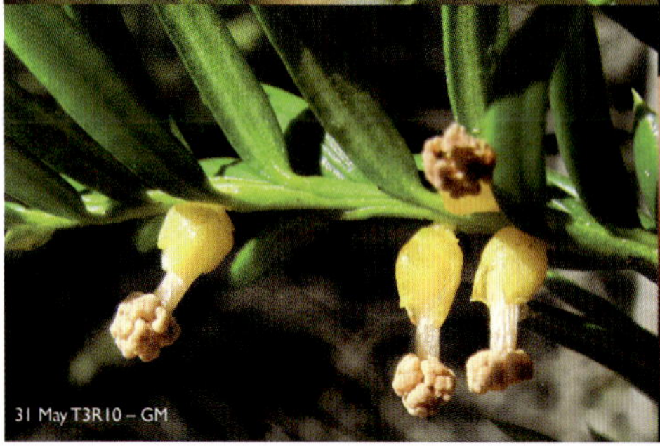

ULMACEAE • ELM FAMILY ▼

American Elm • *Ulmus americana*

Deciduous tree of rich, wet soils, up to 37 m tall. **Flowers** *in loose, drooping fascicles* emerging before leaves, on unequal, long pedicels; petals absent. **Leaves** alternate, *coarsely double-toothed, abruptly pointed*, 8–15 cm long and 4–9 cm wide; *leaf base strongly asymmetric, with one side attached lower on petiole than other, the two halves usually of different shape*. **Fruit** a samara, *winged on all sides*, 8–13 mm long.
OCCURRENCE: Occasional, widely distributed.
NOTES: See *Ostrya virginiana* in the Betulaceae family.

ASPLENIACEAE • SPLEENWORT FAMILY ▼

Maidenhair Spleenwort • *Asplenium trichomanes*

Perennial herb of shaded rocky areas, up to 20 cm tall. **Leaves** with blades once divided, narrow, *tapering at both ends*; fertile leaves erect, otherwise similar to prostrate sterile leaves; leaflets small, rounded, toothed, slightly stalked, reduced in size toward tip of rachis; petioles dark, purplish brown, brittle. **Sori** 2–4 pairs per leaflet; indusia narrow, attached to one side of vein. **Rhizomes** present. **OCCURRENCE:** Rare, with scattered distribution.

BLECHNACEAE • CHAIN FERN FAMILY ▼

Virginia Chain Fern • *Woodwardia virginica*

Perennial herb of acidic wetlands, up to 1.2 m tall. **Leaves** with blades 1.5-times divided; leaflets deeply lobed, *with distinctive netted or chain-like veins*, in 12–23 pairs of decreasing size up rachis; petioles dark purple to black, shiny. **Sori** *long, narrow, and straight*, aligned along midveins. **Rhizomes** present.
OCCURRENCE: Historical record, last documented in 1940, from southern portion of Park.

DENNSTAEDTIACEAE • BRACKEN FAMILY ▼

Eastern Hay-scented Fern • *Dennstaedtia punctilobula*

Perennial herb of spruce-fir forests and old fields, 0.4–1.3 m tall. **Leaves** with blades 2.5-times divided, *tapering to a sharp point, often with a drooping tip*; leaflets with soft hairs on both surfaces; rachis with glandular hairs; petioles dark brown or black near base, slightly pubescent, glandular. **Sori** small, *at margins, in cup-shaped indusia*. **Rhizomes** present.
OCCURRENCE: Occasional, widely distributed.
NOTES: This species forms large colonies, especially in disturbed forests. The distinctive odor of hay or freshly cut grass is present throughout the growing season but strongest in late summer and fall.
OTHER NAMES: Boulder Fern

DENNSTAEDTIACEAE • BRACKEN FAMILY

Bracken Fern • *Pteridium aquilinum*

Perennial herb of forests, meadows, uplands, and wetlands, up to 1.2 m tall. **Leaves** with blades 2.5- to 3-times divided, *broadly triangular*; leaflets longer than wide, *distinctly stalked*; petioles about as long as blade, *grooved*. **Sori** along leafule margins, *rare*.
OCCURRENCE: Common, widely distributed.
NOTES: This fern is among the most widely distributed plants on earth, growing on all continents except Antarctica. Fiddleheads from this species are carcinogenic.
OTHER NAMES: Hogbrake, *Pteridium latiusculum*

DRYOPTERIDACEAE • WOOD FERN FAMILY ▼

Mountain Wood Fern • *Dryopteris campyloptera*

Perennial herb of hardwood forests, up to 1 m tall. **Leaves** *broadly triangular*, 2.5- to 3-times divided; leafules with bristle-tipped teeth; *innermost lower leafule of basal leaflets 2-times width and 3- to 5-times length of offset upper leafule, and longer than or equaling adjacent lower leafule* (see photo below); *rachis without stipitate glands; petioles with tan scales without central dark stripe*. **Sori** round, numerous, between midvein and margin; indusia kidney-shaped.
OCCURRENCE: Common, widely distributed.
NOTES: See *D. carthusiana* and *D. intermedia*.
OTHER NAMES: Spreading Wood Fern, *Dryopteris spinulosa* var. *americana*

DRYOPTERIDACEAE • WOOD FERN FAMILY

Spinulose Wood Fern • *Dryopteris carthusiana*

Perennial herb of hardwood and conifer forests, up to 80 cm tall. **Leaves** with blades 2.5- to 3-times divided; leafules with bristle-tipped teeth; *innermost lower leafule of basal leaflets not much wider and 2-times length of offset upper leafule and longer than or equaling adjacent lower leafule* (see photos); *rachis without stipitate glands*; petioles with pale brown scales. **Sori** small, between midvein and margin; indusia kidney-shaped.
OCCURRENCE: Occasional, widely distributed.
NOTES: See *D. campyloptera* and *D. intermedia*. A hybrid (*Dryopteris ×triploidea*), similar to *D. carthusiana*, has been documented at Baxter State Park. It can be recognized by the stipitate glands on the indusia and leaf blades that are nearly evergreen.
OTHER NAMES: Toothed Wood Fern, *Dryopteris spinulosa*

Clinton's Wood Fern • *Dryopteris clintoniana*

Perennial herb of forested wetlands, 30–60 cm tall. **Leaves** with blades 1.5-times divided; *fertile and sterile leaves similar; leaflets rotated with the plane approximately parallel to the ground*; margins with bristle-tipped teeth; petioles sparsely scaly at base, one-quarter to one-third length of leaf blade. **Sori** kidney-shaped, between midvein and margin.
OCCURRENCE: Rare, with scattered distribution.
NOTES: See *D. cristata*.

DRYOPTERIDACEAE • WOOD FERN FAMILY

Crested Wood Fern • *Dryopteris cristata*

Perennial herb of forested wetlands, up to 70 cm tall. **Leaves** with blades 1.5-times divided, *with narrow, more or less parallel-sided blades; fertile leaves larger and more erect than vegetative leaves; lowest leaflets widely spaced, blunt*; petioles with abundant light brown scales. **Sori** round, numerous, between midvein and margin; indusia kidney-shaped.
OCCURRENCE: Occasional, widely distributed.
NOTES: See. D. clintoniana. A hybrid (*Dryopteris* ×*boottii*), similar to *D. cristata*, has been documented at Baxter State Park. It can be recognized by the stipitate glands on the indusia and leaves with blades 2.5-times divided.

Male Wood Fern • *Dryopteris filix-mas*

Perennial herb of rich, rocky forests, 0.3–1.2 m tall. **Leaves** with blades 1.5- to 2-times divided, gradually tapering to base and tip; leafules with toothed margins; petioles less than one-quarter overall length of blade, *with both broad and hair-like scales*. **Sori** between midvein and margin; indusia kidney-shaped.
OCCURRENCE: Very rare, documented from only one township in Park.
NOTES: See *D. marginalis*. Species is listed as Endangered in Maine.

FERNS AND OTHER SPORE-PRODUCING PLANTS

DRYOPTERIDACEAE • WOOD FERN FAMILY

Fragrant Wood Fern • *Dryopteris fragrans*

Perennial herb of cliffs and rocky ledges, up to 40 cm tall. **Leaves** with blades up to 2.5-times divided, narrowing at both ends, *with aromatic glands; dried leaves persisting at base*; leafules crowded, with rounded teeth; petioles short, with shiny, reddish scales. **Sori** round, numerous, large, between midvein and margin; indusia kidney-shaped, *with glands on margin*.
OCCURRENCE: Very rare, documented from only northern portion of Park.
NOTES: Species is listed as Special Concern in Maine.

DRYOPTERIDACEAE • WOOD FERN FAMILY

Evergreen Wood Fern • *Dryopteris intermedia*

Perennial herb of forests, up to 90 cm tall. **Leaves** with blades 2.5- to 3-times divided, remaining green in winter; *innermost lower leafule of basal leaflets shorter than or equaling adjacent lower leafule* (see photo below right); *rachis with glandular hairs*; petioles with scales mostly at base. **Sori** round, small, between midvein and margin; indusia kidney-shaped.
OCCURRENCE: Common, widely distributed.
NOTES: See *D. campyloptera* and *D. carthusiana*.
OTHER NAMES: Glandular Wood Fern, *Dryopteris spinulosa* var. *intermedia*

Marginal Wood Fern • *Dryopteris marginalis*

Perennial herb of rocky slopes, up to 1.3 m tall. **Leaves** with blades 1.5- to 2-times divided, arching; *leafules with entire margins*; petioles stout, brittle, with many long scales mostly at base. **Sori** *prominent on margins of leafules*; indusia kidney-shaped, without glands.
OCCURRENCE: Uncommon, with scattered distribution.
NOTES: See *Dryopteris filix-mas*.

DRYOPTERIDACEAE • WOOD FERN FAMILY

Christmas Fern • *Polystichum acrostichoides*

Perennial, evergreen herb of moist forests and rocky slopes, 15–70 cm tall. **Leaves** with blades once divided, growing in loose clumps; leaflets lustrous, leathery, *with upward-pointing auricles*; margins with bristle-tipped teeth; fertile leaflets smaller; petioles scaly at base; scales reduced in size and abundance upward. **Sori** covering underside of fertile leaflets; indusia rounded.
OCCURRENCE: Rare, documented from only southern portion of Park.

Braun's Holly Fern • *Polystichum braunii*

Perennial, semi-evergreen herb of rich woods, talus, and rocky slopes, up to 1 m tall. **Leaves** with blades 2-times divided, *lustrous, leathery*, growing in arching, circular clumps; leafules with bristle-tipped teeth, stalked, auricled, often with thread-like scales; *petioles densely covered with both thread-like and broad scales*. **Sori** between midvein and margin; indusia round, *ciliate*.
OCCURRENCE: Rare, with scattered distribution.
OTHER NAMES: *Aspidium braunii*

EQUISETACEAE • HORSETAIL FAMILY ▼

Field Horsetail • *Equisetum arvense*

Perennial herb of wet meadows, roadsides, and woods, 0.1–1 m tall. **Stems** dimorphic; *fertile stems leafless, with strobili 2.5–4 cm long*, emerging in spring and producing spores mid-April to May, withering soon after; sterile stems green, hollow, grooved, emerging later than fertile stems. **Branches** whorled, *not rebranching*. **Leaves** tiny, whorled, fused into a sheath around each node.
OCCURRENCE: Common, widely distributed.

River Horsetail • *Equisetum fluviatile*

Perennial herb of wetlands and shallow fresh water, 0.4–1.2 m tall. **Stems** monomorphic, with both fertile and sterile stems similar, *hollow, thin-walled*, 2.5–9 mm wide; strobili appearing in summer, 1–2 cm long. **Branches** up to 15 cm long, often absent. **Leaves** tiny, whorled, fused into a sheath around each node.
OCCURRENCE: Common, widely distributed.
OTHER NAMES: Water Horsetail

FERNS AND OTHER SPORE-PRODUCING PLANTS

EQUISETACEAE • HORSETAIL FAMILY

Wood Horsetail • *Equisetum sylvaticum*

Perennial herb of wet roadsides, forests, fields, wetlands, and thickets, 25–70 cm tall. **Stems** dimorphic; *fertile stems pale, branching, looking like sterile stems after spores discharge; strobili 1.5–3 cm long, falling off after pollen production*; sterile stems brownish to green, hollow. **Branches** *often rebranching*. **Leaves** tiny, whorled, fused into a sheath around each node.
OCCURRENCE: Common, widely distributed.
OTHER NAMES: Woodland Horsetail

Variegated Scouring-rush • *Equisetum variegatum*

Perennial herb of shores and disturbed areas, 6–48 cm tall. **Stems** unbranched, with 3–12 ridges, *very rough and abrasive*. **Leaves** tiny, whorled, fused into a sheath around each node, *with contrasting black and white color*.
OCCURRENCE: Rare, with scattered distribution.

HUPERZIACEAE • FIRMOSS FAMILY ▼

Mountain Firmoss • *Huperzia appressa*

Perennial, evergreen herb of exposed, rocky sites primarily in alpine areas, 6–10 cm tall. **Leaves** 2–6 mm long, *shorter and more ascending toward apex,* green to yellow-green, pointed, entire, with parallel margins. **Sporangia** yellow, in leaf axils. **Gemmae** with lateral leaves 0.5–1.1 mm wide. **Shoots** clustered, erect, 3–7 mm wide including leaves.
OCCURRENCE: Uncommon, with scattered distribution.
NOTES: Species is listed as Special Concern in Maine. A hybrid (*Huperzia* ×*josephbeitelii*), similar to *H. appressa*, has been documented at Baxter State Park. It can be recognized by its more stocky nature with shoots 7–10 mm wide including leaves and its gemmae with lateral leaves 1–1.5 mm wide.
OTHER NAMES: *Huperzia appalachiana, Lycopodium selago* var. *appressum*

Shining Firmoss • *Huperzia lucidula*

Perennial, evergreen herb of forests and forested wetlands, up to 20 cm tall. **Leaves** *lustrous green,* blunt, *widest above middle,* with 1–8 teeth. **Sporangia** yellow, in leaf axils. **Gemmae** with lateral leaves 1.5–2.5 mm wide. **Shoots** erect to long-trailing, *with conspicuous annual constrictions.*
OCCURRENCE: Common, widely distributed.
OTHER NAMES: *Lycopodium lucidulum*

ISOËTACEAE • QUILLWORT FAMILY ▼

Spiny-spored Quillwort • *Isoëtes echinospora*

Perennial, aquatic herb *of shallow water* of lakes and slow-moving streams, up to 25 cm tall. **Megaspores** white, 0.4–0.6 mm wide, *covered with thin, sharp spines; girdle obscure;* velum covering less than half of sporangium. **Leaves (sporophylls)** bright green to reddish green, pliable, *gradually tapering to tip*.
OCCURRENCE: Occasional, widely distributed.
NOTES: See other species of *Isoëtes*.
OTHER NAMES: *Isoëtes muricata*

Lake Quillwort • *Isoëtes lacustris*

Perennial, aquatic herb of lakes and slow-moving streams, *often in water 1–3 m deep*, up to 25 cm tall. **Megaspores** white, 0.05–0.8 mm wide, *without spines, usually with a net-like raised pattern on surface; girdle visible, usually covered with short, rounded projections;* velum covering less than half of sporangium. **Leaves (sporophylls)** dark green to reddish green, *abruptly tapering to tip*.
OCCURRENCE: Uncommon, with scattered distribution.
NOTES: See other species of *Isoëtes*.
OTHER NAMES: Deep Water Quillwort, *Isoëtes macrospora*

ISOËTACEAE • QUILLWORT FAMILY

Prototype Quillwort • *Isoëtes prototypus*

Perennial, aquatic herb of cold, clear lakes, *submerged in water to 3 m deep*, up to 12 cm tall. **Megaspores** white, 0.4–0.6 mm wide, *without spines, obscurely wrinkled; girdle obscure; velum covering entire sporangium*. **Leaves (sporophylls)** dark green, *rigid, straight*, gradually tapering to tip.
OCCURRENCE: Rare, with scattered distribution.
NOTES: Species is listed as Threatened in Maine. See other species of *Isoëtes*.

Tuckerman's Quillwort • *Isoëtes tuckermanii*

Perennial, aquatic herb *of shallow water* of lakes and slow-moving streams, up to 20 cm tall. **Megaspores** white, 0.4–0.7 mm wide, *without spines, usually with a net-like raised pattern on surface; girdle visible, usually covered with short, rounded projections; velum covering less than half of sporangium*. **Leaves (sporophylls)** olive-green to red-brown, pliable to rigid, gradually tapering to tip, curling at tip.
OCCURRENCE: Historical record, last documented in 1975, from southern portion of Park.
NOTES: See other species of *Isoëtes*.

LYCOPODIACEAE • CLUBMOSS FAMILY ▼

Prickly Tree-Clubmoss • *Dendrolycopodium dendroideum*

Perennial herb of dry forests, 12–30 cm tall. **Leaves** 0.9–1.0 mm wide, pointed at tip, *spreading on main stem below branches*; leaves on lateral branches in alternating pseudowhorls of 2 upper, 2 lower, and 2 lateral ranks, *all roughly equal in size; lateral leaves untwisted.* **Lateral branches** *round in cross section.* **Strobili** *sessile*, 1–14 per stem, 1.2–5.5 cm long. **Upright stems** *prickly.*
OCCURRENCE: Common, widely distributed.
NOTES: See *D. hickeyi* and *D. obscurum.*
OTHER NAMES: *Lycopodium dendroideum*

Hickey's Tree-clubmoss • *Dendrolycopodium hickeyi*

Perennial herb of dry forests, up to 16 cm tall. **Leaves** on main stem pointed at tip, *appressed to ascending*; leaves on lateral branches in pseudowhorls of 1 upper, 1 lower, and 4 lateral ranks, *all roughly equal in size.* **Lateral branches** *round in cross section.* **Strobili** 1–10 per stem, 1.5–6.5 cm long.
OCCURRENCE: Uncommon, widely distributed.
NOTES: See *D. dendroideum* and *D. obscurum.*
OTHER NAMES: *Lycopodium hickeyi*

LYCOPODIACEAE • CLUBMOSS FAMILY

Flat-branched Tree-clubmoss • *Dendrolycopodium obscurum*

Perennial herb of forests and shrubby areas, up to 21 cm tall. **Leaves** 0.5–0.7 mm wide, pointed at tip, appressed to main stem below branches; leaves on lateral branches in pseudowhorls of 1 upper, 1 lower, and 4 lateral ranks; *lower leaves much smaller than upper ones; lateral leaves twisted.* **Lateral branches** *flat to elliptical in cross section.* **Strobili** 1–6 per stem, 1.2–6.0 cm long.
OCCURRENCE: Occasional, widely distributed.
NOTES: See *D. dendroideum* and *D. hickeyi*.
OTHER NAMES: *Lycopodium obscurum*

Northern Ground-cedar • *Diphasiastrum complanatum*

Perennial herb of forests and edges, 8–44 cm tall. **Leaves** in 4 vertical rows, appressed. **Lateral branches** shiny green above, dull green below, flat in cross section, 2–4.4 mm wide, *with abrupt, conspicuous winter bud constrictions.* **Strobili** 1.5–2.5 cm tall, *lacking a sterile tip,* on long, branching stalks.
OCCURRENCE: Uncommon, with scattered distribution.
NOTES: *Diphasiastrum* ×*zeilleri*, a ground-cedar hybrid between *D. complanatum* and *D. tristachyum*, is known from Baxter State Park. The branch width, number of branch divisions, and leaf size of the hybrid are intermediate between these two parent taxa.
OTHER NAMES: *Lycopodium complanatum*

LYCOPODIACEAE • CLUBMOSS FAMILY

Southern Ground-cedar • *Diphasiastrum digitatum*

Perennial herb of dry soils, 15–50 cm tall. **Leaves** in pseudowhorls of 1 upper, 1 lower, and 2 lateral leaves; upper leaf appressed, larger than lower ones; lateral leaves keeled and spreading. **Lateral branches** fan-shaped, *rectangular in cross section, wider than 2 mm, without winter bud constrictions*. **Strobili** 2–4 cm long, *often with slender, sterile tip*; stalks branching, with 2–4 strobili per stalk.
OCCURRENCE: Occasional, widely distributed.
NOTES: See *D. tristachyum*.
OTHER NAMES: *Lycopodium digitatum, Lycopodium flabelliforme*

Sitka Ground-cedar • *Diphasiastrum sitchense*

Perennial herb of alpine and subalpine areas, rarely at lower elevations, up to 12 cm tall. **Leaves** equal in size, in 5 vertical rows, incurved, widest at middle, sharply pointed. **Lateral branches** *round or elliptical in cross section*. **Strobili** *solitary, sessile*, 0.4–3.8 cm long.
OCCURRENCE: Very rare, documented from only one township in Park.
NOTES: Species is listed as Threatened in Maine. *Diphasiastrum ×sabinifolium*, a ground-cedar hybrid between *D. sitchense* and *D. tristachyum*, is known from Baxter State Park. It has somewhat compressed lateral branches with 4 vertical rows of leaves that are adnate (fused) to branches for 50–60% of their length.
OTHER NAMES: *Lycopodium sitchense*

LYCOPODIACEAE • CLUBMOSS FAMILY

Blue Ground-cedar • *Diphasiastrum tristachyum*

Perennial herb of dry forests, edges, and disturbed areas, 15–35 cm tall. **Leaves** in pseudowhorls of 1 upper, 1 lower, and 2 lateral leaves; *upper and lower leaves equal in size; lateral leaves keeled and twice size of upper and lower leaves.* **Lateral branches** *square in cross section,* fan-shaped, with winter bud constrictions, *blue-green, with white, powdery coating.* **Strobili** 10–24 mm long, 3 or 4 per stalk.
OCCURRENCE: Uncommon, widely distributed.
NOTES: See *D. digitatum*.
OTHER NAMES: *Lycopodium tristachyum*

Northern Bog-clubmoss • *Lycopodiella inundata*

Perennial herb of bogs, wet sand, and gravel, up to 10 cm tall. **Leaves** narrow, ascending or spreading, *seldom toothed.* **Strobili** *1 per stalk, bushy,* 1–7 cm long. **Upright stems** *solitary, unbranched.* **Horizontal shoots** usually flat to the ground.
OCCURRENCE: Occasional, widely distributed.
OTHER NAMES: *Lycopodium inundatum*

FERNS AND OTHER SPORE-PRODUCING PLANTS

LYCOPODIACEAE • CLUBMOSS FAMILY

Common Clubmoss • *Lycopodium clavatum*

Perennial herb of dry fields and open woods, 10–25 cm tall. **Leaves** 4–6 mm long, *with a long colorless hair at tip.* **Strobili** *1–5 per stalk (commonly more than 2), ~8 cm long, on long, branched stalks.* **Lateral branches** *round in cross section.* **Upright stems** bristly, *with 2 or 3 spreading to ascending branches.*
OCCURRENCE: Common, widely distributed.
NOTES: See *L. lagopus*.

One-cone Clubmoss • *Lycopodium lagopus*

Perennial herb of fields and open woods, 10–25 cm tall. **Leaves** 3–5 mm long, *with a long colorless hair at tip.* **Strobili** *1 or rarely 2 per stalk, ~8 cm long.* **Lateral branches** *round in cross section.* **Upright stems** bristly, *with 2 or 3 erect branches.*
OCCURRENCE: Uncommon, with scattered distribution.
NOTES: See *L. clavatum*.

LYCOPODIACEAE • CLUBMOSS FAMILY

Common Interrupted-clubmoss • *Spinulum annotinum*

Perennial herb of coniferous forests, up to 28 cm tall. **Leaves** 5–10 mm long near middle of seasonal growth, obscurely to evidently toothed, *with a minute bristle at tip*. **Strobili** *sessile, usually 1 per erect stem*, 1.7–4.3 cm long. **Upright stems** *with evident growth constrictions, branching near base.*
OCCURRENCE: Common, widely distributed.
NOTES: A clubmoss hybrid between *S. annotinum* and *S. canadense* is known from Baxter State Park. It is intermediate in character between parent taxa.
OTHER NAMES: *Lycopodium annotinum*

Northern Interrupted-clubmoss • *Spinulum canadense*

Perennial herb of alpine and subalpine areas, cliffs, and ridges, sometimes at low elevations, up to 27 cm tall. **Leaves** 3–5.9 mm long near middle of seasonal growth, obscurely toothed to entire. **Strobili** 1–3 per erect stem, 8–21 mm long. **Upright stems** *with evident growth constrictions.*
OCCURRENCE: Rare, with scattered distribution.
OTHER NAMES: *Lycopodium canadense*

FERNS AND OTHER SPORE-PRODUCING PLANTS

ONOCLEACEAE • FIDDLEHEAD FERN FAMILY ▼

Fiddlehead Fern • *Matteuccia struthiopteris*

Perennial herb of floodplains and riparian forests, up to 1.3 m tall. **Sterile leaves** with blades 1.5-times divided, arching, widest above middle, in a circular clump; *leaflets gradually reduced in size toward base of blade; rachis grooved*; petioles black at maturity, flattened at base, with deep, rounded groove. **Fertile leaves** green, maturing to brown, arising in center of clump, much shorter than sterile leaves.
OCCURRENCE: Occasional, widely distributed.
OTHER NAMES: Ostrich Fern

Sensitive Fern • *Onoclea sensibilis*

Perennial herb of open and forested wetlands, up to 1.3 m tall. **Sterile leaves** with blades lobed nearly to rachis, green, triangular, with netted veins, *withering at first frost*; margins of lobes entire to undulate; rachis winged; petioles longer than the blade, yellowish. **Fertile leaves** *erect, appearing as clusters of green beads, becoming brown and persisting through winter*. **Rhizomes** present.
OCCURRENCE: Common, widely distributed.

OPHIOGLOSSACEAE • ADDER'S-TONGUE FERN FAMILY ▼

Narrow Triangle Moonwort • *Botrychium angustisegmentum*

Perennial herb of bogs, fens, forests, and fields, up to 30 cm tall. **Leaves** divided into a sterile portion (trophophore) and a reproductive portion (sporophore); trophophores dark green, shiny, *sessile*, 1–5 cm long, with three main axes; *sporophores always present, reflexed in bud*, with sporangia 0.7–1.0 mm in diameter.
OCCURRENCE: Very rare, documented from only northern portion of Park.
NOTES: See *B. matricariifolium*.

Daisy-leaved Moonwort • *Botrychium matricariifolium*

Perennial herb of forests, edges, and fields, up to 30 cm tall. **Leaves** divided into a sterile portion (trophophore) and a reproductive portion (sporophore); trophophores oblong to ovate, with one main axis, *1.5–6 cm long, lobed to 3 times divided, on a short stalk; sporophores always present*, with tip curved over when plant is in bud.
OCCURRENCE: Very rare, documented from only one township in Park.
NOTES: See *B. angustisegmentum*.

OPHIOGLOSSACEAE • ADDER'S-TONGUE FERN FAMILY

Leathery Grapefern • *Botrychium multifidum*

Perennial herb of forest edges and open areas, up to 40 cm tall. **Leaves** divided into a sterile portion (trophophore) and a reproductive portion (sporophore); trophophores broadly triangular, leathery, up to 20 cm wide and 30 cm long, *remaining green into winter and often persisting into next summer*, diverging from sporophore near ground; leafules ovate, crowded, with entire to crenulate margins; *sporophores often absent*, 8–40 cm long.
OCCURRENCE: Uncommon, with scattered distribution.

Swamp Moonwort • *Botrychium tenebrosum*

Perennial herb of forests, forested wetlands, and streambanks. **Leaves** divided into a sterile portion (trophophore) and a reproductive portion (sporophore); trophophores slender, simple or divided into lobes or leaflets; *stalk of trophophore usually diverging at or above mid-height of plant*; lowest pair of lobes or leaflets, if present, roughly equal in size to adjacent pair. Sporophores erect in bud.
OCCURRENCE: Very rare, documented from only one township in Park.

OPHIOGLOSSACEAE • ADDER'S-TONGUE FERN FAMILY

Rattlesnake Fern • *Botrychium virginianum*

Perennial herb of forests, 5–75 cm tall. **Leaves** divided into a sterile portion (trophophore) and a reproductive portion (sporophore); trophophores sessile, broadly triangular, sparsely pilose, 7–20 cm long and 10–30 cm wide; basal leaflets often large, diverging near junction with sporophore, *giving a 3-parted appearance*; sporophores erect, branched, 6–15 cm long.
OCCURRENCE: Uncommon, widely distributed.
OTHER NAMES: Common Grapefern

OSMUNDACEAE • ROYAL FERN FAMILY ▼

Interrupted Fern • *Osmunda claytoniana*

Perennial herb of forests and wetlands, up to 1.5 m tall. **Sterile leaves** with blades 1.5-times divided, widest at middle; *leaflets with few or no tufts of hairs at base*; petioles smooth, green, *pubescent early, becoming glabrous*. **Fertile leaves** with blade 1.5-times divided, taller than sterile leaves, *with 2–5 pairs of fertile leaflets near midpoint of rachis*. **Sori** absent; sporangia short-stalked.
OCCURRENCE: Occasional, widely distributed.
NOTES: See *Osmundastrum cinnamomeum*. Fiddleheads from this species carcinogenic.

OSMUNDACEAE • ROYAL FERN FAMILY

Royal Fern • *Osmunda regalis*

Perennial herb of wetlands, up to 1.5 m tall. **Sterile leaves** with blades 2-times divided; leafules ovate, stalked, with entire to finely serrate margins; petioles smooth, reddish at base. **Fertile leaves** similar to sterile leaves, *with fertile leaflets at branch tips*. **Sori** absent; sporangia short-stalked.
OCCURRENCE: Occasional, widely distributed.
NOTES: Fiddleheads from this species carcinogenic.
OTHER NAMES: Flowering Fern

Cinnamon Fern • *Osmundastrum cinnamomeum*

Perennial herb of forests and wetlands, up to 1.5 m tall. **Sterile leaves** with blades 1.5-times divided; *leaflets with tufts of white-brown hairs at base*; petioles slightly shorter than blade, smooth, green, *with at least some long hairs*. **Fertile leaves** erect, arising in center of clump, *cinnamon-colored at maturity, narrow, without expanded leaflets*. **Sori** absent; sporangia large, short-stalked.
OCCURRENCE: Common, widely distributed.
NOTES: See *Osmunda claytoniana*. Fiddleheads from this species carcinogenic.
OTHER NAMES: *Osmunda cinnamomea*

372 THE PLANTS OF BAXTER STATE PARK

POLYPODIACEAE • FERN FAMILY

Appalachian Polypody • *Polypodium appalachianum*

Perennial herb of rocks and cliffs, up to 40 cm tall. **Leaves** with blades lobed but not fully cut to rachis, *widest just above base; lobes with narrow, pointed tips*; petioles slender, round, light green. **Sori** round, prominent; indusia absent. **Rhizomes** present.
OCCURRENCE: Uncommon, with scattered distribution.
NOTES: See *P. virginianum*.

Rock Polypody • *Polypodium virginianum*

Perennial, evergreen herb of rocks and cliffs, usually under forest canopy, up to 40 cm tall. **Leaves** with blades lobed but not fully cut to rachis, *widest near middle; lobes with blunt, rounded tips*; petioles slender, round, light green. **Sori** round, prominent; indusia absent.
OCCURRENCE: Occasional, widely distributed.
NOTES: See *P. appalachianum*.

PTERIDACEAE • MAIDENHAIR FERN FAMILY ▼

Northern Maidenhair Fern • *Adiantum pedatum*

Perennial herb of rich, hardwood forests, up to 75 cm tall. **Leaves** *divided at apex into 2 spreading rachises; leaflets elongated, smooth, divided into numerous leafules; leafules alternate, fan-shaped, 12–22 mm long and 5–9 mm wide;* sterile and fertile leaves similar; petioles long and smooth, *black or purplish brown*, with scales at base. **Sori** *elongate; indusia formed from reflexed margin of leafules.*
OCCURRENCE: Very rare, documented from only one township in Park.

THELYPTERIDACEAE • MARSH FERN FAMILY ▼

New York Fern • *Parathelypteris noveboracensis*

Perennial herb of forests and open areas, up to 70 cm tall. **Leaves** with blades 1.5-times divided; *leaflets gradually tapering to base and tip, without red to orange resin glands; lowest leaflets much shorter than middle ones; petioles shorter than blade, often darkened.* **Sori** few, round, small, near margins; indusia less than 1 mm wide, pale, with long hairs.
OCCURRENCE: Common, widely distributed.
NOTES: See *P. simulata* and *Thelypteris palustris*.
OTHER NAMES: *Dryopteris noveboracensis, Thelypteris noveboracensis*

THELYPTERIDACEAE • MARSH FERN FAMILY

Massachusetts Fern • *Parathelypteris simulata*

Perennial herb of acidic bogs, swamps, and marshes, up to 80 cm tall. **Leaves** with blades 1.5- to 2-times divided; fertile leaves often taller than sterile leaves; *leaflets with minute, red to orange resin glands; lowest leaflets nearly same length as middle ones*; petioles pale green to straw-colored.
OCCURRENCE: Rare, with scattered distribution.
NOTES: See *P. noveboracensis* and *Thelypteris palustris*.
OTHER NAMES: *Thelypteris simulata*

Long Beech Fern • *Phegopteris connectilis*

Perennial herb of wet woods, up to 50 cm tall. **Leaves** with blades 1.5-times divided, *triangular*, often pubescent; *lowest leaflets distinctly angled down*; petioles pubescent, scaly. **Sori** small and round, near margins; *indusia absent*.
OCCURRENCE: Common, widely distributed.
OTHER NAMES: *Dryopteris phegopteris, Thelypteris phegopteris*

THELYPTERIDACEAE • MARSH FERN FAMILY

Marsh Fern • *Thelypteris palustris*

Perennial herb of open and forested wetlands, up to 60 cm tall. **Leaves** with blades 1.5-times divided, without glands, pale green, *often twisted*; leafules with forked veins; *margins of fertile segments rolled toward lower surface; petioles longer than blade.* **Sori** numerous, round, near midvein; indusia often pubescent.

OCCURRENCE: Uncommon, with scattered distribution.

NOTES: See *Parathelypteris noveboracensis* and *P. simulata*.

OTHER NAMES: *Dryopteris thelypteris*

WOODSIACEAE • LADY FERN FAMILY ▼

Northern Lady Fern • *Athyrium angustum*

Perennial herb of wet woods and wetlands, up to 1 m tall. **Leaves** with blades 2.5-times divided, broadest near middle, *usually drooping at tip, in circular clusters; leafules with toothed margins*; sterile and fertile leaves similar; *petioles smooth, with dark brown scales, without glandular hairs.* **Sori** *elongate, arched;* indusia attached along one side, pubescent.
OCCURRENCE: Common, widely distributed.
OTHER NAMES: *Asplenium filix-femina*

Fragile Fern • *Cystopteris fragilis*

Perennial herb of cliffs and ledges, up to 40 cm tall. **Leaves** with blades 1.5- to 2.5-times divided; fertile and sterile leaves similar; leafules with toothed margins, *with veins extending to tip of teeth; leafules on basal leaflets sessile and perpendicular to main axes of leaflets; petioles brittle,* dark at base, green above, sparsely scaly at base. **Sori** round; indusia ovate to lanceolate.
OCCURRENCE: Very rare, documented from only northern portion of Park.
NOTES: See *C. tenuis.*

WOODSIACEAE • LADY FERN FAMILY

Mackay's Fragile Fern • *Cystopteris tenuis*

Perennial herb of cliffs, ledges, and talus, up to 40 cm tall. **Leaves** with blades 1.5- to 2.5-times divided; fertile and sterile leaves similar; margins of leafules dentate to deeply incised, *with veins extending to both teeth and notches; leafules on basal leaflets short-stalked and angled toward tip of leaflets; petioles brittle*, dark at base, green above, sparsely scaly at base. **Sori** round; indusia ovate to cup-shaped.
OCCURRENCE: Very rare, documented from only northern portion of Park.
NOTES: See *C. fragilis*.

Silvery False Spleenwort • *Deparia acrostichoides*

Perennial herb of rich, hardwood forests, up to 1.2 m tall. **Leaves** with blades 1.5-times divided, pubescent along veins, tapering at both ends; *lowest pair of leaflets downward-pointing*; fertile leaves taller and more erect than sterile ones; *petioles short, dark red-brown and swollen at base*, with long, white hairs and light brown scales. **Sori** long, narrow.
OCCURRENCE: Very rare, documented from only northern portion of Park.
OTHER NAMES: Silver Glade Fern, *Athyrium thelypteroides*

WOODSIACEAE • LADY FERN FAMILY

Northern Oak Fern • *Gymnocarpium dryopteris*

Perennial herb of upland and wetland forests, up to 50 cm tall. **Leaves** with blade 2.5-times divided, *broadly triangular*, upper surface without glandular hairs; *lower leaflets at right angles to petiole and almost parallel to ground; petioles longer than blade, dark at base*. **Sori** small, rounded, near margin; *indusia absent*.
OCCURRENCE: Common, widely distributed.

Smooth Cliff Fern • *Woodsia glabella*

Perennial herb of cliffs and talus on high-pH bedrock, up to 15 cm tall. **Leaves** with blades 1.5-times divided, *glabrous, narrowly linear to linear-lanceolate*, 3–15 cm long and 1–1.5 cm wide; *basal leaflets fan-shaped; petioles glabrous*, green or straw-colored; often persistent into next growing season. **Sori** round; *indusia composed of filaments*.
OCCURRENCE: Very rare, documented from only one township in Park.
NOTES: Species is listed as Threatened in Maine.

WOODSIACEAE • LADY FERN FAMILY

Rusty Cliff Fern • *Woodsia ilvensis*

Perennial herb of dry, exposed rocks and cliffs, up to 30 cm tall. **Leaves** with blades 1.5-times divided, the blades 5–15 cm long and wider than 1 cm, *with abundant brown hairs and scales on leaflets and rachis*; petioles stout, brittle, dark brown, pubescent, often persistent into next growing season. **Sori** round, numerous; *indusia composed of many filaments*.
OCCURRENCE: Rare, with scattered distribution.

CYPERACEAE • SEDGE FAMILY ▼

Tufted Hair-sedge • *Bulbostylis capillaris*

Annual herb of dry, sandy, disturbed soils such as roadsides or gravel pits, *up to 30 cm tall*. **Inflorescence** *with lowest subtending bract bristle-like*. **Spikelets** relatively large, *purplish to black, clustered, on short peduncles*. **Leaves** mostly basal, minutely pubescent, *thread-like*. **Stems** densely tufted, round in cross section, *much longer than leaves, stiff, hair-like*. **Achenes** *with a tiny bump (tubercle) at top*.
OCCURRENCE: Rare, with scattered distribution.
OTHER NAMES: Vagabond

White-tinged Sedge • *Carex albicans*

Perennial herb of forests and open areas, 5–45 cm tall. **Spikes** borne on long aerial stems; *internode between two lowest spikes less than 7 mm*. **Leaves** *0.8–1.5 mm wide*. **Stems** tufted. **Carpellate scales** *longer than or equaling perigynium*. **Perigynia** pubescent, *the body (excluding beak and basal stalk) longer than wide*.
OCCURRENCE: Very rare, documented from only one township in Park.
NOTES: Section *Acrocystis*. See *C. communis*, *C. deflexa*, *C. lucorum*, and *C. pensylvanica*.
OTHER NAMES: *Carex emmonsii, Carex artitecta*

CYPERACEAE • SEDGE FAMILY

Appalachian Sedge • *Carex appalachica*

Perennial herb of sandy or rocky soils in mixed woods, up to 60 cm tall. **Spikes** not densely crowded; distance between two lowest spikes at least 2-times as long as lowest spike; *stigmas coiled 1- to 3-times.* **Leaves** *0.9–1.5 mm wide*, fine, thread-like, reaching tip of spikes; sheaths tightly clasping. **Stems** loosely tufted. **Perigynia** *ascending to slightly spreading*, spongy at base.
OCCURRENCE: Uncommon, with scattered distribution.
NOTES: Section *Phaestoglochin*. See *C. radiata* and *C. rosea*.

Water Sedge • *Carex aquatilis*

Perennial herb of shallow water and open marshes, *up to 1.2 m tall*. **Inflorescence** *longer than lowest subtending bract*. **Carpellate spikes** erect to ascending. **Leaves** *2.5–5 mm wide*, often with whitish waxy coating. **Stems** *not tufted*. **Carpellate scales** without awns. **Perigynia** *without veins or with obscure veins on faces*. **Achenes** glossy. **Rhizomes** *long*.
OCCURRENCE: Uncommon, with scattered distribution.
NOTES: Section *Phacocystis*. See *C. lenticularis, C. nigra*, and *C. stricta*.

Northern Cluster Sedge • *Carex arcta*

Perennial herb of coniferous forest wetlands, up to 80 cm tall. **Inflorescence** *congested, 1.5–4 cm long*; lowest subtending bract less than 3 cm long. **Spikes** 2–15 per stem, sessile, similar looking, appressed along stem, *overlapping and clustered*. **Leaves** 2.5–4 mm wide; sheaths with conspicuous purple dots. **Stems** densely tufted. **Perigynia** *widest near base*; margin of beak finely toothed.
OCCURRENCE: Uncommon, with scattered distribution.
NOTES: Section *Glareosae*.
OTHER NAMES: Bear Sedge

Drooping Woodland Sedge • *Carex arctata*

Perennial herb of forested uplands and woodland edges, 0.2–1 m tall. **Spikes** long, narrow, drooping, 2.5–8 cm long; peduncles up to 3 cm long, shorter than spike; terminal spike usually unisexual, staminate. **Leaves** 3–10 mm wide; *basal sheath reddish purple*. **Stems** densely tufted. **Perigynia** *3–5 mm long*, borne on a short stipe, *abruptly tapering to a long beak*. **Achenes** *sessile, distinctly 3-sided*.
OCCURRENCE: Common, widely distributed.
NOTES: Section *Hymenochlaenae*. See *C. debilis* and *C. gracillima*.

Prickly Bog Sedge • *Carex atlantica*

Perennial herb of bogs, swamps, and peatlands, 0.1–1.1 m tall. **Spikes** with obvious staminate portion at base of terminal one. **Leaves** 0.4–4 mm wide, *the largest wider than 2.8 mm*. **Stems** tufted. **Perigynia** *spreading to reflexed, the body nearly circular in outline, up to 1.7-times as long as wide, widest near base, gradually tapering or concavely tapering to beak from widest point, not forming shoulder*; beak sparsely and finely toothed.
OCCURRENCE: Very rare, documented from only one township in Park.
NOTES: Section *Stellulatae*. See *C. echinata*, *C. interior*, and *C. wiegandii*.

Scabrous Black Sedge • *Carex atratiformis*

Perennial herb of calcareous shores, alpine seeps, and streambanks, up to 70 cm tall. **Spikes** *spreading or drooping, on a peduncle longer than 10 mm*; carpellate spikes cylindrical. **Leaves** 2.5–5 mm wide. **Stems** loosely tufted. **Carpellate scales** *without awns*, longer than and as wide as perigynium. **Achenes** filling less than half of a perigynium.
OCCURRENCE: Very rare, documented from only one township in Park.
NOTES: Section *Racemosae*. See *C. magellanica*. Species is listed as Special Concern in Maine.

CYPERACEAE • SEDGE FAMILY

Bebb's Sedge • *Carex bebbii*

Perennial herb of wet meadows, swamps, and shorelines, often associated with calcareous habitats, up to 90 cm tall. **Spikes** *rounded in outline*, usually reddish brown, *tightly crowded*. **Leaves** 2–3 mm wide; *sheaths tight along stem*. **Stems** densely tufted. **Carpellate scales** *shorter and narrower than perigynium*. **Perigynia** small, 2.5–3.8 mm long and *1–2 mm wide*.
OCCURRENCE: Uncommon, with scattered distribution.
NOTES: Section *Cyperoideae*. See *C. crawfordii*, *C. oronensis*, and *C. tribuloides*.

Date Unknown – MA

Bigelow's Sedge • *Carex bigelowii*

Perennial herb of *exposed, alpine summits*, up to 30 cm tall. **Spikes** erect, narrow-cylindric; *lowest subtending bract shorter than inflorescence*. **Leaves** 1.5–4 mm wide; sheaths concave at apex. **Stems** colonial. **Carpellate scales** black, *without awns*. **Perigynia** ascending, green or mottled, smooth, 1.8–3.5 mm long; *veins inconspicuous*. **Rhizomes** present.
OCCURRENCE: Uncommon, with scattered distribution.
NOTES: Section *Phacocystis*. See *C. nigra*. Species is listed as Special Concern in Maine.

18 June – AH   27 June – GM

Billings' Sedge • *Carex billingsii*

Perennial herb of *open*, acidic wetlands, up to 36 cm tall. **Inflorescence** *14–32 mm long*; lowest subtending bract thread-like, longer than or equaling inflorescence. **Spikes** *2–3 per stem*; *terminal spike with 1–3 perigynia*; lowest spike removed from upper spikes by more than 2 cm. **Leaves** 0.3–0.8 mm wide, delicate, thread-like; *margins rolled inward toward upper surface*. **Stems** loosely tufted, slender, delicate.
OCCURRENCE: Occasional, widely distributed.
NOTES: Section *Glareosae*. See *C. trisperma*.
OTHER NAMES: *Carex trisperma* var. *billingsii*.

10 September Russell Pond area – GM

CYPERACEAE • SEDGE FAMILY

Brome-like Sedge • *Carex bromoides*

Perennial herb of deciduous forested wetlands, floodplains, and streambanks, up to 90 cm tall. **Inflorescence** *with lowest subtending bract inconspicuous, shorter than lowest spike.* **Spikes** sessile, similar looking, with staminate flowers below perigynia (gynecandrous). **Leaves** *1.1–3 mm wide*, flat. **Stems** tufted. **Carpellate scales** *1.1–1.6 mm wide.* **Perigynia** appressed to ascending, *0.8–1.3 mm wide*, 4- to 6-times as long as wide, with 3–6 conspicuous veins on outer surface.
OCCURRENCE: Very rare, documented from only one township in Park.
NOTES: Section *Deweyanae*. See *C. deweyana*.

Brownish Sedge • *Carex brunnescens*

Perennial herb of forested wetlands, seepy ledges, and summits, up to 90 cm tall. **Inflorescence** longer than 15 mm; lowest subtending bract less than 3 cm long. **Spikes** 5–10 per stem, sessile, similar looking, ascending, not congested. **Leaves** 1–2.5 mm wide. **Stems** densely tufted, *delicate*. **Perigynia** *5–15 per spike, with a distinct slit on outer (abaxial) side of beak.*
OCCURRENCE: Common, widely distributed.
NOTES: Section *Glareosae*. See *C. canescens*.

Hoary Sedge • *Carex canescens*

Perennial herb of bogs, swamps, and shallow water, up to 90 cm tall. **Inflorescence** longer than 15 mm; lowest subtending bract less than 3 cm long. **Spikes** 5–8 per stem, sessile, similar looking, ascending, not congested. **Leaves** 2–4 mm wide. **Stems** tufted, *firm*. **Perigynia** 10–30 per spike, *without a distinct slit on outer (abaxial) side of beak.*
OCCURRENCE: Common, widely distributed.
NOTES: Section *Glareosae*. See *C. brunnescens*.

CYPERACEAE • SEDGE FAMILY

Rope-root Sedge • *Carex chordorrhiza*

Perennial herb of *open wetlands with deep organic soil*, up to 35 cm tall. **Spikes** sessile, similar looking, densely crowded. **Leaves** 0.4–3 mm wide. **Stems** *colonial; erect shoots arising from axils of prostrate stems.* **Rhizomes** short. **Stolons** *present.*
OCCURRENCE: Very rare, documented from only one township in Park.
NOTES: Section *Chordorrhizae.*

Fibrous-rooted Sedge • *Carex communis*

Perennial herb of clearings and forested areas, 10–60 cm tall. **Spikes** borne on long aerial stems. **Leaves** *3–5 mm wide.* **Stems** densely tufted. **Perigynia** pubescent, *the body (excluding beak and basal stalk) as long as wide.* **Rhizomes** *absent.*
OCCURRENCE: Common, widely distributed.
NOTES: Section *Acrocystis*. See *C. albicans, C. deflexa, C. lucorum,* and *C. pensylvanica.*

Open-field Sedge • *Carex conoidea*

Perennial herb of wet meadows and shorelines, up to 75 cm tall. **Inflorescence** *with lowest subtending bract with prolonged, loose sheath.* **Spikes** on short, ascending, *rough peduncles.* **Leaves** 3–4 mm wide, glabrous. **Stems** densely tufted. **Carpellate scales** *with rough awn up to 2.7 mm long.* **Perigynia** plump, *with numerous, conspicuous, impressed veins.*
OCCURRENCE: Very rare, documented from only one township in Park.
NOTES: Section *Griseae*. See *C. pallescens.*
OTHER NAMES: *Carex katahdinensis*

CYPERACEAE • SEDGE FAMILY

Crawford's Sedge • *Carex crawfordii*

Perennial herb of wet meadows, shorelines, and swamps, up to 65 cm tall. **Inflorescence** slightly overtopping leaves, *very tightly crowded*. **Spikes** *8–10 mm long and 4.5–6.5 mm wide*. **Leaves** *3–4 mm wide*. **Stems** densely tufted. **Carpellate scales** not glossy, *shorter than perigynium*. **Perigynia** lance-shaped, very narrow, *up to 1.3 mm wide*.
OCCURRENCE: Occasional, widely distributed.
NOTES: Section *Cyperoideae*. See *C. oronensis, C. scoparia,* and *C. tincta*.

2 July – DC

Fringed Sedge • *Carex crinita*

Perennial herb of wet woods, thickets, and swales, 0.4–1.3 m tall. **Inflorescence** arching with drooping spikes; lowest subtending bract longer than inflorescence. **Leaves** 4–10.5 mm wide; *basal sheaths smooth*. **Stems** densely tufted. **Carpellate scales** *truncate to notched at apex*, with rough awn nearly as long as scale body. **Perigynia** *widest above middle, abruptly tapering to small beak*.
OCCURRENCE: Common, widely distributed.
NOTES: Section *Phacocystis*. See *C. gynandra*.
OTHER NAMES: Long-hair Sedge

25 June Trout Brook area – HC

Northeastern Sedge • *Carex cryptolepis*

Perennial herb of sandy shorelines and peaty wetlands, up to 50 cm tall. **Inflorescence** with crowded, short-cylindrical to spherical spikes. **Leaves** 1.5–2.9 mm wide. **Stems** tufted. **Carpellate scales** *same color as perigynium, often concealed by perigynium at maturity*. **Perigynia** yellowish green, reflexed, spreading; *beak bent, smooth*.
OCCURRENCE: Occasional, with scattered distribution.
NOTES: Section *Ceratocystis*. See *C. flava* and *C. viridula*.

13 August Katahdin Lake – ML   20 June Abol Pond – ML

CYPERACEAE • SEDGE FAMILY

White-edged Sedge • *Carex debilis*

Perennial herb of forested uplands and wet fields, 0.2–1 m tall. **Spikes** up to 8 cm long, narrow, drooping, on long peduncles; *terminal spike often entirely staminate*. **Leaves** 2–7 mm wide; *basal sheaths dark reddish purple*. **Stems** densely tufted. **Perigynia** *5–6.2 mm long*, without a short stipe, *gradually tapering to beak*. **Achenes** obscurely 3-sided, short-stalked.
OCCURRENCE: Common, widely distributed.
NOTES: Section *Hymenochlaenae*. See *C. arctata* and *C. gracillima*.

Northern Sedge • *Carex deflexa*

Perennial herb of forest edges, gaps, and rocky slopes, up to 30 cm tall. **Inflorescence** *with densely clustered spikes; lowest subtending bract longer than or equaling inflorescence*. **Spikes** borne on both long, aerial stems and short peduncles sometimes hidden by leaf bases. **Leaves** 1–2.6 mm wide, *overtopping or equaling flowering stem*. **Stems** loosely to densely tufted. **Carpellate scales** *approximately half as long as perigynium*. **Perigynia** pubescent, *the body (excluding beak and basal stalk) longer than wide*.
OCCURRENCE: Occasional, widely distributed.
NOTES: Section *Acrocystis*. See *C. albicans*, *C. communis*, *C. lucorum*, and *C. pensylvanica*.

Round-fruited Short-scaled Sedge • *Carex deweyana*

Perennial herb of forests and forest edges, up to 90 cm tall. **Inflorescence** *with lowest subtending bract conspicuous, longer than lowest spike*. **Spikes** sessile, all similar looking, with staminate flowers below perigynia (gynecandrous). **Leaves** 2.4–4.2 mm wide, flat. **Stems** tufted. **Carpellate scales** *1.6–2.2 mm wide*. **Perigynia** appressed to ascending, *1.3–1.5 mm wide*, 3- to 4-times as long as wide, with obscure veins on outer face.
OCCURRENCE: Uncommon, with scattered distribution.
NOTES: Section *Deweyanae*. See *C. bromoides*.

Lesser Tussock Sedge • *Carex diandra*

Perennial herb of calcareous wetlands and swamps, up to 90 cm tall. **Inflorescence** erect, 2–5 cm long; lowest subtending bract short, bristle-like. **Spikes** densely crowded, overlapping. **Leaves** 1–2.5 mm wide; *sheaths whitish with red dots*. **Stems** firm. **Perigynia** *dark brown, 2.3–2.5 mm long*. **OCCURRENCE:** Very rare, documented from only one township in Park. **NOTES:** Section *Heleoglochin*. See *C. stipata* and *C. vulpinoidea*.

Slender Woodland Sedge • *Carex digitalis*

Perennial herb of mixed upland forests, up to 52 cm tall. **Spikes** 2–4 per stem, with 3–9 perigynia per spike; *carpellate spikes on long, slender peduncles borne near base of plant*. **Leaves** 1.5–4.5 mm wide; *basal sheaths white to light brown*. **Stems** densely tufted. **Achenes** *sharply triangular*. **OCCURRENCE:** Very rare, documented from only one township in Park. **NOTES:** Section *Careyanae*. See *C. laxiflora*.

Soft-leaved Sedge • *Carex disperma*

Perennial herb of bogs and wet woods, up to 60 cm tall. **Inflorescence** 1.5–2.5 cm long; *lowest subtending bract small, scale-like*. **Spikes** *2–4 per stem*, with 1–4 perigynia per spike. **Leaves** 0.7–1.5 mm wide, *delicate*, flat. **Stems** loosely tufted, *slender, weak*. **Perigynia** *plump, not tightly appressed to stem*, smooth, hairless; beak minute, up to 0.3 mm long. **OCCURRENCE:** Occasional, widely distributed. **NOTES:** Section *Dispermae*. See *C. billingsii*, *C. leptalea*, and *C. trisperma*.

CYPERACEAE • SEDGE FAMILY

Star Sedge • *Carex echinata*

Perennial herb of bogs, swamps, wet meadows, and ditches, 10–50 cm tall. **Spikes** terminal with obvious staminate portion at base. **Leaves** *0.7–2.7 mm wide*. **Stems** densely tufted. **Perigynia** *0.8–2.1 mm wide, 1.8- to 3.2-times as long as wide; beak 1–1.6 mm long*.
OCCURRENCE: Common, widely distributed.
NOTES: Section *Stellulatae*. See *C. atlantica, C. interior*, and *C. wiegandii*.
OTHER NAMES: *Carex angustior, Carex josselynii*

Meager Sedge • *Carex exilis*

Perennial herb of bogs and wet soils, up to 80 cm tall. **Spikes** *solitary, rarely with 1–3 smaller, accessory spikes*. **Leaves** *0.4–1.5 mm wide, stiff, wiry; margins rolled inward toward upper surface*. **Stems** tufted. **Perigynia** *strongly spreading to reflexed*.
OCCURRENCE: Very rare, documented from only one township in Park.
NOTES: Section *Stellulatae*.

Yellow-green Sedge • *Carex flava*

Perennial herb of shorelines, wet meadows, and seepy ledges, up to 75 cm tall. **Inflorescence** with short-cylindrical to spherical spikes. **Leaves** *1.6–4.7 mm wide*. **Stems** tufted. **Carpellate scales** *reddish brown to copper-brown, distinctly contrasting in color with perigynium*. **Perigynia** yellowish green, *reflexed; beak scabrous, bent, 1.3–2.7 mm long*.
OCCURRENCE: Occasional, widely distributed.
NOTES: Section *Ceratocystis*. See *C. cryptolepis* and *C. viridula*.

CYPERACEAE • SEDGE FAMILY

Straw Sedge • *Carex foenea*

Perennial herb of calcareous wetlands and swamps, up to 90 cm tall. **Inflorescence** erect, 2–5 cm long; lowest subtending bract short, bristle-like. **Spikes** densely crowded, overlapping. **Leaves** 1–2.5 mm wide; *sheaths whitish with red dots*. **Stems** firm. **Perigynia** dark brown, 2.3–2.5 mm long. **OCCURRENCE:** Very rare, documented from only one township in Park. **NOTES:** Section *Cyperoideae*. See *C. stipata* and *C. vulpinoidea*.

Northern Long Sedge • *Carex folliculata*

28 June – AH

Perennial herb of forested wetlands and swales, *up to 1.75 m tall*. **Inflorescence** 15–85 cm long. **Leaves** *6–18 mm wide, not overtopping inflorescence*. **Stems** loosely tufted, *long-arching to prostrate*. **Carpellate scales** *(excluding awn) two-thirds as long as perigynium*. **Perigynia** *1.8–3.3 mm wide, 4- to 7-times as long as wide*, slightly inflated. **OCCURRENCE:** Occasional, widely distributed. **NOTES:** Section *Rostrales*. See *C. intumescens* and *C. michauxiana*.

24 August – DC | 10 September Upper Togue Pond – ML

Graceful Sedge • *Carex gracillima*

Perennial herb of forests, roadsides, and wetlands, up to 90 cm tall. **Spikes** long, narrow, drooping, up to 7 cm long; peduncles up to 4 cm long; terminal spike usually with a few perigynia at apex. **Leaves** 3–9 mm wide; *basal sheath reddish brown*. **Stems** densely tufted. **Perigynia** 2–3.7 mm long, *blunt at tip, beakless*. **OCCURRENCE:** Common, widely distributed. **NOTES:** Section *Hymenochlaenae*. See *C. arctata* and *C. debilis*.

3 July – DC | 29 May – DC

CYPERACEAE • SEDGE FAMILY

Nodding Sedge • *Carex gynandra*

Perennial herb of wet woods, clearings, and ditches, up to 1.4 m tall. **Inflorescence** arching with drooping spikes; lowest subtending bract longer than inflorescence. **Leaves** 4–10.5 mm wide; *basal sheaths rough*. **Stems** tufted. **Carpellate scales** *gradually tapering to awn*; awn rough, nearly as long as scale body. **Perigynia** *widest below middle, gradually tapering to small beak*.
OCCURRENCE: Common, widely distributed.
NOTES: Section *Phacocystis*. See *C. crinita*.

Hayden's Sedge • *Carex haydenii*

Perennial herb of open marshes, wet meadows, and fields, up to 1.1 m tall. **Spikes** *ascending to spreading, not appressed to stem*. **Leaves** 3–4 mm wide; *basal sheaths smooth, hairless*. **Stems** tufted. **Carpellate scales** without awns, *longer than perigynium*. **Perigynia** *slightly inflated, spreading, widest above middle*.
OCCURRENCE: Rare, with scattered distribution.
NOTES: Section *Phacocystis*. See *C. stricta*.

Pubescent Sedge • *Carex hirtifolia*

Perennial herb of meadows, rich woods, and floodplain forests, up to 60 cm tall. **Inflorescence** with narrow-cylindric spikes. **Spikes** 2–5 per stem; lateral spikes on short peduncles. **Leaves** 5–8 mm wide, *densely pubescent*; basal sheaths reddish. **Stems** loosely tufted, *pubescent*. **Perigynia** *pubescent*.
OCCURRENCE: Very rare, documented from only one township in Park.
NOTES: Section *Hirtifoliae*. Species is listed as Special Concern in Maine.

CYPERACEAE • SEDGE FAMILY

Porcupine Sedge • *Carex hystericina*

Perennial herb of rich soils of shorelines, seeps, fens, and marshes, up to 1 m tall. **Spikes** 1–6 cm long and 10–15 mm wide; *lateral spikes on peduncles.* **Leaves** 2.5–8.5 mm wide; *basal sheaths reddish purple.* **Stems** densely to loosely tufted. **Perigynia** *soft*, spreading, slightly inflated, *4.5–7.3 mm long and 1.4–2.1 mm wide*; with 13–21 conspicuous veins.
OCCURRENCE: Historical record, last documented in 1982, from northern portion of Park.
NOTES: Section *Vesicariae*. See *C. lurida* and *C. pseudocyperus*.

Inland Sedge • *Carex interior*

Perennial herb of swamps, fens, and seeps, up to 1.1 m tall. **Spikes** terminal with obvious staminate portion at base. **Leaves** *0.6–2.4 mm wide.* **Stems** tufted. **Perigynia** *1.1–1.8 mm wide*, 1.4- to 2.2-times as long as wide; *beak less than 1 mm long.*
OCCURRENCE: Very rare, documented from only one township in Park.
NOTES: Section *Stellulatae*. See *C. atlantica* and *C. echinata*.

Greater Bladder Sedge • *Carex intumescens*

Perennial herb of open and forested wetlands, *up to 80 cm tall.* **Inflorescence** 2–15 cm long. **Leaves** *3.5–8 mm wide, overtopping inflorescence; basal sheaths reddish purple.* **Stems** tufted. **Perigynia** *10–16.5 mm long and 4–5.5 mm wide, greatly inflated.*
OCCURRENCE: Common, widely distributed.
NOTES: Section *Lupulinae*. See *C. folliculata*.

Lakeside Sedge • *Carex lacustris*

Perennial herb of wetlands, up to 1.3 m tall. **Inflorescence** 17–60 cm long. **Leaves** 8.5–15 mm wide, *M-shaped in cross section*; *basal sheaths reddish purple, strongly shredding*; ligules much longer than wide. **Stems** colonial. **Perigynia** *ascending, slightly inflated*, smooth, hairless, *tapering gradually to inconspicuous beak*. **Rhizomes** present.
OCCURRENCE: Uncommon, with scattered distribution.
NOTES: Section *Paludosae*. See *C. utriculata* and *C. vesicaria*.

Woolly-fruited Sedge • *Carex lasiocarpa*

Perennial herb of shallow water and open fens, up to 1.2 m tall. **Inflorescence** 6–20 cm long. **Leaves** *narrow, 0.2–2 mm wide*, wispy, U-shaped in cross section or with margins rolled inward toward upper surface. **Stems** colonial. **Perigynia** *densely pubescent*, 3–4.3 mm long; beak with 2 teeth at apex. **Rhizomes** present.
OCCURRENCE: Common, widely distributed.
NOTES: Section *Paludosae*.

Broad Loose-flowered Sedge • *Carex laxiflora*

Perennial herb of deciduous and mixed forests, ledges, and outcrops, up to 47 cm tall. **Spikes** 9–60 mm long and up to 4 mm wide, loosely flowered. **Leaves** 5–26 mm wide, not reaching tip of inflorescence; *basal sheaths brownish*, without any red or purple. **Stems** densely tufted. **Perigynia** *3.2–4.1 mm long and 1.2–1.6 mm wide, not tightly overlapping, with more than 20 conspicuous veins*; beak short, straight or slightly bent.
OCCURRENCE: Rare, with scattered distribution.
NOTES: Section *Laxiflorae*. See *C. leptonervia* and *C. ormostachya*.

CYPERACEAE • SEDGE FAMILY

Lake Shore Sedge • *Carex lenticularis*

Perennial herb of shorelines of ponds and streams, *up to 60 cm tall*. **Inflorescence** erect; *lowest subtending bract longer than inflorescence*. **Spikes** erect to ascending. **Leaves** *1.5–3 mm wide*. **Stems** densely tufted. **Carpellate scales** without awns. **Perigynia** green; *beak straight*. **Achenes** *not glossy*.
OCCURRENCE: Occasional, widely distributed.
NOTES: Section *Phacocystis*. See *C. aquatilis*, *C. nigra*, and *C. torta*.

Oval Sedge • **Carex leporina*

Perennial herb of disturbed meadows, roadsides, ditches, and fields, up to 85 cm tall. **Inflorescence** erect, fairly congested. **Leaves** 2–4 mm wide; basal sheaths smooth. **Stems** densely tufted. **Carpellate scales** *reddish brown*, 3.4–5 mm long, mostly covering perigynium. **Perigynia** 3.4–4.7 mm long and 1.3–2.1 mm wide; *beak round in cross section, unwinged, smooth*.
OCCURRENCE: Very rare, documented from only one township in Park.
NOTES: Section *Cyperoideae*; See *C. oronensis* and *C. tincta*.

Bristle-stalk Sedge • *Carex leptalea*

Perennial herb of wet woods and fields, up to 70 cm tall. **Inflorescence** 4–15 mm long; lowest subtending bract small, bristle-like. **Spikes** *solitary*, with tightly appressed perigynia. **Leaves** 0.4–1.3 mm wide, *fine, delicate*. **Stems** loosely tufted, mat-forming. **Perigynia** *tightly appressed to stem*, 2.5–4.9 mm long and 0.8–1.3 mm wide.
OCCURRENCE: Occasional, widely distributed.
NOTES: Section *Leptocephalae*. See *C. disperma*.

Nerveless Woodland Sedge • *Carex leptonervia*

Perennial herb of wet forests, clearings, and thickets, up to 44 cm tall. **Spikes** up to 22 mm long and 2.8–4 mm wide, loosely flowered; *carpellate spikes overlapping*. **Leaves** 3–10 mm wide; *basal sheaths brown*. **Stems** densely tufted. **Perigynia** *2.5–3.2 mm long* and 1–1.5 mm wide, not tightly overlapping, *with only 2 or 3 distinct veins*; beak short, curved.
OCCURRENCE: Uncommon, with scattered distribution.
NOTES: Section *Laxiflorae*. See *C. laxiflora* and *C. ormostachya*.

Mud Sedge • *Carex limosa*

Perennial herb of bogs and pond margins, up to 60 cm tall. **Inflorescence** longer than lowest subtending bract. **Spikes** drooping, cylindrical; *staminate spikes 10–30 mm long*. **Leaves** 1–2.5 mm wide; *margins rolled inward toward upper surface*. **Stems** rough, angled. **Carpellate scales** 2–3.4 mm wide, shorter than and as wide as perigynium, blunt at apex. **Rhizomes** short.
OCCURRENCE: Uncommon, with scattered distribution.
NOTES: Section *Limosae*. See *C. magellanica* and *C. rariflora*.

Blue Ridge Sedge • *Carex lucorum*

Perennial herb of upland woods and clearings, up to 55 cm tall. **Inflorescence** longer than lowest subtending bract. **Spikes** borne on long, aerial stems. **Leaves** 1–3 mm wide; blades poorly developed along upper portion of stem. **Stems** loosely tufted. **Perigynia** *(excluding beak and basal stalk) as long as wide, weakly pubescent*; beak 0.9–1.6 mm long, more than half as long as body of perigyna. **Rhizomes** long.
OCCURRENCE: Very rare, documented from only one township in Park.
NOTES: Section *Acrocystis*; See *C. albicans, C. communis, C. deflexa,* and *C. pensylvanica*.

CYPERACEAE • SEDGE FAMILY

Hop Sedge • *Carex lupulina*

Perennial herb of forested wetlands, fens, and vernal pools, up to 1 m tall. **Inflorescence** with lowest subtending bract leaf-like, *with a long closed sheath up to 15 cm long*. **Spikes** longer than wide, *typically wider than 2 cm*. **Leaves** 6–10 mm wide; *basal sheaths red to brown*. **Stems** loosely tufted. **Perigynia** *greatly inflated, ascending*.
OCCURRENCE: Very rare, documented from only one township in Park.
NOTES: Section *Lupulinae*. See *C. lurida, C. retrorsa* and *C. vesicaria*.

Sallow Sedge • *Carex lurida*

Perennial herb of wetlands, up to 95 cm tall. **Inflorescence** 2.7–18 cm long. **Carpellate spikes** 1–4 per stem, the lowest often spreading on a short peduncle. **Leaves** 4.5–13 mm wide, overtopping inflorescence. **Stems** densely to loosely tufted. **Carpellate scales** *with a long, rough awn*. **Perigynia** spreading, inflated, *2–3.5 mm wide, with 7–12 prominent veins*; beak 0.7- to 0.9-times as long as body.
OCCURRENCE: Uncommon, with scattered distribution.
NOTES: Section *Vesicariae*. See *C. hystericina, C. retrorsa,* and *C. utriculata*.

Boreal Bog Sedge • *Carex magellanica*

Perennial herb of open bogs and forested peatlands, up to 70 cm tall. **Inflorescence** shorter than lowest subtending bract. **Spikes** drooping, short-cylindrical; *staminate spikes 4–15 mm long*. **Leaves** 1–4 mm wide, flat. **Carpellate scales** *1.1–2 mm wide, narrower and longer than perigynium, acuminate to awned at apex*. **Rhizomes** short.
OCCURRENCE: Occasional, widely distributed.
NOTES: Section *Limosae*. See *C. atratiformis, C. limosa,* and *C. rariflora*.
OTHER NAMES: *Carex paupercula*

CYPERACEAE • SEDGE FAMILY

Michaux's Sedge • *Carex michauxiana*

Perennial herb of open wetlands, peatlands, and fens, *up to 70 cm tall*. **Inflorescence** 1.8–18 cm long. **Leaves** *1.6–3.5 mm wide; basal sheaths not reddish*. **Stems** loosely tufted. **Carpellate scales** *without awns, one-third to two-thirds as long as perigynium*. **Perigynia** *8.7–12.1 mm long, 4- to 7-times as long as wide, slightly inflated*.
OCCURRENCE: Occasional, widely distributed.
NOTES: Section *Rostrales*. See *C. folliculata* and *C. intumescens*.

Muhlenberg's Sedge • *Carex muehlenbergii*

Perennial herb of disturbed soils and forests, up to 90 cm tall. **Spikes** all similar looking, densely crowded, *the lowest not separated from other spikes; distance between two lowest spikes less than 2-times as long as lowest spike*. **Leaves** *2–4 mm wide*; sheaths tightly clasping stem. **Stems** tufted. **Carpellate scales** greater than two-thirds as long as perigynium. **Perigynia** usually with 5–9 veins on inner or upper (adaxial) face.
OCCURRENCE: Historical record, last documented in 1901, from southern portion of Park.
NOTES: Section *Phaestoglochin*. Species is listed as Endangered in Maine.

Smooth Black Sedge • *Carex nigra*

Perennial herb of wet meadows and ditches, up to 1.1 m tall. **Inflorescence** as long as lowest subtending bract. **Spikes** *with lowest ascending*. **Leaves** *2–3 mm wide; basal sheaths not shredding into horizontal and vertical fibers (ladder fibrillose)*. **Carpellate scales** *black to purplish black*, without awns. **Perigynia** green, turning purplish black at maturity, with numerous minute, smooth bumps (papillae); *beak straight*.
OCCURRENCE: Historical record, last documented in 1985, from southern portion of Park.
NOTES: Section *Phacocystis*. See *C. haydenii*, *C. stricta*, and *C. torta*.
OTHER NAMES: Goodenough's Sedge

Greater Straw Sedge • *Carex normalis*

Perennial herb of rich woods, meadows, thickets, and ditches, up to 1.4 m tall. **Inflorescence** *compact, stiff*. **Spikes** *usually round at base*, without prolonged staminate portion. **Leaves** *2.2–6 mm wide; sheaths loosely attached*, greenish white, smooth. **Stems** tufted. **Carpellate scales** *whitish green*, 2.1–3.3 mm long, much shorter and narrower than perigynium. **Perigynia** 2.9–3.8 mm long and *1.3–2.3 mm wide, with 4 or more conspicuous veins on each face*.
OCCURRENCE: Very rare, documented from only one township in Park.
NOTES: Section *Cyperoideae*. See *C. brevior*, *C. tenera*, and *C. tincta*.

New England Sedge • *Carex novae-angliae*

Perennial herb of forests, usually under deciduous trees, 5–40 cm tall. **Spikes** borne on long, aerial stems; *lowest spike on short peduncle separated from upper spikes by at least 7 mm*. **Leaves** 0.7–1.5 mm wide, *delicate, pale green*. **Stems** loosely tufted, *weak, very slender*. **Perigynia** (excluding beak and basal stalk) longer than wide, white, nearly translucent, weakly pubescent.
OCCURRENCE: Uncommon, with scattered distribution.
NOTES: Section *Acrocystis*.

Few-seeded Sedge • *Carex oligosperma*

Perennial herb of *open peatlands*, up to 90 cm tall. **Inflorescence** 3–20 cm long. **Leaves** *wiry, thread-like; margins rolled inward toward upper surface*. **Stems** loosely tufted. **Carpellate scales** *without awns*. **Perigynia** *shiny, inflated*; beak small, up to 0.9 mm long, with minute teeth.
OCCURRENCE: Occasional, widely distributed.
NOTES: Section *Vesicariae*.

CYPERACEAE • SEDGE FAMILY

Necklace Spike Sedge • *Carex ormostachya*

Perennial herb of deciduous and mixed forests, up to 44 cm tall. **Spikes** up to 33 mm long and up to 4 mm wide, loosely flowered; *carpellate spikes well separated from each other, usually with no overlap.* **Leaves** 3.5–8 mm wide, overtopping inflorescence; *basal sheaths red.* **Stems** densely tufted. **Perigynia** not tightly overlapping, 2.2–3.3 mm long and 1.2–1.6 mm wide, *with 20 or more conspicuous veins.*
OCCURRENCE: Very rare, documented from only one township in Park.
NOTES: Section *Laxiflorae.* See *C. laxiflora, C. leptonervia.*

Orono Sedge • *Carex oronensis*

Perennial herb of fields and thickets, up to 1 m tall. **Inflorescence** stiff, erect, fairly congested. **Spikes** 3–7 mm wide. **Leaves** 2–4 mm wide. **Stems** loosely tufted. **Carpellate scales** *dark, coppery brown, 3.5–3.7 mm long* and 1.5 mm wide, usually as long and as wide as perigynium. **Perigynia** 2.9–4.3 mm long and *0.9–1.2 mm wide,* green, glossy, *with chestnut-colored blotch on outer (abaxial) face; beak flattened.*
OCCURRENCE: Rare, with scattered distribution.
NOTES: Section *Cyperoideae.* See *C. leporina* and *C. tincta. Carex oronensis* is endemic to Maine. Species is listed as Threatened in Maine.

Pale Sedge • *Carex pallescens*

Perennial herb of roadsides, forest edges, and open fields, up to 80 cm tall. **Inflorescence** *with lowest subtending bract without prolonged, loose sheath.* **Spikes** on short peduncles. **Leaves** *2–3 mm wide, finely pubescent.* **Stems** tufted. **Carpellate scales** *acute to acuminate at apex.* **Perigynia** plump, *without impressed veins.*
OCCURRENCE: Occasional, widely distributed.
NOTES: Section *Porocystis.* See *C. conoidea.*

CYPERACEAE • SEDGE FAMILY

Few-flowered Sedge • *Carex pauciflora*

Perennial herb of bogs and fens, up to 40 cm tall. **Inflorescence** lacking subtending bract. **Spikes** terminal, solitary, with staminate flowers at apex. **Leaves** 0.5–1.6 mm wide. **Stems** *solitary or loosely clumped*. **Perigynia** *narrow, strongly reflexed*.
OCCURRENCE: Occasional, widely distributed.
NOTES: Section *Leucoglochin*.

Long-stalked Sedge • *Carex pedunculata*

Perennial herb of rich forests, up to 30 cm tall. **Inflorescence** with lowest subtending bract bladeless, the bract with a prolonged sheath. **Spikes** *2–4 per stem, with long peduncles usually hidden in extended sheaths of bladeless bracts, with 2–5 perigynia per spike; some carpellate spikes borne on elongate peduncles from base of plant*. **Leaves** 2–4 mm wide, hairless; *basal sheaths reddish purple*. **Stems** tufted. **Perigynia** *often minutely pubescent*.
OCCURRENCE: Uncommon, with scattered distribution.
NOTES: Section *Clandestinae*. See *C. lucorum* and *C. penslyvanica*.

Pennsylvania Sedge • *Carex pensylvanica*

Perennial herb of sandy soils in hardwood forests and fields, up to 45 cm tall. **Inflorescence** longer than lowest subtending bract. **Spikes** borne on long, aerial stems. **Leaves** *1–3 mm wide*. **Stems** tufted. **Perigynia** *(excluding beak and basal stalk) as long as wide, weakly pubescent; beak 0.5–0.9 mm long, less than half as long as body of perigynium*. **Rhizomes** *long*.
OCCURRENCE: Rare, with scattered distribution.
NOTES: Section *Acrocystis*. See *C. albicans, C. communis, C. deflexa,* and *C. lucorum*.

CYPERACEAE • SEDGE FAMILY

Necklace Sedge • *Carex projecta*

Perennial herb of damp meadows, wet forests, swamps, and shorelines, up to 90 cm tall. **Inflorescence** *drooping, with 7–15 spikes.* **Spikes** becoming progressively more clustered toward tip; internode between two lowest spikes 5–12 mm long. **Leaves** 3–7 mm wide; *sheaths loose, winged.* **Stems** tufted, *with well-developed, abundant vegetative stems.* **Carpellate scales** shorter and narrower than perigynium. **Perigynia** *spreading.*
OCCURRENCE: Occasional, widely distributed.
NOTES: Section *Cyperoideae*. See *C. tenera* and *C. tribuloides*.

7 July Trout Brook – ML

Cypress-like Sedge • *Carex pseudocyperus*

Perennial herb of swamps, bogs, and slow-moving shallow water, up to 1 m tall. **Inflorescence** 4–15 cm long, drooping. **Carpellate spikes** stiff, cylindrical, 2.5–7.5 cm long and 8–12 mm wide. **Leaves** 4–10 mm wide; *basal sheaths mostly pale brown.* **Stems** loosely tufted. **Carpellate scales** *with a long, rough awn.* **Perigynia** spreading to reflexed, *firm, leathery,* 3.4–6.1 mm long and 1–1.7 mm wide; *beak with 2 sharp, minute, straight teeth.*
OCCURRENCE: Occasional, with scattered distribution.
NOTES: Section *Vesicariae*. See *C. hysterincina* and *C. lurida*.

14 August Katahdin Lake – JL   25 June Trout Brook area – HC

Eastern Star Sedge • *Carex radiata*

Perennial herb of wet woods and slopes, up to 80 cm tall. **Inflorescence** 3–7 cm long and 5–7.5 mm wide. **Spikes** sessile, similar looking, short; *distance between two lowest spikes at least 2-times as long as lowest spike.* **Leaves** 1.3–1.9 mm wide; sheaths tightly clasping stem. **Stems** tufted. **Perigynia** 3–8 per spike, spreading to reflexed, with a spongy base 1–1.5 mm long.
OCCURRENCE: Very rare, documented from only one township in Park.
NOTES: Section *Phaestoglochin*. See *C. appalachica* and *C. rosea*.

7 June – MA   7 June – DC

CYPERACEAE • SEDGE FAMILY

Loose-flowered Alpine Sedge • *Carex rariflora*

Perennial herb of *open, alpine wetlands and pond shores*, up to 35 cm tall. **Spikes** drooping, short-cylindrical. **Leaves** 1–2 mm wide; *margins rolled inward toward upper surface.* **Carpellate scales** purplish brown to black, *as long and as wide as perigynium.* **Perigynia** bluish green, *beakless.* **Rhizomes** short.
OCCURRENCE: Historical record, last documented in 1862, from southern portion of Park.
NOTES: Section *Limosae.* See *C. limosa* and *C. magellanica.* Species is listed as historical in Maine.

Retrorse Sedge • *Carex retrorsa*

Perennial herb of open wetlands and shores, up to 1 m tall. **Inflorescence** *with lowest subtending bract 3- to 9-times as long as inflorescence.* **Spikes** *densely clustered*, 15–20 mm wide. **Stems** tufted. **Carpellate scales** *without awns.* **Perigynia** *reflexed, at least on lower half of spikes*, inflated; beak 2.1–4.5 mm long.
OCCURRENCE: Uncommon, with scattered distribution.
NOTES: Section *Vesicariae.* See *C. lupulina.*

Rosy Sedge • *Carex rosea*

Perennial herb of dry to moist forests, up to 90 cm tall. **Inflorescence** *2–7 cm long and 5–8 mm wide.* **Spikes** sessile, similar looking, short; *distance between two lowest spikes at least 2-times as long as lowest spike.* **Leaves** *1.8–2.6 mm wide*; sheaths tightly clasping stem. **Stems** tufted. **Perigynia** *7–14 per spike, spreading to reflexed, with spongy base 0.8–1.3 mm long.*
OCCURRENCE: Uncommon, with scattered distribution.
NOTES: Section *Phaestoglochin.* See *C. appalachica* and *C. radiata.*
OTHER NAMES: *Carex convoluta*

CYPERACEAE • SEDGE FAMILY

Beaked Sedge • *Carex rostrata*

Perennial herb of fens, bogs, and shorelines, up to 90 cm tall. **Spikes** cylindrical, erect. **Leaves** *1.5–4.5 mm wide, with minute, dense, smooth bumps (papillae) on upper surface, U-shaped in cross section; margins rolled inward toward upper surface.* **Stems** not tufted, colonial, *spongy at base.* **Carpellate scales** without awns. **Perigynia** *glabrous,* inflated, spreading, 3.6–5.8 mm long and 1.7–2.8 mm wide. **Rhizomes** *long.*
OCCURRENCE: Uncommon, with scattered distribution.
NOTES: Section *Vesicariae.* See *C. lasiocarpa, C. utriculata,* and *C. vesicaria.* Species is listed as Special Concern in Maine.

27 July – DC   27 July – DC

Russet Sedge • *Carex saxatilis*

Perennial herb of *subalpine pond shores and fens,* up to 90 cm. **Carpellate spikes** 1–3 per stem. **Leaves** 2–4 mm wide, V-shaped in cross section, usually with margins rolled toward lower surface. **Stems** loosely tufted. **Carpellate scales** purplish black. **Perigynia** shiny, often purplish black, with obscure veins, *not conspicuously inflated, abruptly tapering to beak, 0.2–0.8 mm long.*
OCCURRENCE: Very rare, documented from only one township in Park.
NOTES: Section *Vesicariae.* Species is listed as Endangered in Maine. A hybrid (*C. ×stenolepis*), similar to *C. saxatilis,* has been documented at Baxter State Park. It can be recognized by its leaf blades 2–3 mm wide with margins rolled inward toward upper surface and scarce mature achenes.

11 September – MA   17 August – AH

Eastern Rough Sedge • *Carex scabrata*

Perennial herb of wet forests, streams, and springs, up to 90 cm tall. **Spikes** cylindrical, on ascending peduncles; peduncles progressively reduced toward top of stem. **Leaves** 6–8 mm wide, *very rough.* **Stems** loosely tufted, colonial. **Perigynia** *rough-textured.* **Rhizomes** present.
OCCURRENCE: Common, widely distributed.
NOTES: Section *Anomalae.* See *C. lacustris, C. utriculata,* and *C. versicaria.*

10 June – ML   4 June – AH

CYPERACEAE • SEDGE FAMILY

Canadian Single-spike Sedge • *Carex scirpoidea*

Perennial herb of *calcareous ledges and seeps in boreal and subalpine habitats*, up to 35 cm tall. **Spikes** usually solitary, narrow-cylindric, *unisexual, with staminate and carpellate flowers on separate plants.* **Leaves** 1.5–4 mm wide; *basal sheaths reddish brown.* **Stems** tufted. **Carpellate scales** reddish brown to purple. **Perigynia** *pubescent, not spreading in spike*; beak minute.
OCCURRENCE: Very rare, documented from only one township in Park.
NOTES: Section *Scirpinae*. See *C. exilis*. Species is listed as Special Concern in Maine.

Pointed Broom Sedge • *Carex scoparia*

Perennial herb of wetlands, forests, and disturbed areas, up to 1 m tall. **Inflorescence** erect to nodding, *crowded or distant, golden-brown, shiny, tawny at maturity.* **Spikes** tapering to both ends. **Leaves** 1.4–3.5 mm wide, 3–5 per fertile stem. **Stems** densely tufted. **Carpellate scales** *acuminate to short-awned*, shorter and narrower than perigynium. **Perigynia** ascending, narrow, lanceolate, *1.2–2 mm wide.*
OCCURRENCE: Uncommon, with scattered distribution.
NOTES: Section *Cyperoideae*. See *C. crawfordii* and *C. tincta*.

Dry Land Sedge • *Carex siccata*

Perennial herb of fields and rocky outcrops, up to 90 cm tall. **Inflorescence** erect, *1–5 cm long.* **Spikes** tapering to both ends. **Leaves** 1–3 mm wide. **Stems** *not tufted, sometimes colonial.* **Carpellate scales** *shorter than or equaling perigynium.* **Perigynia** 1.4–2.4 mm wide. **Rhizomes** long.
OCCURRENCE: Very rare, documented from only one township in Park.
NOTES: Section *Ammoglochin*. See *C. foenea*.

CYPERACEAE • SEDGE FAMILY

Long-beaked Sedge • *Carex sprengelii*

Perennial herb of hardwood forests, riparian habitats, and floodplains, up to 90 cm tall. **Spikes** *drooping, narrow-cylindric,* up to 3.5 cm long and 1 cm wide; terminal spike usually unisexual, staminate; peduncles of carpellate spikes usually longer than or equaling spikes. **Leaves** 3–4 mm wide; *basal sheath brown, often fibrous.* **Stems** loosely tufted. **Perigynia** *globose, shiny, very abruptly tapering to beak;* beak as long as body of perigynium.
OCCURRENCE: Very rare, documented from only one township in Park.
NOTES: Section *Hymenochlaenae.*
OTHER NAMES: *Carex longirostris*

Awl-fruited Sedge • *Carex stipata*

Perennial herb of wet meadows, fens, bogs, and shorelines, up to 1.2 m tall. **Inflorescence** branched at least in lower spikes. **Spikes** sessile, similar looking, ascending, yellowish green, prickly looking, crowded. **Leaves** 4–8 mm wide; blades well developed; ligules long, acute; *sheaths thin, brittle.* **Stems** tufted, spongy, 3-sided in cross section, winged, easily compressed. **Perigynia** truncate at base, spongy.
OCCURRENCE: Common, widely distributed.
NOTES: Section *Vulpinae.* See *C. diandra* and *C. laevivaginata.*

Tussock Sedge • *Carex stricta*

Perennial herb of open wetlands, shorelines, and bogs, up to 1.5 m tall. **Spikes** *erect, appressed.* **Leaves** 4–6 mm wide; basal sheaths rough, shredding into horizontal and vertical fibers (ladder fibrillose), reddish brown. **Stems** densely tufted. **Carpellate scales** *brown, shorter than perigynium.* **Perigynia** flattened, ascending, acute at apex, widest below middle. **Rhizomes** present.
OCCURRENCE: Common, widely distributed.
NOTES: Section *Phacocystis.* See *C. haydenii, C. nigra,* and *C. torta.*

CYPERACEAE • SEDGE FAMILY

Delicate Quill Sedge • *Carex tenera*

Perennial herb of meadows, roadsides, and open forests, up to 1 m tall. **Inflorescence** *with 3–8 spikes, arching or drooping, especially early in season.* **Spikes** *small, 4–10 mm long,* golden-brown and greenish, the lowest well separated. **Leaves** *narrow, 1.3–2.5 mm wide.* **Stems** densely tufted. **Carpellate scales** slightly shorter than or equaling perigynium. **Perigynia** *wing-margined.*
OCCURRENCE: Rare, with scattered distribution.
NOTES: Section *Cyperoideae.* See *C. brunnescens, C. canescens, C. foenea,* and *C. normalis.*

Sparse-flowered Sedge • *Carex tenuiflora*

Perennial herb of *calcareous, white-cedar forests,* up to 50 cm tall. **Inflorescence** *6–12 mm long.* **Spikes** 2–4 per stem. **Leaves** 0.5–2 mm wide, pale green or with a whitish, waxy coating. **Stems** loosely tufted, slender, overtopping leaves. **Perigynia** *grayish green, beakless,* obscurely veined.
OCCURRENCE: Very rare, documented from only one township in Park.
NOTES: Section *Glareosae.* Species is listed as Special Concern in Maine.

Tinged Sedge • *Carex tincta*

Perennial herb of fields, ditches, roadsides, and woodland edges, up to 85 cm tall. **Inflorescence** *relatively short,* stiff, erect, crowded, *dark green or dark brown when mature.* **Spikes** 3.6–6.5 mm wide. **Leaves** 1.5–3.5 wide. **Stems** loosely tufted, *relatively long, often slightly arching.* **Carpellate scales** *reddish brown to dark brown.* **Perigynia** 1.5–2.4 mm wide, *not glossy, with conspicuous veins on both surfaces,* brown at maturity, contrasting in color with shorter, narrower carpellate scales; *beak flat in cross section.*
OCCURRENCE: Rare, with scattered distribution.
NOTES: Section *Cyperoideae.* See *C. leporina* and *C. oronensis.*

CYPERACEAE • SEDGE FAMILY

Twisted Sedge • *Carex torta*

Perennial herb of rocky lakeshores and stream beds, up to 75 cm tall. **Inflorescence** with lowest subtending bract shorter than inflorescence. **Spikes** narrow-cylindric; *lowest spike spreading to drooping*, on short peduncle. **Leaves** 3–5 mm wide. **Stems** tufted. **Carpellate scales** *purplish black, strongly contrasting with green perigynium*. **Perigynia** green; *beak bent or twisted*.
OCCURRENCE: Rare, with scattered distribution.
NOTES: Section *Phacocystis*. See *C. lenticularis* and *C. nigra*.

Blunt Broom Sedge • *Carex tribuloides*

Perennial herb of damp ditches, streambanks, wet meadows, and woods, up to 1.1 m tall. **Inflorescence** *long, erect, overtopping leaves, congested*. **Spikes** 6–15 per stem. **Leaves** 3.2–7 mm wide; *sheaths loosely attached to stem, sharply angled toward apex*. **Stems** densely tufted. **Carpellate scales** half as long as perigynium. **Perigynia** narrow, *tightly ascending*, somewhat wedge-shaped toward base.
OCCURRENCE: Rare, with scattered distribution.
NOTES: Section *Cyperoideae*. See *C. projecta*.

Three-seeded Sedge • *Carex trisperma*

Perennial herb of *shaded* bogs and wet woods, up to 65 cm tall. **Inflorescence** *23–55 mm long*, with lowest subtending bract thread-like and longer than or equaling inflorescence. **Spikes** *3–4 per stem; terminal spike with 2–6 perigynia*; lowest spike removed from upper spikes by more than 2 cm. **Leaves** *1–2 mm wide*, soft, delicate, *flat*. **Stems** loosely tufted.
OCCURRENCE: Common, widely distributed.
NOTES: Section *Glareosae*. See *C. billingsii*.

CYPERACEAE • SEDGE FAMILY

Parasol Sedge • *Carex umbellata*

Perennial herb of sandy soils, rocky ledges, and deciduous upland forests, up to 8 cm tall. **Inflorescence** *usually longer than lowest subtending bract.* **Spikes** *borne on long, aerial stems and short peduncles, sometimes hidden under leaves.* **Leaves** 1–2.3 mm wide. **Stems** densely tufted. **Carpellate scales** *nearly equaling or longer than perigynium.* **Perigynia** *as long as or shorter than subtending carpellate scale, weakly pubescent*; beak 0.4–1 mm long.
OCCURRENCE: Rare, with scattered distribution.
NOTES: Section *Acrocystis*. See *C. deflexa*.

Swollen-beaked Sedge • *Carex utriculata*

Perennial herb of swamps, wet meadows, ditches, and pond margins, up to 1 m tall. **Spikes** 1–15 cm long and 10–20 mm wide, erect. **Leaves** *mostly 4.5–12 mm wide, flat to V-shaped in cross section; without minute, smooth bumps (papillae) on upper surface.* **Stems** colonial, spongy at base. **Perigynia** spreading, 4–8.6 mm long and 1.7–3 mm wide. **Rhizomes** present.
OCCURRENCE: Common, widely distributed.
NOTES: Section *Vesicariae*. See *C. lacustris, C. rostrata*, and *C. vesicaria*.

Lesser Bladder Sedge • *Carex vesicaria*

Perennial herb of wet meadows, shorelines, and swamps, up to 1 m tall. **Leaves** 1.8–6.5 mm wide; basal sheaths brown to reddish purple. **Stems** *densely tufted, very rough below inflorescence.* **Carpellate scales** without awns. **Perigynia** *inflated, thin,* 1.7–3.5 mm wide, thin walled, ascending, gradually tapering to beak.
OCCURRENCE: Common, widely distributed.
NOTES: Section *Vesicariae*. See *C. lacustris, C. rostrata*, and *C. utriculata*.

CYPERACEAE • SEDGE FAMILY

Little Green Sedge • *Carex viridula*

Perennial herb of shorelines, up to 40 cm tall. **Spikes** stout, 2–6 per stem; *lower perigynia spreading in spikes, not reflexed.* **Leaves** 1–3.1 mm wide. **Stems** tufted. **Perigynia** yellowish green, nearly filled by achene; *beaks straight or slightly bent, 0.3–1.3 mm long.*
OCCURRENCE: Rare, with scattered distribution.
NOTES: Section *Ceratocystis*. See *C. cryptolepis* and *C. flava*.

Common Fox Sedge • *Carex vulpinoidea*

Perennial herb of wet open fields, ditches, and roadsides, up to 1 m tall. **Inflorescence** 7–10 cm long. **Spikes** sessile, similar looking, ascending, crowded. **Leaves** 2–6 mm wide, overtopping inflorescence. **Stems** loosely to densely tufted, *firm, round.* **Perigynia** *yellowish brown; abruptly tapering to short beak; beak roughly one-third as long as body of perigynium, up to 1.2 mm long.*
OCCURRENCE: Rare, with scattered distribution.
NOTES: Section *Multiflorae*. See *C. diandra* and *C. stipata*.

Wiegand's Sedge • *Carex wiegandii*

Perennial herb of bogs and forested peatlands, up to 1.1 m tall. **Spikes** 4–6 per stem; terminal spike with obvious staminate portion. **Leaves** *2.8–5 mm wide.* **Stems** tufted. **Perigynia** *1.6- to 2.5-times as long as wide,* spreading to reflexed, *broadly ovate,* concavely tapering to finely toothed beak.
OCCURRENCE: Historical record, last documented in 1900, from southern portion of Park.
NOTES: Section *Stellulatae*. See *C. atlantica, C. echinata,* and *C. interior*.

CYPERACEAE • SEDGE FAMILY

Smooth Saw-sedge • *Cladium mariscoides*

Perennial herb of swamps, marshes, and shorelines, up to 1 m tall. **Spikelets** brown to chestnut brown, ascending in dense clusters. **Leaves** 2–3.5 mm wide, *channeled near base, becoming round in cross section near tip.* **Stems** stiff, solitary or a few together, slender. **Achenes** *hardened, conspicuously pointed,* dull brown, 2.5–3 mm long.
OCCURRENCE: Very rare, documented from only one township in Park.

Toothed Flatsedge • *Cyperus dentatus*

Perennial herb of sandy, gravelly, and peaty shorelines, up to 50 cm tall. **Inflorescence** with spikelets in spreading-ascending clusters. **Spikelets** flattened, often replaced by leafy, vegetative plantlets with elongate scales. **Stigmas** 3, 1.5 mm long. **Leaves** 2–5 mm wide, numerous. **Stems** tufted or solitary. **Achenes** reddish brown, triangular in cross section. **Rhizomes** present. **Stolons** present.
OCCURRENCE: Very rare, documented from only one township in Park.

Umbrella Flatsedge • *Cyperus diandrus*

Annual herb of sandy, gravelly, and peaty shorelines, up to 50 cm tall. **Inflorescence** with spikelets in spreading-ascending clusters. **Spikelets** flattened, *never replaced by leafy, vegetative plantlets with elongate scales.* **Stigmas** 2, 2.2–3.1 mm long. **Leaves** 1.5–3 mm wide. **Stems** tufted or solitary. **Achenes** *brown, lens-shaped in cross section.* **Rhizomes** *absent.* **Stolons** *absent.*
OCCURRENCE: Historical record, last documented in 1982, from northern portion of Park.

CYPERACEAE • SEDGE FAMILY

Three-way Sedge • *Dulichium arundinaceum*

Perennial herb of swamps, marshes, and shorelines, up to 1 m tall. **Inflorescence** with spikelets emerging from axils of leaves. **Spikelets** 10–30 mm long; floral scales oriented in 2 vertical rows. **Leaves** *5–15 cm long and 3.5–8 mm wide, numerous, oriented in 3 vertical rows along stem.*
OCCURRENCE: Common, widely distributed.

Needle Spikesedge • *Eleocharis acicularis*

Perennial herb of wetlands and muddy shorelines, *3–30 cm tall.* **Spikelets** 2–8 mm long and *1–2 mm wide, scarcely wider than stem*; lowest scale at base of spikelet with achene. **Leaf sheaths** reddish at base, blunt to acute at apex. **Stems** *extremely thin, up to 0.5 mm wide.* **Achenes** with no perianth bristles or sometimes with up to 4 bristles; tubercles at tip of achene 0.1–0.2 mm long and 0.1–0.3 mm wide, without constriction at base. **Rhizomes** thread-like.
OCCURRENCE: Common, widely distributed.
NOTES: Species often forms submerged mats without flowering or fruiting.

Red-footed Spikesedge • *Eleocharis erythropoda*

Perennial herb of wetlands and shorelines, up to 80 cm tall. **Spikelets** 3–18 mm long and *2–3 mm wide*; lowest scale at base of spikelet lacking achene. **Leaf sheaths** reddish purple, occasionally with a tiny tooth-like projection up to 0.1 mm long at apex. **Stems** *up to 1.4 mm wide, forming large mats.* **Achenes** with no perianth bristles or sometimes with up to 4 bristles; tubercles at tip of achene 0.3–0.7 mm long and 0.2–0.6 mm wide, constricted at base. **Rhizomes** present.
OCCURRENCE: Very rare, documented from only one township in Park.
NOTES: See *E. palustris*.

CYPERACEAE • SEDGE FAMILY

Blunt Spikesedge • *Eleocharis obtusa*

Annual herb of wetlands and shorelines, 3–50 cm tall. **Spikelets** *5–13 mm long and 3–4 mm wide*; lowest scale at base of spikelet lacking achene. **Leaf sheaths** purplish, with a tiny tooth-like projection up to 0.3 mm long at apex. **Stems** *tufted, not forming large mats, usually all stems in tuft of similar height*. **Achenes** with 6 or 7 perianth bristles, the bristles exceeding tip of tubercle; tubercles at tip of achene 0.2–0.4 mm long and 0.5–0.9 mm wide, without constriction at base. **Rhizomes** absent.
OCCURRENCE: Rare, with scattered distribution.
NOTES: See *E. ovata*.

Ovoid Spikesedge • *Eleocharis ovata*

Annual herb of wetlands and muddy shorelines, up to 35 cm tall. **Spikelets** *2–8 mm long and 2–4 mm wide*; lowest scale at base of spikelet lacking achene. **Leaf sheaths** purplish, with a tiny tooth-like projection up to 0.2 mm long at apex. **Stems** *tufted, not forming large mats, usually all stems in tuft of unequal height*. **Achenes** with 6 or 7 perianth bristles, the bristles exceeding tip of tubercle; tubercles at tip of achene 0.3–0.5 mm long and 0.3–0.5 mm wide, without constriction at base. **Rhizomes** absent.
OCCURRENCE: Rare, with scattered distribution.

Common Spikesedge • *Eleocharis palustris*

Perennial herb of shorelines, up to 1 m tall. **Spikelets** 5–25 mm long and *3–7 mm wide*; lowest scale at base of spikelet lacking achene. **Leaf sheaths** black or red, without a tiny tooth-like projection at apex. **Stems** *up to 5 mm wide, forming large mats*. **Achenes** with 4 perianth bristles or rarely none; tubercles at tip of achene 0.3–0.7 mm long and 0.3–0.7 mm wide, constricted at base. **Rhizomes** present.
OCCURRENCE: Occasional, widely distributed.

Robbins' Spikesedge • *Eleocharis robbinsii*

Perennial herb of *shallow water*, up to 70 cm tall. **Spikelets** 9–33 mm long and 1.5–3 mm wide; lowest scale at base of spikelet with achene. **Leaf sheaths** dull brown, blunt to pointed at apex. **Stems** *not tufted, forming large mats, often producing many elongate, hair-like, sterile, floating stems; dimorphic, reproductive stems firm and triangular, vegetative stems weak and thread-like.* **Achenes** with 6 or 7 perianth bristles, the bristles exceeding tip of tubercle; tubercles at tip of achene 0.5–1.1 mm long and 0.3–0.7 mm wide, barely constricted at base. **Rhizomes** present.
OCCURRENCE: Uncommon, with scattered distribution.
NOTES: See *Schoenoplectus subterminalis*.

Tall Cottonsedge • *Eriophorum angustifolium*

Perennial herb of peatlands, up to 1 m tall. **Spikelets** 2–10 per stem, *at least some on drooping peduncles; bracts below inflorescence 2 or 3, usually longer than inflorescence,* often blackish at base; *perianth bristles white.* **Leaves** 3.5–6 mm wide, flat at least basally, *developing a reddish purple tinge near base.*
OCCURRENCE: Uncommon, with scattered distribution.
NOTES: See *E. virginicum*.
OTHER NAMES: Tall Cotton-grass

Slender Cottonsedge • *Eriophorum gracile*

Perennial herb of open bogs and fens, up to 60 cm tall. **Spikelets** 2–5 per stem, *at least some on drooping peduncles; bract below inflorescence solitary, usually shorter than inflorescence, dark gray to black at base*; perianth bristles white. **Leaves** 1–2 mm wide, channeled; *uppermost leaf on stem 1–4 cm long, the blade shorter than sheath.*
OCCURRENCE: Very rare, documented from only one township in Park.
NOTES: See *E. tenellum*.

CYPERACEAE • SEDGE FAMILY

Few-nerved Cottonsedge • *Eriophorum tenellum*

Perennial herb of open bogs and fens, 30–90 cm tall. **Spikelets** 2–7 per stem, *at least some on drooping peduncles; bract below inflorescence solitary, usually shorter than inflorescence, green to reddish brown at base*; perianth bristles creamy white. **Leaves** 1–1.5 mm wide, channeled; *uppermost leaf on stem 3–25 cm long, the blade longer than sheath.*
OCCURRENCE: Occasional, widely distributed.
NOTES: See *E. gracile*.
OTHER NAMES: Five-nerved Cotton-grass

Tussock Cottonsedge • *Eriophorum vaginatum*

Perennial herb of open peatlands and alpine bogs, 10–60 cm tall. **Spikelets** *solitary, erect to spreading, not drooping; bracts below inflorescence bladeless, black to dark gray with white margins.* **Leaves** ~1 mm wide, thread-like. **Stems** *densely tufted, often forming dense tussocks.*
OCCURRENCE: Occasional, widely distributed.
OTHER NAMES: Harestail, *Eriophorum spissum*

Tawny Cottonsedge • *Eriophorum virginicum*

Perennial herb of open peatlands, 0.4–1.2 m tall. **Spikelets** 2–10 per stem, *erect to spreading, not drooping*; bracts below inflorescence 2–5, longer than inflorescence; *perianth bristles light brown.* **Leaves** 1.5–4 mm wide, flat at least basally.
OCCURRENCE: Common, widely distributed.
OTHER NAMES: Tawny Cotton-grass

CYPERACEAE • SEDGE FAMILY

White Beaksedge • *Rhynchospora alba*

Perennial herb of bogs and fens, up to 50 cm tall. **Spikelets** *whitish, becoming pale brown, in wide clusters.* **Leaves** *1–1.5 mm wide, flat at base.* **Achenes** lens-shaped, 1.5–1.8 mm long (excluding tubercle) and 0.9–1.3 mm wide; perianth bristles 10–12, retrorsely barbed, exceeding tip of tubercle; tubercles narrowly triangular, 0.5–1.2 mm long.
OCCURRENCE: Common, widely distributed.
OTHER NAMES: White Beak-rush

Brown Beaksedge • *Rhynchospora fusca*

Perennial herb of wetlands and shorelines, up to 50 cm tall. **Spikelets** *reddish brown to deep brown, in wide clusters.* **Leaves** *up to 1.5 mm wide, thread-like.* **Achenes** lens-shaped, 1–1.5 mm long (excluding tubercle) and 1–1.1 mm wide; perianth bristles 5 or 6, antrorsely barbed, longer than or equaling tubercle; tubercles triangular, 1–1.3 mm long.
OCCURRENCE: Rare, with scattered distribution.
OTHER NAMES: Brown Beak-rush

Hard-stemmed Bulrush • *Schoenoplectus acutus*

Perennial herb of shallow water of fens and shorelines, 1–3 m tall. **Spikelets** 6–18 mm long and 3–4 mm wide, *in clusters of 2–8 at end of spreading to ascending peduncles;* floral scales 3–4 mm long, *with awn 0.5–2 mm long.* **Leaves** 3 or 4 per stem, short bladed or bladeless. **Stems** *stout, erect, round, very firm, with numerous cross partitions.*
OCCURRENCE: Historical record, last documented in 1982, from northern portion of Park.
NOTES: See *S. tabernaemontani.*
OTHER NAMES: *Scirpus acutus*

CYPERACEAE • SEDGE FAMILY

Water Bulrush • *Schoenoplectus subterminalis*

Perennial, aquatic herb of shallow, quiet water, up to 1.5 m tall. **Spikelets** 5–15 mm long and 3–7 mm wide, *1 per stem, sessile*. **Leaves** *numerous, hair-like*, arising from near base, *trailing just below surface*. **Stems** firm, emergent, *round in cross section*.
OCCURRENCE: Occasional, with scattered distribution.
NOTES: See *Eleocharis robbinsii*. Tubers of this species are an important food for waterfowl during the fall.
OTHER NAMES: *Scirpus subterminalis*

11 August Lower Fowler Pond – GM

Soft-stemmed Bulrush • *Schoenoplectus tabernaemontani*

Perennial herb of shallow water along shorelines, 1–3 m tall. **Spikelets** 3–17 mm long and 2.5–4 mm wide, *usually solitary at end of spreading or ascending peduncles*; floral scales 2–3.5 mm long, *with awn up to 0.8 mm long*. **Leaves** 1 or 2 per stem, short bladed. **Stems** *soft, easily crushed between fingers, with few, large cross partitions*.
OCCURRENCE: Uncommon, with scattered distribution.
NOTES: See *S. acutus*.
OTHER NAMES: *Scirpus validus*

15 August T3R8 – GM   25 July – DC

Black-girdled Woolsedge • *Scirpus atrocinctus*

Perennial herb of wet fields, marshes, and shorelines, 0.3–1.8 m tall. **Inflorescence** with branches ascending to drooping; *bracts subtending inflorescence black at base*. **Spikelets** 4–7 mm long and 2–2.7 mm wide, *usually solitary at tip of pedicels*. **Leaves** 3–5 mm wide, 4–7 per stem. **Achenes** *with smooth, strongly contorted perianth bristles, the bristles much longer than achene; mature fruits present June through early July*.
OCCURRENCE: Common, widely distributed.
NOTES: See *S. cyperinus* and *S. pedicellatus*.
OTHER NAMES: Black-girdled Wool-grass

26 July Kidney Pond – GM

CYPERACEAE • SEDGE FAMILY

Common Woolsedge • *Scirpus cyperinus*

Perennial herb of swamps, meadows, and shorelines, 0.3–1.8 m tall. **Inflorescence** with branches ascending to drooping; *bracts subtending inflorescence reddish brown to black at base.* **Spikelets** 3.5–8 mm long and *2.5–3.5 mm wide, in dense clusters at tip of pedicels.* **Leaves** *4–10 mm wide,* 5–10 per stem. **Achenes** *with smooth, strongly contorted perianth bristles, the bristles much longer than achene;* mature fruits present mid-July through September.
OCCURRENCE: Common, widely distributed.
NOTES: See *S. atrocinctus* and *S. pedicellatus*.
OTHER NAMES: Common Wool-grass

Mosquito Bulrush • *Scirpus hattorianus*

Perennial herb of marshes, meadows, and ditches, up to 1.8 m tall. **Inflorescence** with branches ascending to spreading; *bracts subtending inflorescence green to brown at base.* **Spikelets** 2–3.5 mm long and *1.3–2.5 mm wide, in dense clusters at tip of pedicels.* **Leaves** *5–9 mm wide,* 3–9 per stem; *sheaths of lower leaves green or brown.* **Achenes** *with 5 or 6 retrorsely barbed perianth bristles, the bristles straight or merely curved and 0.5- to 1.5-times as long as the achene.*
OCCURRENCE: Occasional, widely distributed.

Barber-pole Bulrush • *Scirpus microcarpus*

Perennial herb of wetlands, up to 0.9 m tall. **Inflorescence** with branches ascending to spreading; *bracts subtending inflorescence green to black at base.* **Spikelets** 2–8 mm long and *1–3.5 mm wide, in dense clusters at tip of pedicels.* **Leaves** *5–15 mm wide,* 4–11 per stem; *sheaths of lower leaves red.* **Achenes** *with 3–6 retrorsely barbed perianth bristles, the bristles straight or merely curved and up to 1.5-times as long as the achene.*
OCCURRENCE: Uncommon, with scattered distribution.
OTHER NAMES: Red-tinged Bulrush, *Scirpus rubrotinctus*

CYPERACEAE • SEDGE FAMILY

Stalked Woolsedge • *Scirpus pedicellatus*

Perennial herb of wet meadows and shorelines, up to 2 m tall. **Inflorescence** with branches ascending to drooping; *bracts subtending inflorescence light brown to blackish brown at base.* **Spikelets** 3–9 mm long and *2–3 mm wide, solitary at tip of pedicels.* **Leaves** *5–9 mm wide,* 8 per stem. **Achenes** *with smooth, strongly contorted perianth bristles, the bristles much longer than the achene;* mature fruits present mid-July through September.
OCCURRENCE: Occasional, widely distributed.
NOTES: See *S. atrocinctus* and *S. cyperinus.*
OTHER NAMES: Pedicellate Wool-grass

Alpine Clubsedge • *Trichophorum alpinum*

Perennial herb of seepy ledges, fens, and shorelines, 10–40 cm tall. **Spikelets** *1 per stem,* terminal, brown. **Leaves** 0.4–0.5 mm wide. **Stems** *triangular in cross section, scabrous on angles.* **Achenes** with 6 white, flat perianth *bristles 1–3 cm long at maturity and clearly surpassing spikelet.*
OCCURRENCE: Uncommon, widely distributed.
OTHER NAMES: Alpine Club-rush, *Scirpus hudsonianus*

Tufted Clubsedge • *Trichophorum cespitosum*

Perennial herb of peatlands, fens, and alpine areas, 5–45 cm tall. **Spikelets** 1 per stem, terminal, brown. **Leaves** 0.3–0.4 mm wide. **Stems** *round in cross section, smooth.* **Achenes** *with 3–6 brown perianth bristles ~2 mm long, the bristles hidden by scales.*
OCCURRENCE: Uncommon, with scattered distribution.
OTHER NAMES: Deer's Hair, Deer Grass, *Scirpus cespitosus*

JUNCACEAE • RUSH FAMILY ▼

Joint-leaved Rush • *Juncus articulatus*

Perennial herb of bogs, wet meadows, ditches, and shorelines, 10–60 cm tall. **Flowers** in a terminal inflorescence 3.5–8 cm long with 3–30 compact clusters of flowers; primary bract erect; *compact clusters 6–8 mm wide, with 3–10 flowers each*; outer tepals acute to acuminate at apex. **Leaves** 3.5–12 cm long and 0.5–1.1 mm wide, with blades, round in cross section *with internal cross partitions, the lowest not overtopping inflorescence*; auricles present. **Stems** loosely tufted, often decumbent at base. **Capsules** 2.8–4 mm long, with a beak. **Seeds** *merely pointed at each end, 0.3–0.9 mm long*.
OCCURRENCE: Very rare, documented from only one township in Park.

Small-headed Rush • *Juncus brachycephalus*

Perennial herb of marshes, meadows, and shorelines, 20–70 cm tall. **Flowers** in a terminal inflorescence 5–25 cm long with 5–80 compact clusters of flowers; primary bract erect; *compact clusters 2–5 mm wide, with 2–6 flowers each; outer tepals obtuse to nearly acute.* **Leaves** 0.2–12 cm long and 0.5–2 mm wide, with blades, round in cross section, *with internal cross partitions*; auricles present. **Stems** tufted. **Capsules** 2.4–3.8 mm long, with a beak. **Seeds** *tailed at each end, 0.8–1.2 mm long including tails*.
OCCURRENCE: Rare, with scattered distribution.

Short-tailed Rush • *Juncus brevicaudatus*

Perennial herb of marshes, wet meadows, and shorelines, 14–55 cm tall. **Flowers** in a terminal inflorescence 1–12 cm long with 2–35 compact clusters of flowers; primary bract erect; compact clusters 2–9 mm wide, *with 2–5 flowers each*; outer tepals acuminate to acute. **Leaves** 1.5–25 cm long and 0.5–2.5 mm wide, with blades, round in cross section, *with internal cross partitions*; auricles present. **Stems** densely tufted. **Capsules** 3.2–4.8 mm long, with a beak. **Seeds** *tailed usually from only one end, 0.7–1.2 mm long including tail; tail less than one-quarter length of seed*.
OCCURRENCE: Common, widely distributed.
NOTES: See *J. canadensis*.

JUNCACEAE • RUSH FAMILY

Toad Rush • *Juncus bufonius*

Annual herb of wet soils along shorelines and roadsides, 5–40 cm tall. **Flowers** *solitary, pedicellate, in a diffuse inflorescence occupying over 50 percent of total plant height*; primary bract shorter than inflorescence. **Leaves** 3–13 cm long and 0.3–1.1 mm wide, with blades, *very fine, flat*, without internal cross partitions; *auricles absent*. **Stems** tufted. **Capsules** 2.7–4 mm long. **Seeds** *without tails, 0.2–0.5 mm long*.
OCCURRENCE: Rare, with scattered distribution.
NOTES: See *J. pelocarpus* and *J. tenuis*.

Canada Rush • *Juncus canadensis*

Perennial herb of swamps, marshes, wet shorelines, and ditches, 0.3–1 m tall. **Flowers** in a terminal inflorescence 2–20 cm long with 3–50 compact clusters of flowers; primary bract erect; compact clusters 3–10 mm wide, *with 5–50 flowers each*; outer tepals acuminate. **Leaves** 7–22 cm long and 1.2–3 mm wide, with blades, round in cross section, *with internal cross partitions*; auricles present. **Stems** tufted. **Capsules** 3.3–4.5 mm long, abruptly narrowing to a short beak. **Seeds** *tailed at each end, 1.1–1.9 mm long including tails; each tail over half as long as seed body*.
OCCURRENCE: Occasional, widely distributed.
NOTES: See *J. brevicaudatus*.

Common Soft Rush • *Juncus effusus*

Perennial herb of swamps, thickets, and pool margins, up to 1.3 m tall. **Flowers** *in a compact, lateral appearing inflorescence*; bract subtending inflorescence solitary, round in cross section, appearing to be a continuation of stem; stamens 3; *tepals 2.2–3.4 mm long, occasionally shorter than capsule at maturity*. **Leaves** *bladeless, consisting of sheath only*; basal sheaths up to 20 cm long, *pale brown, darkening to dark brown or dark reddish brown at base, but without purple-black*; cauline leaves absent. **Stems** densely tufted, lustrous, smooth below inflorescence, with 30–60 tiny, obscure, compressible ridges. **Capsules** 1.5–3.2 mm long. **Seeds** 0.4–0.5 mm long.
OCCURRENCE: Occasional, widely distributed.
NOTES: See *J. filiformis* and *J. pylaei*.

JUNCACEAE • RUSH FAMILY

Thread Rush • *Juncus filiformis*

Perennial herb of shorelines, ridges, and ledges, 2–35 cm tall. **Flowers** in a compact, lateral appearing inflorescence; bract subtending inflorescence solitary, *very long*, round in cross section, appearing to be a continuation of stem; stamens 6; tepals 2.5–4.2 mm long, longer than capsule at maturity. **Leaves** bladeless, consisting of sheath only; cauline leaves absent. **Stems** *not tufted*. **Capsules** 2.5–3 mm long. **Seeds** 0.5–0.6 mm long.
OCCURRENCE: Occasional, widely distributed.
NOTES: See *J. effusus* and *J. pylaei*.

Bayonet Rush • *Juncus militaris*

Perennial, aquatic herb of shallow water and wet shorelines, up to 1.5 m tall. **Flowers** in a terminal inflorescence 4–15 cm long with 20–100 compact clusters of flowers; primary bract erect; compact clusters 6–8 mm wide, *with 5–13 flowers each*; outer tepals acuminate. **Leaves** with 2 blades on stem, *round in cross section, dimorphic; rhizome leaves submerged, capillary, with a reddish tinge, with internal cross partitions; cauline leaves emergent and thicker; lowest cauline leaf overtopping inflorescence.* **Stems** emerging along rhizomes. **Capsules** 2.3–3.3 mm long, with a sharp beak. **Seeds** *pointed at each end, 0.5–0.6 mm long.*
OCCURRENCE: Rare, with scattered distribution.

Brown-fruited Rush • *Juncus pelocarpus*

Perennial herb of damp shorelines and boggy areas, occasionally submerged, 3–70 cm tall. **Flowers** *solitary or paired at nodes, often replaced by vegetative bulbils*, in a diffuse inflorescence, the inflorescence 2–25 cm long, with spreading to erect branches; primary bract erect; outer tepals obtuse at apex. **Leaves** 1.5–11 cm long and 0.8–1.1 mm wide, with blades, round in cross section, with obscure internal cross partitions; *auricles present*. **Capsules** *1.5–3.5 mm long, with a beak*. **Seeds** *without tails, merely pointed at each end, 0.3–0.5 mm long.*
OCCURRENCE: Occasional, widely distributed.
NOTES: See *J. bufonius* and *J. tenuis*.

JUNCACEAE • RUSH FAMILY

Pylae's Soft Rush • *Juncus pylaei*

Perennial herb of swamps, thickets, and pool margins, up to 1 m tall. **Flowers** *in a compact, lateral appearing inflorescence*; bract subtending inflorescence solitary, round in cross section, appearing to be a continuation of stem; stamens 3; *tepals 2.5–3.8 mm long, longer than capsule at maturity.* **Leaves** *bladeless, consisting of sheath only*; basal sheaths up to 20 cm long, *purple-black at base; cauline leaves absent.* **Stems** densely tufted, dull, *rough below inflorescence, with 10–30 firm ridges.* **Capsules** 1.5–3.2 mm long. **Seeds** 0.4–0.5 mm long.
OCCURRENCE: Occasional, widely distributed.
NOTES: See *J. effusus* and *J. filiformis*.
OTHER NAMES: *Juncus effusus* var. *pylaei*

Moor Rush • *Juncus stygius*

Perennial herb of bogs and pools, 0.3–1 m tall. **Flowers** in a terminal inflorescence *with 1–3 compact clusters of flowers*; primary bract equaling or slightly surpassing inflorescence; *compact clusters with 1–3 flowers each*; outer tepals acute. **Leaves** 8–18 cm long and 0.5–1 mm wide, with blades, round in cross section or slightly flattened; auricles present. **Stems** loosely tufted. **Capsules** 5.5–9 mm long. **Seeds** *tailed at each end, 2.8–3.5 mm long including tails.*
OCCURRENCE: Very rare, documented from only one township in Park.
NOTES: Species is listed as Special Concern in Maine.

Path Rush • *Juncus tenuis*

Perennial herb of fields, roadsides, and open areas, 15–50 cm tall. **Flowers** *in a terminal inflorescence occupying less than 30% of total plant height*; flowers solitary on a pedicel; *primary bract usually surpassing inflorescence.* **Leaves** basal only, flat, 3–12 cm long and 0.5–1 mm wide, *entire, without internal cross partitions*; auricles present, entire. **Stems** tufted. **Capsules** 3.8–4.7 mm long, usually without a beak. **Seeds** *without tails, merely pointed at each end,* 0.5–0.7 mm long.
OCCURRENCE: Common, widely distributed.
NOTES: See *J. bufonius* and *J. pelocarpus*.

JUNCACEAE • RUSH FAMILY

Hairy Wood Rush • *Luzula acuminata*

Perennial herb of woods and open areas, 10–40 cm tall. **Flowers** in a terminal inflorescence, *solitary or rarely paired at end of branches, on a drooping or curving pedicel; primary bract not reaching tip of inflorescence*; tepals pale to dark brown, with clear margins, 3–4.5 mm long. **Leaves** *sparsely pubescent, with blunt tips*; cauline leaves 2–4. **Stems** loosely tufted. **Capsules** longer than tepals.
OCCURRENCE: Occasional, widely distributed.
NOTES: See *L. parviflora*.
OTHER NAMES: *Luzula saltuensis*

Northern Wood Rush • *Luzula confusa*

Perennial herb of wet hillsides and alpine areas, 3–28 cm tall. **Flowers** in a terminal, *arching* inflorescence *with 1–3 main branches; primary bract usually inconspicuous; compact clusters with 1–5 flowers each*; tepals dark brown with clear apex, 1.6–2.6 mm long. **Leaves** *reddish, glabrous*, pointed at tip, with margins rolled inward toward upper surface; *cauline leaves 2 or 3, often reaching or exceeding inflorescence*. **Stems** tufted. **Capsules** equaling or shorter than tepals.
OCCURRENCE: Very rare, documented from only southern portion of Park.
NOTES: See *L. spicata*. Species is listed as Endangered in Maine.
OTHER NAMES: *Luzula hyperborea*

Common Wood Rush • *Luzula multiflora*

Perennial herb of open woods, fields, and meadows, 10–40 cm tall. **Flowers** in a terminal inflorescence *with branches erect, usually straight; primary bract usually longer than inflorescence; compact clusters with 8–16 flowers each*; tepals pale brown to black with clear margins, 2–4 mm long. **Leaves** *flat, with blunt tips*; cauline leaves reaching or exceeding inflorescence. **Stems** densely to loosely tufted. **Capsules** equaling or shorter than tepals.
OCCURRENCE: Occasional, widely distributed.

JUNCACEAE • RUSH FAMILY

Small-flowered Wood Rush • *Luzula parviflora*

Perennial herb of shorelines and alpine areas, 0.3–1 m tall. **Flowers** *in small clusters of 2–4 per branch tip; main branches spreading, often arching*; primary bract inconspicuous or leaf-like; tepals pale brown to brown, *1.8–2.5 mm long*. **Leaves** *pilose on upper portions of sheaths*; cauline leaves 3–6. **Stems** loosely tufted, *reddish at base and lower nodes*. **Capsules** *equaling or longer than tepals*.
OCCURRENCE: Uncommon, with scattered distribution.
NOTES: See *L. acuminata*.

Spiked Wood Rush • *Luzula spicata*

Perennial herb of alpine areas, 3–33 cm tall. **Flowers** *in dense, nodding inflorescence with spike-like clusters; primary bract conspicuous, usually longer than inflorescence*; tepals brown with clear margins, 2–2.5 mm long. **Leaves** pointed at tip, with margins rolled inward toward upper surface, *densely pubescent on upper portions of sheaths*; cauline leaves 2 or 3. **Stems** densely tufted. **Capsules** *usually shorter than tepals*.
OCCURRENCE: Very rare, documented from only southern portion of Park.
NOTES: See *L. confusa*. Species is listed as Threatened in Maine.

Highland-rush • *Oreojuncus trifidus*

Perennial herb of *alpine areas*, 10–40 cm tall. **Flowers** *in a terminal inflorescence occupying less than 30% of total plant height*; flowers solitary on a pedicel; primary bract usually longer than inflorescence. **Leaves** with blades, *minutely toothed, without internal cross partitions; auricles present, with ragged margins*. **Stems** tufted. **Capsules** 2.2–3.5 mm long. **Seeds** without tails, 0.9–1.4 mm long.
OCCURRENCE: Occasional, widely distributed.
OTHER NAMES: *Juncus trifidus*

POACEAE • GRASS FAMILY ▼

Dog Bentgrass • *Agrostis canina*

Perennial herb of *fields and meadows*, 20–70 cm tall. **Inflorescence** an open panicle 3–10 cm long; *lower panicle branches with spikelets confined to distal third*. **Spikelets** with only 1 floret; glumes or lemmas or both folded, *the glumes 1.7–3 mm long*. **Florets** with lemmas not firmer than glumes, *the lemmas with an awn visible beyond glumes; paleas less than one-third as long as associated lemma; anthers 1–1.5 mm long*. **Leaves** flat or margins rolled inward toward upper surface, 1–10 cm long and 1–3 mm wide; ligules of upper leaves 1–4 mm long. **Stems** occasionally rooting at lower nodes, with 2–4 nodes. **Rhizomes** absent. **Stolons** *present, up to 25 cm long*.
OCCURRENCE: Very rare, documented from only one township in Park.
NOTES: See *A. mertensii*.

Rhode Island Bentgrass • *Agrostis capillaris*

Perennial herb of roadsides and disturbed areas, 10–75 cm tall. **Inflorescence** an open panicle 3–20 cm long; *lower panicle branches with spikelets confined to distal half*. **Spikelets** with only 1 floret; glumes or lemmas or both folded. **Florets** with lemmas not firmer than glumes; *paleas greater than half as long as associated lemma*. **Leaves** flat, 3–10 cm long and 1–5 mm wide, *usually less than 4 mm wide*; ligules of upper leaves 0.3–2 mm long. **Stems** *not rooting at lower nodes*, with 2–5 nodes. **Rhizomes** *present, up to 5 cm long*. **Stolons** *present, up to 5 cm long*.
OCCURRENCE: Occasional, widely distributed.
NOTES: See *A. gigantea* and *A. stolonifera*.
OTHER NAMES: *Agrostis tenuis*

Redtop Bentgrass • *Agrostis gigantea*

Perennial herb of fields, roadsides, and disturbed areas, up to 1 m tall. **Inflorescence** an open panicle 8–25 cm long; *lower panicle branches with spikelets nearly to base on at least some branches*. **Spikelets** with only 1 floret; glumes or lemmas or both folded. **Florets** with lemmas not firmer than glumes; *paleas greater than half as long as associated lemma*. **Leaves** flat, 4–10 cm long and 3–8 mm wide, *usually wider than 4 mm*; ligules of upper leaves 2–7 mm long. **Stems** occasionally rooting at lower nodes, with 4–7 nodes. **Rhizomes** *present, up to 25 cm long*. **Stolons** absent.
OCCURRENCE: Very rare, documented from only northern portion of Park.
NOTES: See *A. capillaris* and *A. stolonifera*.

POACEAE • GRASS FAMILY

Northern Bentgrass • *Agrostis mertensii*

Perennial herb of *alpine and subalpine areas*, 10–40 cm tall. **Inflorescence** an open panicle 3–10 cm long; *lower panicle branches with spikelets confined to distal half*. **Spikelets** with only 1 floret; glumes or lemmas or both folded, *the glumes 2.5–3.8 mm long*. **Florets** with lemmas not firmer than glumes, *the lemmas with an awn visible beyond glumes; paleas less than one-third as long as associated lemma; anthers 0.5–0.8 mm long*. **Leaves** flat or margins rolled inward toward upper surface, 2.5–13 cm long and 0.5–3 mm wide; ligules of upper leaves 0.7–3.3 mm long. **Stems** not rooting at lower nodes, with 2–4 nodes. **Rhizomes** absent. **Stolons** absent.
OCCURRENCE: Rare, with scattered distribution.
NOTES: See *A. canina* and *A. scabra*. Species is listed as Threatened in Maine.

Autumn Bentgrass • *Agrostis perennans*

Perennial herb of roadsides, fields, and forests, 20–80 cm tall. **Inflorescence** an open panicle 10–25 cm long; panicle branches forking near or below middle, soon spreading; *lower panicle branches with spikelets aggregated toward tip*. **Spikelets** with only 1 floret; glumes or lemmas or both folded. **Florets** with lemmas not firmer than glumes, *the lemmas with an awn not visible beyond glumes; paleas less than one-third as long as associated lemma*. **Leaves** flat, 6–20 cm long and *2–5 mm wide*; ligules 1.5–7.3 mm long. **Stems** sometimes rooting at lower nodes, with 3–10 nodes. **Rhizomes** absent. **Stolons** absent.
OCCURRENCE: Occasional, widely distributed.
NOTES: See *A. scabra*.

Rough Bentgrass • *Agrostis scabra*

Annual or perennial herb of wet and dry areas, 15–90 cm tall. **Inflorescence** a diffuse panicle 8–25 cm long; panicle branches abundantly scabrous, usually forking beyond middle. **Spikelets** with only 1 floret; glumes or lemmas or both folded. **Florets** with lemmas not firmer than glumes; *paleas less than one-third as long as associated lemma*. **Leaves** flat or margins rolled inward toward upper surface, 4–14 cm long and *1–2 mm wide*; ligules 0.7–5 mm long. **Stems** rough, with 1–3 nodes. **Rhizomes** absent. **Stolons** absent.
OCCURRENCE: Common, widely distributed.
NOTES: See *A. mertensii* and *A. perennans*.
OTHER NAMES: Hairgrass, Fly-away Grass, Ticklegrass

POACEAE • GRASS FAMILY

Creeping Bentgrass • *Agrostis stolonifera*

Perennial herb of wet areas, 15–60 cm tall. **Inflorescence** a narrow, dense panicle 4–20 cm long; *lower panicle branches with spikelets nearly to base on at least some branches*. **Spikelets** with only 1 floret; glumes or lemmas or both folded. **Florets** with lemmas not firmer than glumes; *paleas greater than half as long as associated lemma*. **Leaves** flat, 2–10 cm long and 2–6 mm wide, *usually wider than 4 mm; ligules of upper leaves 3–7.5 mm long.* **Stems** *prostrate at base, usually rooting at lower nodes*, with 4–7 nodes. **Rhizomes** absent. **Stolons** *present, up to 1 m long.*
OCCURRENCE: Very rare, with scattered distribution.
NOTES: See *A. capillaris* and *A. gigantea*.
OTHER NAMES: *Agrostis alba*

Field Meadow-foxtail • *Alopecurus pratensis*

Perennial herb of roadsides, fields, and shorelines, 0.3–1.1 m tall, *flowering in May and fruiting in June.* **Inflorescence** a spike-like panicle, 3.5–9 cm long and 6–10 mm wide, with very short branches, the branches not clearly visible. **Spikelets** with only 1 floret; glumes folded, *4–6 mm long, acute at apex, without an awn.* **Florets** with *lemmas 4–6 mm long, the lemmas with an awn 5–10.5 mm long.* **Leaves** flat, 6–40 cm long and 1.9–8 mm wide; ligules 1.5–3 mm long. **Stems** erect. **Rhizomes** present, short.
OCCURRENCE: Rare, documented from only northern portion of Park.
NOTES: See *Phleum pratense*.

Alpine Sweet Grass • *Anthoxanthum monticola*

Perennial herb of *alpine areas*, 20–55 cm tall. **Inflorescence** a *compact panicle 1–8.5 cm long and 1.2–2 cm wide, with 10–20 spikelets; lowest branches of panicle wide spreading to drooping.* **Spikelets** 5–8 mm long, with 3 or 4 florets, *the lowest two florets staminate; glumes nearly equal in length, 5–7 mm long*, usually concealing all florets within. **Florets** with staminate lemmas awned, the uppermost awn *4.5–7 mm long.* **Leaves** flat or folded, 1–12 cm long and 1–3 mm wide; ligules 0.2–1.5 mm long. **Rhizomes** present, *rarely longer than 2 cm.*
OCCURRENCE: Very rare, documented from only one township in Park.
NOTES: See *A. nitens*. Species is listed as Threatened in Maine.
OTHER NAMES: *Hierochloe alpina, Hierochloe monticola*

POACEAE • GRASS FAMILY

Vanilla Sweet Grass • *Anthoxanthum nitens*

Perennial herb of wet meadows, marshes, and roadsides, 15–50 cm tall. **Inflorescence** an *open panicle 4–9 cm long and 2–5 cm wide, with 8–100 spikelets; lowest branches of panicle wide spreading to drooping*. **Spikelets** 3–7.5 mm long, with 3 or 4 florets, *the lowest two florets staminate; glumes nearly equal in length, 3.5–6 mm long*, usually concealing all florets within. **Florets** *with staminate lemmas without awns*. **Leaves** flat or rolled, sweetly scented, 10–30 cm long and 2–8 mm wide; ligules 0.5–6.5 mm long. **Rhizomes** present, *longer than 2 cm*.
OCCURRENCE: Very rare, documented from only one township in Park.
NOTES: See *A. monticola*.
OTHER NAMES: *Hierochloe odorata*

Large Sweet Grass • **Anthoxanthum odoratum*

Perennial herb of fields, roadsides, and disturbed sites, 25–60 cm tall. **Inflorescence** a *dense, spike-like, narrow panicle 4–14 cm long; lowest branches of panicle appressed to appressed-ascending*. **Spikelets** 7–9.5 mm long, with 3 or 4 florets, *the lowest two florets sterile; glumes distinctly unequal in length, the lower glume 3–4 mm long and the upper glume 8–10 mm long*, usually concealing all florets within. **Florets** *with staminate lemmas awned, the uppermost awn 4–9 mm long*. **Leaves** flat, 1–31 cm long and 3–10 mm wide; ligules 2–7 mm long.
OCCURRENCE: Uncommon, with scattered distribution.
OTHER NAMES: Sweet Vernal Grass

Tall Oat Grass • **Arrhenatherum elatius*

Perennial herb of wet fields, roadsides, and disturbed sites, 0.5–1.4 m tall. **Inflorescence** a narrow panicle 7–30 cm long and 1–6 cm wide, shiny, with short, bunched branches. **Spikelets** 7–11 mm long, *with 2 florets, the upper floret carpellate or bisexual and the lower floret staminate*; glumes distinctly unequal in length, usually concealing all florets within. **Florets** with awn of lower lemmas 10–20 mm long, the awn bent near middle; awn of upper lemmas 0–6 mm long; *paleas violet*. **Leaves** flat, scabrous, 4–8 mm wide; ligules 0.5–2 mm long.
OCCURRENCE: Very rare, documented from only one township in Park.

POACEAE • GRASS FAMILY

Northern Long-awned Wood Grass • *Brachyelytrum aristosum*

Perennial herb of moist forests, 40–80 cm tall. **Inflorescence** a narrow panicle 9.5–17.5 cm long, with few flowers; *lowest branches of panicle appressed*. **Spikelets** 23–36 mm long including awn, *with only 1 floret; glumes flat, arched, or rounded, not folded, not concealing all florets within, the lower glume absent or up to 0.4 mm long, the upper glume 0.6–1.7 mm long*. **Florets** with lemmas 8–10 mm long, *scabrous on nerves*, the lemmas with an awn 17–24 mm long. **Leaves** pubescent, 8.6–13 cm long and 8–16 mm wide; margins scabrous. **Rhizomes** *present*.
OCCURRENCE: Common, widely distributed.
OTHER NAMES: *Brachyelytrum septentrionale*

Fringed Brome • *Bromus ciliatus*

Perennial herb of wet forests, thickets, slopes, and shorelines, 0.5–1.2 m tall. **Inflorescence** an open, *nodding* panicle 10–20 cm long, with drooping lower branches. **Spikelets** 15–25 mm long, with 4–9 florets; glumes not concealing all florets within, the lower glume 1(–3)-veined, the upper glume 3-veined. **Florets** with lemmas 9.5–14 mm long, *the lemmas with an awn 3–5 mm long and arising less than 1.5 mm below apex;* sheaths with united edges at least in basal half. **Leaves** flat, usually glabrous beneath, pilose above, 13–25 cm long and 4–10 mm wide; ligules 0.4–1.4 mm long. **Stems** pubescent on upper nodes, with 4–7 nodes. **Rhizomes** *absent*.
OCCURRENCE: Occasional, widely distributed.
NOTES: See *B. secalinus* and *B. tectorum*.

Smooth Brome • **Bromus inermis*

Perennial herb of fields and roadsides, 0.5–1.3 m tall. **Inflorescence** an open, *erect* panicle 10–20 cm long, with spreading lower branches. **Spikelets** 20–40 mm long, with 8–10 florets; glumes not concealing all florets within, the lower glume 1(–3)-veined, the upper glume 3-veined. **Florets** *with lemmas 9–13 mm long, the lemmas without an awn or with an awn up to 3 mm long, the awns, if present, arising less than 1.5 mm below lemma apex*. **Leaves** flat, usually glabrous, 11–35 cm long and 5–15 mm wide; ligules up to 3 mm long; sheaths with united edges at least in basal half. **Stems** usually glabrous, with 3–5 nodes. **Rhizomes** *present, usually long*
OCCURRENCE: Uncommon, with scattered distribution.

Cheat Brome • *Bromus tectorum*

Annual herb of roadsides and disturbed sites, 20–70 cm tall. **Inflorescence** an open, nodding panicle 5–20 cm long, with drooping lower branches. **Spikelets** 10–20 mm long, *widest near apex*, with 4–8 florets; glumes not concealing all florets within, *the lower glume 1(–3)-veined, the upper glume 3(–5)-veined*. **Florets** with *lemmas 9–12 mm long, the lemmas with an awn 10–18 mm long, arising more than 1.5 mm below lemma apex*. **Leaves** soft-pubescent on both surfaces, up to 16 cm long and 1–6 cm wide; ligules 2–3 mm long; sheaths with united edges at least in basal half. **Stems** with fine hairs below inflorescence.
OCCURRENCE: Historical record, last documented in 1985, from one township in Park.
NOTES: See *B. ciliatus* and *B. secalinus*.
OTHER NAMES: Downy Chess, Junegrass

Canada Reed Grass • *Calamagrostis canadensis*

Perennial herb of wet meadows, swamps, and bogs, 0.5–1.5 m tall. **Inflorescence** *an open, loose, nodding panicle 20–40 mm wide*, with longer branches 2.7–6 cm long. **Spikelets** *2–5 mm long*, with only 1 floret; glumes or lemmas or both folded. **Florets** with lemmas not firmer than glumes; *lemmas with moderate to dense, uniformly distributed pubescence on callus, the hairs greater than 90% as long as lemma; awn of lemmas smooth at least in basal half, delicate, often difficult to distinguish from callus hairs*. **Leaves** scabrous on both sides, flat, *4–8 mm wide*; ligules 3–8 mm long; sheaths glabrous to pubescent. **Stems** *often branched above*, with 3–8 nodes.
OCCURRENCE: Occasional, widely distributed.
NOTES: See *C. stricta* and *Phalaris arundinacea*.
OTHER NAMES: Canada Bluejoint

Neglected Reed Grass • *Calamagrostis stricta*

Perennial herb of alpine areas, rocky slopes, and shorelines, 0.3–1 m tall. **Inflorescence** *a contracted, dense panicle 10–20 mm wide*, with longer branches 1.4–5 cm long. **Spikelets** 3–4 mm long, with only 1 floret; glumes or lemmas or both folded. **Florets** with lemmas not firmer than glumes; *lemmas with scant to moderate pubescence on callus, the hairs not uniformly distributed but concentrated in unequal lateral tufts less than 75% as long as lemma; awn of lemmas minutely scabrous throughout its length, delicate to stout, usually distinguishable from callus hairs*. **Leaves** flat, stiff, 1.5–5 mm wide; ligules 1.5–5 mm long; sheaths smooth. **Stems** *generally not branched*, with 2–8 nodes.
OCCURRENCE: Historical record, last documented in 1927, from one township in Park.
NOTES: See *C. canadensis*. Species is listed as Endangered in Maine.

POACEAE • GRASS FAMILY

Sweet Wood-reed • *Cinna arundinacea*

Perennial herb of wet woods, thickets, and clearings, 0.3–1.8 m tall. **Inflorescence** *an open, stiff panicle, not drooping.* **Spikelets** *3.5–7.5 mm long*, with only 1 floret; glumes or lemmas or both folded; *glumes unequal in length, the lower glume 3.5–5 mm long with 1 vein, the upper glume 4–6 mm long with 3 veins.* **Florets** on stipes 0.2–0.7 mm long; lemmas 3.5–5 mm long, not firmer than glumes, *with an awn 0.2–0.5 mm long.* **Leaves** *15–40 cm long* and 3–19 mm wide; *ligules deeply tinged with reddish brown, 2–8 mm long.*
OCCURRENCE: Very rare, documented from only one township in Park.
NOTES: See *C. latifolia*.

21 July – DC

Slender Wood-reed • *Cinna latifolia*

Perennial herb of wet woods, thickets, and clearings, 0.2–1.9 m tall. **Inflorescence** *an open, lax panicle.* **Spikelets** *2–4 mm long*, with only 1 floret; glumes or lemmas or both folded; *glumes nearly equal in length, 2.5–4 mm long, each with 1 vein.* **Florets** on a stipe 0.1–0.5 mm long; lemmas 1.8–3.8 mm long, not firmer than glumes, *with an awn 0.5–2.5 mm long.* **Leaves** *6–25 cm long* and 1–20 mm wide; *ligules colorless or nearly so, 2–8 mm long.*
OCCURRENCE: Occasional, widely distributed.
NOTES: See *C. arundinacea*.

10 July Burnt Mountain – AD   22 July – GM

Orchard Grass • **Dactylis glomerata*

Perennial herb of wet fields, roadsides, and disturbed sites, 0.5–2.1 m tall. **Inflorescence** an open panicle 4–20 cm long; *branches and main axis bearing numerous short-stalked spikelets in dense, one-sided clusters.* **Spikelets** 5–8 mm long, with 2–6 florets; glumes 3–5 mm long, not concealing all florets within. **Florets** with lemmas 4–8 mm long, usually ciliate on keel. **Leaves** *dark green, mostly basal, glabrous, flat to folded, 4–8 mm wide; sheaths closed for at least half their length, keeled;* ligules 3–11 mm long.
OCCURRENCE: Uncommon, with scattered distribution.

13 June – DC   24 April – AH

POACEAE • GRASS FAMILY

Flattened Oatgrass • *Danthonia compressa*

Perennial herb of *semi-shaded* forests and meadows, 40–80 cm tall. **Inflorescence** a lax panicle 5–10 cm long with 6–17 spikelets, *the lowest branches usually flexible and divergent after anthesis*. **Spikelets** 10–16 mm long, with 3–12 florets; glumes usually concealing all florets within; *pedicel on lowest inflorescence branch longer than or equaling spikelets*. **Florets** with pilose lemmas 2.5–5 mm long; awns 6–10 mm long, partially coiled, emerging between 2 terminal teeth of lemma, *the teeth 2–4 mm long*. **Leaves** *not curling at maturity*, mostly basal, usually flat, up to 30 cm long and *2–4 mm wide*; ligules with a band of white hairs; sheaths glabrous. **Stems** very slender, *flat*, overtopping leaves.
OCCURRENCE: Uncommon, widely distributed.
NOTES: See *D. spicata*.

Poverty Oatgrass • *Danthonia spicata*

Perennial herb of *open sunny* areas, 10–70 cm tall. **Inflorescence** a dense, crowded panicle 2–5 cm long with 5–10 spikelets, *the lowest branches usually stiff and appressed to strongly ascending after anthesis*. **Spikelets** 7–15 mm long, with 3–12 florets; glumes usually concealing all florets within; *pedicel on lowest inflorescence branch shorter than or equaling length of spikelets*. **Florets** with pilose lemmas 2.5–5 mm long; awns 5–8 mm long, partially coiled, emerging from between 2 terminal teeth of lemma, *the teeth 0.5–2 mm long*. **Leaves** *curled at maturity*, usually with margins rolled inward toward upper surface, 6–15 cm long and *0.8–3 mm wide*; ligules with a band of whitw hairs; sheaths pilose or glabrous.
OCCURRENCE: Common, widely distributed.
NOTES: See *D. compressa*.

Tufted Hair Grass • *Deschampsia cespitosa*

Perennial herb of bogs, wet meadows, and shorelines, 0.3–1.2 m tall. **Inflorescence** an open to contracted panicle 8–30 cm long; lower branches in bundles of 2–5. **Spikelets** purplish or silvery, 2.5–7.6 mm long, with 2 florets. **Florets** *with glabrous, shiny lemmas 2–5 mm long, the lemmas with an awn 1–8 mm long, the awn shorter than to slightly exceeding lemma*. **Leaves** flat to rolled, 5–30 cm long and *1–4 mm wide*; ligules 2–13 mm long.
OCCURRENCE: Historical record, last documented in 1901, from one township in Park.
NOTES: See *Vahlodea atropurpurea*.

POACEAE • GRASS FAMILY

Wavy Hair Grass • *Deschampsia flexuosa*

Perennial herb of dry rocky slopes and forests, 0.3–0.8 m tall. **Inflorescence** an open panicle 5–15 cm long, nodding before flowering, becoming erect and widely spreading. **Spikelets** 4–7 mm long, with 2 florets. **Florets** *with lemmas 3.5–5 mm long, minutely scabrous-pubescent, dull, the lemmas with an awn 3.7–7 mm long, the awns exceeding lemma by 1–3 mm.* **Leaves** with margins rolled inward toward upper surface, 12–25 cm long and *0.3–0.5 mm wide*; ligules 1.5–3.6 mm long.
OCCURRENCE: Uncommon, with scattered distribution.

Northern Rosette-panicgrass • *Dichanthelium boreale*

Perennial herb of thickets, fields, wet meadows, and shorelines, 18–75 cm tall. **Inflorescence** 5–11 cm long and *3–8 cm wide*. **Spikelets** *2–2.2 mm long*; glumes and lemmas flat, arched, or rounded, not folded; lower glumes 0.5–1 mm long. **Cauline leaves** 3–5, 5–11 cm long and *5–13 mm wide; upper surface glabrous; lower surface usually glabrous, rarely pubescent; ligules composed entirely of hairs up to 0.5 mm long*, barely protruding above apex of sheath; sheaths with sparsely ciliate margins, the lowest pubescent, uppermost glabrous.
OCCURRENCE: Uncommon, widely distributed.
NOTES: See *D. columbianum, D. implicatum, D. lanuginosum,* and *D. lindheimeri.*
OTHER NAMES: *Panicum boreale*

Deer-tongue Rosette-panicgrass • *Dichanthelium clandestinum*

Perennial herb of damp or sandy soils in woodlands, thickets, and banks, 50–140 cm tall. **Inflorescence** 8–16 cm long and *4–12 mm wide*. **Spikelets** *2.4–3.6 mm long*; glumes and lemmas flat, arched, or rounded, not folded; lower glumes less than half as long as the spikelet. **Cauline leaves** 5–10, 10–25 cm long and *15–30 mm wide; ligules membranous, with ciliate apex, 0.4–0.9 mm long*; sheaths with coarse, stiff hairs, with ciliate margins, with pale spots at summit, not overlapping on stem.
OCCURRENCE: Very rare, documented from only one township in Park.

POACEAE • GRASS FAMILY

Sand Rosette-panicgrass • *Dichanthelium columbianum*

Perennial herb of sandy woods or clearings, 10–80 cm tall. **Inflorescence** 3–12 cm long and 1–8 cm wide. **Spikelets** *1.5–1.9 mm long*; glumes and lemmas flat, arched, or rounded, not folded; lower glumes 0.4–1 mm long. **Cauline leaves** 4–7, 3–7 cm long and *3–7 mm wide*; upper surface glabrous or with sparse long hairs at base; lower surface densely pubescent to nearly glabrous; ligules composed entirely of hairs 1.5–5 mm long; ligules composed entirely of hairs 1–1.5 mm long, protruding above apex of sheath; sheaths densely pubescent with minute hairs and sometimes also scattered longer hairs.
OCCURRENCE: Uncommon, with scattered distribution.
NOTES: See *D. boreale, D. implicatum*.
OTHER NAMES: *Panicum columbianum, Dichanthelium acuminatum* ssp. *columbianum*

Tangled Rosette-panicgrass • *Dichanthelium implicatum*

Perennial herb of dry to wet areas, 10–50 cm tall. **Inflorescence** 3–12 cm long and 1–8 cm wide. **Spikelets** *1.3–1.6 mm long*; glumes and lemmas flat, arched, or rounded, not folded; lower glumes 0.4–1 mm long. **Cauline leaves** 4–7, 2–9 cm long and *2–6.5 mm wide*; upper surface pubescent with hairs 3–6 mm long; lower surface densely pubescent with minute hairs; ligules composed entirely of hairs 1.5–5 mm long, protruding above apex of sheath; sheaths without dense minute hairs, pubescent with long hairs, the hairs with a small blister-like base.
OCCURRENCE: Occasional, widely distributed.
NOTES: See *D. boreale, D. columbianum, D. lanuginosum*, and *D. lindheimeri*.
OTHER NAMES: *Panicum lanuginosum* var. *implicatum, Panicum acuminatum* var. *implicatum, Dichanthelium acuminatum* ssp. *implicatum*

Woolly Rosette-panicgrass • *Dichanthelium lanuginosum*

Perennial herb of dry to wet areas, 15–75 cm tall. **Inflorescence** 3–12 cm long and 1–8 cm wide. **Spikelets** *1.5–1.9 mm long*; glumes and lemmas flat, arched, or rounded, not folded; lower glumes 0.4–1 mm long. **Cauline leaves** 4–7, 5–12 cm long and *6–12 mm wide*; upper surface glabrous or pubescent with hairs up to 3 mm long; lower surface usually minutely pubescent; ligules composed entirely of hairs 1.5–5 mm long, protruding above apex of sheath; sheaths without dense minute hairs, pubescent with long hairs, the hairs with a small blister-like base.
OCCURRENCE: Uncommon, with scattered distribution.
NOTES: See *D. boreale, D. columbianum, D. implicatum*, and *D. lindheimeri*.
OTHER NAMES: *Dichanthelium lanuginosum* var. *fasciculatum, Panicum lanuginosum* var. *fasciculatum, Dichanthelium acuminatum* ssp. *fasciculatum*

POACEAE • GRASS FAMILY

Lindheimer's Rosette-panicgrass • *Dichanthelium lindheimeri*

Perennial herb of dry to wet areas, 10–80 cm tall. **Inflorescence** 3.5–7 cm long and 1–3.5 cm wide. **Spikelets** *1.3–1.6 mm long*; glumes and lemmas flat, arched, or rounded, not folded; lower glumes 0.4–1 mm long. **Cauline leaves** 4–7, 4–9 cm long and *4–8 mm wide*; upper surface glabrous; lower surface usually glabrous or rarely puberulent; ligules composed entirely of hairs 1.5–5 mm long, protruding above apex of sheath; sheaths without dense minute hairs, glabrous or nearly so; margins ciliate with long hairs.
OCCURRENCE: Uncommon, with scattered distribution.
NOTES: See *D. boreale*, *D. columbianum*, *D. implicatum*, and *D. lanuginosum*.
OTHER NAMES: *Panicum lanuginosum* var. *lindheimeri*, *Panicum lanuginosum* var. *septentrionale*, *Dichanthelium acuminatum* ssp. *lindheimeri*

Smooth Crabgrass • **Digitaria ischaemum*

Annual herb of disturbed sites, 20–55 cm tall. **Inflorescence** a terminal or axillary panicle, with 2–7 spike-like primary branches 6–15 cm long. **Spikelets** 1.2–2.3 mm long, becoming dark brown to purplish black, *with only 1 floret; glumes and lemmas flat, arched, or rounded, not folded; glumes not concealing all florets within*, the lower glume absent or minute and the upper glume over half as long as entire spikelet. **Florets** *with lemmas lacking awns*. **Leaves** smooth, glabrous, except for a few long hairs at junction of blade and sheath; ligules 2–3 mm long.
OCCURRENCE: Rare, with scattered distribution.

Creeping Wild-rye • **Elymus repens*

Perennial herb of bogs, wet meadows, and shorelines, 0.3–1.2 m tall. **Inflorescence** an open to contracted panicle 8–30 cm long. **Spikelets** purplish or silvery, 2.5–7.6 mm long, with 2 florets, *disarticulating below glumes*. **Florets** *with glabrous, shiny lemmas 2–5 mm long*, the lemmas with an awn 1–8 mm long, the awns shorter than to slightly exceeding the lemma. **Leaves** flat to rolled, 5–30 cm long and *1–4 mm wide*; ligules 2–13 mm long.
OCCURRENCE: Historical record, last documented in 1901, from one township in Park.
NOTES: See *Vahlodea atropurpurea*.

POACEAE • GRASS FAMILY

Slender Wild-rye • *Elymus trachycaulus*

Perennial herb of fields, bogs, shorelines, and forests, 0.3–1.5 m tall. **Inflorescence** a solitary, terminal spike 4–25 cm long and 0.4–1 cm wide; *middle internodes usually 7–9 mm long*. **Spikelets** 9–17 mm long, 1 at all or most nodes, with 3–9 florets, *disarticulating above glumes*. **Florets** with lemmas 6–13 mm long, the lemmas usually with an awn up to 40 mm long; *anthers 1.2–2.5 mm long*. **Leaves** flat or margins rolled inward toward upper surface, 2–5 mm wide; ligules 0.2–0.8 mm long. **Stems** usually tufted. **Rhizomes** short, sometimes absent.
OCCURRENCE: Rare, with scattered distribution.
NOTES: See *E. repens*.
OTHER NAMES: Slender Wheatgrass, *Agropyron trachycaulum*

Common Eastern Wild-rye • *Elymus virginicus*

Perennial herb of wet woods, thickets, meadows, and shorelines, 0.5–1 m tall. **Inflorescence** a solitary, terminal, rigidly erect spike 4–16 cm long and 1–2.2 cm wide, the base often included in apex of more or less inflated, uppermost sheath. **Spikelets** 2 at all or most nodes, with 3 or 4 florets, disarticulating below glumes. **Florets** with lemmas 6–10 mm long, the lemmas with long straight awns 8–20 mm long. **Leaves** flat or margins rolled inward toward upper surface, 2–14 mm wide, with ligules less than 1 mm long. **Stems** tufted. **Rhizomes** absent.
OCCURRENCE: Rare, with scattered distribution.
OTHER NAMES: Virginia Lyme Grass, Terrel Grass

Fine-leaved Sheep Fescue • **Festuca filiformis*

Perennial herb of dry soils in mixed forests, fields, and disturbed sites, 18–40 cm tall. **Inflorescence** a contracted panicle 1–6 cm long, with 1 or 2 branches per node. **Spikelets** *3–6 mm long*, with 2–6 florets; glumes not concealing all florets within. **Florets** with lemmas 2.3–4 mm long, *the lemmas without an awn*. **Leaves** *0.2–0.4 mm wide*, with margins rolled inward toward upper surface; ligules 0.1–0.4 mm long; *lower sheaths closed for less than 30% of their length when young, usually glabrous in apical half, not shredding, whitish brown to light brown*.
OCCURRENCE: Very rare, documented from only one township in Park.
NOTES: Species has potential to spread aggressively in some natural habitats (particularly dry, open sites and ledges).
OTHER NAMES: *Festuca capillata*

POACEAE • GRASS FAMILY

Proliferous Fescue • *Festuca prolifera*

Perennial herb of *alpine areas*, 20–41 cm tall. **Inflorescence** an open to compact panicle 5–12 cm long, with 1 or 2 branches per node. **Spikelets** *producing vegetative bulbils*, varying in length with stage of proliferation, with 2 or more florets; glumes not concealing all florets within. **Florets** *with upper lemmas modified into leafy bracts.* **Leaves** 0.3–0.8 mm wide, with margins rolled inward toward upper surface; ligules 0.1–0.4 mm long; lower sheaths closed for about 75% of their length when young, turning into loose, reddish brown fibers with age. **Stems** *usually mat-forming*, loosely tufted.
OCCURRENCE: Very rare, documented from only one township in Park.
NOTES: Species is listed as Endangered in Maine.
OTHER NAMES: *Festuca rubra* var. *prolifera*

Red Fescue • **Festuca rubra*

Perennial herb of disturbed sites and shorelines, 0.1–1.2 m tall. **Inflorescence** an open panicle, 3.5–25 cm long, with 1–3 branches per node. **Spikelets** *7–17 mm long*, with 3–10 florets; glumes not concealing all florets within. **Florets** with lemmas 4–9.5 mm long, *the lemmas with an awn 0.4–4.5 mm long emerging from tip.* **Leaves** folded or margins rolled inward toward upper surface, *0.3–2.5 mm wide*; ligules less than 0.5 mm long; *lower sheaths closed for about 75% of their length when young, usually minutely pubescent in apical half, turning into loose, reddish brown fibers with age.* **Stems** loosely to densely tufted, usually decumbent at base. **Rhizomes** usually present.
OCCURRENCE: Uncommon, with scattered distribution.

Northern Manna Grass • *Glyceria borealis*

Perennial, *aquatic* herb of shallow water and wet areas, 0.6–1 m tall. **Inflorescence** an open to closed panicle 18–40 cm long and 0.5–2 cm wide, with appressed to strongly ascending branches; branches usually 1–3 per node; *internodes of rachillae 1–4 mm long.* **Spikelets** *9–22 mm long and 0.8–2.5 mm wide*, with 8–12 florets; glumes not concealing all florets within. **Florets** with lemmas 2.7–5.4 mm long. **Leaves** limp, floating, 2–7 mm wide, with pubescent surface when growing in water; sheaths with united edges at least in basal half, *conspicuously compressed.* **Stems** weak, decumbent, rooting at lower nodes.
OCCURRENCE: Occasional, widely distributed.
NOTES: See *Torreyochloa pallida*.

POACEAE • GRASS FAMILY

Rattlesnake Manna Grass • *Glyceria canadensis*

Perennial herb of wetlands, wet woods, and shorelines, 0.6–1.5 m tall. **Inflorescence** *an open panicle 10–30 cm long and 10–20 cm wide with drooping branches*; internodes of rachillae less than 1 mm long. **Spikelets** *5–8 mm long and 3–5 mm wide, with 4–10 florets; upper glume acute at apex*, not concealing all florets within. **Florets** with lemmas 2.4–4 mm long, the lemmas distinctly pointed at apex. **Leaves** firm, 3–8 mm wide; sheaths with united edges at least in basal half, not conspicuously compressed; ligules 2–6 mm long.
OCCURRENCE: Occasional, widely distributed.
NOTES: See *G. grandis*, *G. laxa* and *G. striata*.

American Manna Grass • *Glyceria grandis*

Perennial herb of wet meadows, streamsides, marshes, and shallow water, 0.5–1.5 m tall. **Inflorescence** *an open panicle 16–42 cm long and 12–20 cm wide with drooping branches*; internodes of rachillae less than 1 mm long. **Spikelets** *3.2–10 mm long and 2–3 mm wide, with 4–10 florets; upper glume acute at apex*, not concealing all florets within. **Florets** with lemmas 1.8–3 mm long, *the lemmas blunt at apex*. **Leaves** 4.5–15 mm wide; sheaths with united edges at least in basal half, not conspicuously compressed; ligules 1–5 mm long.
OCCURRENCE: Uncommon, with scattered distribution.
NOTES: See *G. canadensis* and *G. striata*.

Flaccid Manna Grass • *Glyceria laxa*

Perennial herb of swamps, bogs, and wet woods, 0.6–1.5 m tall. **Inflorescence** *an open panicle 10–30 cm long and 10–20 cm wide with drooping branches*; internodes of rachillae less than 1 mm long. **Spikelets** *3–5 mm long and 3–5 mm wide, with 3–5 florets; upper glume usually blunt at apex*, not concealing all florets within. **Florets** with lemmas 1.8–2.5 mm long, the lemmas distinctly pointed at apex. **Leaves** firm, 3–8 mm wide; sheaths with united edges at least in basal half, not conspicuously compressed; ligules 2–6 mm long.
OCCURRENCE: Uncommon, with scattered distribution.
NOTES: See *G. canadensis*.

POACEAE • GRASS FAMILY

Northeastern Manna Grass • *Glyceria melicaria*

Perennial herb of swamps and wet soils, 0.5–1 m tall. **Inflorescence** *a dense, nodding panicle 15–25 cm long and 0.8–1.5 cm wide, with strictly ascending to erect branches, with 1–3 branches at each node*; internodes of rachillae less than 1 mm long. **Spikelets** *3.5–5 mm long and 1–2.5 mm wide, with 3 or 4 florets; upper glume pointed at apex*, not concealing all florets within. **Florets** *with lemmas 1.9–2.8 mm long, the lemmas pointed at apex.* **Leaves** 2–7 mm wide; sheaths with united edges at least in basal half, not conspicuously compressed; *ligules 0.2–0.9 mm long.*
OCCURRENCE: Uncommon, widely distributed.
NOTES: See *G. obtusa*.

Atlantic Manna Grass • *Glyceria obtusa*

Perennial herb of wet woods, bogs, and shallow water, 0.6–1 m tall. **Inflorescence** *a dense, erect panicle 5–15 cm long and 2.5–6 cm wide, with strictly ascending to erect branches, with 3–8 branches at each node*; internodes of rachillae less than 1 mm long. **Spikelets** 4–7 mm long and 2.5–4 mm wide, with 4–7 florets; upper glume blunt at tip, not concealing all florets within. **Florets** *with lemmas 3–3.9 mm long, the lemmas blunt at apex.* **Leaves** 2–8 mm wide; sheaths with united edges at least in basal half, not conspicuously compressed; *ligules 0.5–0.8 mm long.*
OCCURRENCE: Very rare, documented from only one township in Park.
NOTES: See *G. melicaria*.

Fowl Manna Grass • *Glyceria striata*

Perennial herb of bogs, shorelines, and wetlands, 20–80 cm tall. **Inflorescence** *an open, nodding panicle usually 12–17 cm long and 2.5–21 cm wide, with ascending to drooping branches*; internodes of rachillae less than 1 mm long. **Spikelets** *1.8–4 mm long and 1.2–2.9 mm wide, with 3–7 florets*; upper glume rounded to pointed at apex, not concealing all florets within. **Florets** *with lemmas 1.2–2 mm long, the lemmas pointed at apex.* **Leaves** flat, 2–6 mm wide; sheaths with united edges at least in basal half, not conspicuously compressed; *ligules 1–4 mm long.*
OCCURRENCE: Occasional, widely distributed.
NOTES: See *G. canadensis, G. grandis*, and *Torreyochloa pallida*.

POACEAE • GRASS FAMILY

Rice Cut Grass • *Leersia oryzoides*

Perennial herb of wet meadows, ditches, and shorelines, up to 1.5 m tall. **Inflorescence** *an open panicle 10–30 cm long, diffusely branched, with branches clearly visible.* **Spikelets** 4.2–6.5 mm long and 1.3–1.7 mm wide, *with only 1 floret; glumes absent.* **Florets** *with lemmas ciliate on keel and margins.* **Leaves** 7–30 cm long and 5–15 mm wide, very harsh and scabrous on surfaces and margins; ligules 0.5–1 mm long. **Stems** ascending to sprawling, decumbent at base, rooting at nodes, *the nodes densely pubescent with spreading, white hairs.* **Rhizomes** long, scaly.
OCCURRENCE: Occasional, with scattered distribution.

Millet Grass • *Milium effusum*

Perennial herb of forests, 0.5–1.4 m tall. **Inflorescence** a panicle 10–27 cm long, *with spreading or drooping branches, the branches with spikelets primarily near ends.* **Spikelets** *with only 1 floret; glumes flat, arched, or rounded, not folded, usually concealing all florets within.* **Leaves** 5–26 cm long and 8–17 mm wide; ligules 3–9 mm long. **Rhizomes** present.
OCCURRENCE: Uncommon, with scattered distribution.

Spike Muhly • *Muhlenbergia glomerata*

Perennial herb of wet meadows, bogs, and shorelines, 0.3–1.2 m tall. **Inflorescence** *a slender, contracted panicle 1.5–12 cm long and 0.3–1.8 cm wide, often strongly tinged with purple.* **Spikelets** *3–8 mm long*, with only 1 floret, *sessile or on a pedicel up to 1 mm long; glumes 3–8 mm long including awn.* **Florets** with pubescent lemmas 1.9–3.1 mm long, *the lemmas without an awn.* **Leaves** flat, scabrous, 7–15 cm long and 2–6 mm wide; ligules 0.2–0.6 mm long.
OCCURRENCE: Uncommon, with scattered distribution.

POACEAE • GRASS FAMILY

Woodland Muhly • *Muhlenbergia sylvatica*

Perennial herb of forests, rocky ledges, and shorelines, 0.4–1.1 m tall. **Inflorescence** *a slender, contracted panicle* 6–21 cm long and 0.2–1 cm wide. **Spikelets** 2.2–3.7 mm long, with only 1 floret, *with pedicels 0.8–3.5 mm long; glumes 1.8–3 mm long including awn.* **Florets** with pubescent lemmas 2.2–3.7 mm long, the lemmas with an *awn 5–18 mm long.* **Leaves** flat, scabrous, 5–18 cm long and 3–7 mm wide; *ligules 1–2.5 mm long.*
OCCURRENCE: Historical record, last documented in 1953, from one township in Park.

Bog Muhly • *Muhlenbergia uniflora*

Perennial herb of wet meadows, bogs, and shorelines, 5–45 cm tall. **Inflorescence** *an open, diffuse panicle* 2–20 cm long and 2.5–6 cm wide. **Spikelets** 1.3–2.1 mm long, with only 1 or occasionally 2 florets, *with pedicels 3–7 mm long; glumes 0.4–1.3 mm long, without awns.* **Florets** with lemmas 1.2–2 mm long, the lemmas without an awn. **Leaves** flat, smooth or scabrous, 1–15 cm long and 1–2 mm wide.
OCCURRENCE: Uncommon, with scattered distribution.

White-grained Rice Grass • *Oryzopsis asperifolia*

Perennial herb of forest understory, 25–65 cm tall. **Inflorescence** *a contracted panicle* 5–12 cm long. **Spikelets** 5–7.5 mm long, with only 1 floret. **Florets** with sparsely pubescent lemmas, the lemmas with an awn 7–15 mm long. **Leaves** flat, evergreen, erect, scabrous, 4–12 mm wide, the lowest 4–9 mm wide; ligules 0.2–0.7 mm long. **Stems** erect or bending at lowest node.
OCCURRENCE: Uncommon, with scattered distribution.
NOTES: Seeds look like grains of rice, hence common name "ricegrass."

Tuckerman's Panicgrass • *Panicum tuckermanii*

Annual herb of shorelines, 8–50 cm tall. **Inforescence** an open, diffuse panicle 7–27 cm long and 4–24 cm wide. **Spikelets** *1.4–1.7 mm long, with only 1 floret*; glumes and lemmas flat, arched, or rounded, not folded; glumes not concealing all florets within. **Florets** with lemmas 1.6–1.9 mm long. **Leaves** 2–6 mm wide, usually erect; *basal leaves not forming a rosette, similar in shape and length to cauline leaves.* **Stems** *decumbent at base.*
OCCURRENCE: Very rare, documented from only one township in Park.

Reed Canary Grass • *Phalaris arundinacea*

Perennial herb of shorelines of lakes, ponds, and streams, 0.6–2.5 m tall. **Inflorescence** a dense, spike-like panicle 5–40 cm long and *1–2 cm wide*, with branches clearly visible, the longer branches up to 9 cm long. **Spikelets** 4–6 mm long, *with 3 florets; terminal floret reproductive, the lower florets sterile*; glumes folded. **Florets** *with lemmas glabrous on lower portion and pubescent distally.* **Leaves** flat, 10–30 cm long and *5–20 mm wide*; ligules 4–10 mm long. **Stems** *not branched above base.*
OCCURRENCE: Uncommon, with scattered distribution.
NOTES: See *Calamagrostis canadensis*. Plants in Maine may represent a mixture of native and nonnative forms.

Mountain Timothy • *Phleum alpinum*

Perennial herb of *alpine* meadows and streambanks, 15–50 cm long. **Inflorescence** a spike-like *panicle 1–6 cm long, 1.5- to 3-times as long as wide*, with branches not clearly visible. **Spikelets** with only 1 floret; glumes folded, 2.5–4.5 mm long, *with awns 1.5–2.5 mm long*. **Leaves** flat, up to 17 cm long and 4–7 mm wide; *sheath of uppermost leaf inflated*; ligules 1–4 mm long. **Stems** *smooth below inflorescence, not enlarged or bulbous at lower nodes.*
OCCURRENCE: Very rare, documented from only one township in Park.
NOTES: See *P. pratense*. Species is listed as Threatened in Maine.

POACEAE • GRASS FAMILY

Common Timothy • *Phleum pratense*

Perennial herb of fields and roadsides, up to 1.5 m tall, *flowering in June and fruiting in July and August*. **Inflorescence** a spike-like *panicle 5–10 cm long, 5- to 20-times as long as wide*, with branches not clearly visible. **Spikelets** with only 1 floret; glumes folded, *3–4 mm long, with awns 1–1.5 mm long*. **Florets** with *lemmas 1.7–2 mm long, the lemmas without an awn*. **Leaves** flat, up to 45 cm long and 4–8 mm wide; *sheath of uppermost leaf not inflated*; ligules 2–4 mm long. **Stems** minutely roughened below inflorescence, usually enlarged or bulbous at lower nodes.
OCCURRENCE: Occasional, widely distributed.
NOTES: See *Alopecurus pratensis* and *P. alpinum*.

8 July – AH   12 July Nesowadnehunk Field – AD

Common Reed • *Phragmites australis*

Perennial herb of wetlands, ditches, and brackish marshes, *1–4 m tall*. **Inflorescence** an open to dense panicle 15–35 cm long and 8–20 cm wide, often purple when young and straw-colored at maturity; branches ascending and densely flowered. **Spikelets** *with 3–10 florets; glumes not concealing all florets within*. **Leaves** flat, 15–40 cm long and 2–4 cm wide; ligules 0.4–0.9 mm long. **Stems** *with middle and upper internodes dull and tan during growing season*. **Rhizomes** *wider than 15 mm*.
OCCURRENCE: Very rare, documented from only northern portion of Park.
NOTES: Species has potential to spread aggressively in some natural habitats. A native and non-invasive species, *Phragmites americanus*, can be recognized by having stems with middle and upper internodes shiny and red-brown to dark red-brown during growing season, ligules 1–1.7 mm long, and rhizomes less than 15 mm wide.
OTHER NAMES: *Phragmites communis*

30 August – DC

Grove Blue Grass • *Poa alsodes*

Perennial herb of moist forests, 0.3–1.2 m tall. **Inflorescence** an open, erect or nodding panicle 11.4–36 cm long, *with 3–5 sparsely scabrous branches at each node*. **Spikelets** 3.5–6.7 mm long, with 2–4 florets; glumes not concealing all florets within, the upper glume widest at or near base. **Florets** *with lemmas 2.7–4.2 mm long, the lemmas glabrous on marginal veins*; callus with cobwebby hairs; anthers 0.4–0.8 mm long. **Leaves** flat, lax, 0.8–4.1 mm wide, prow-shaped at apex; ligules 0.1–1.7 mm long. **Rhizomes** absent. **Stolons** absent.
OCCURRENCE: Very rare, documented from only northern portion of Park.
NOTES: See *P. saltuensis*.

16 May – AH

POACEAE • GRASS FAMILY

Annual Blue Grass • *Poa annua*

Annual herb of disturbed sites, 2–20 cm tall. **Inflorescence** an open, erect panicle, 1–7 cm long, with 1 or 2 smooth branches at each node. **Spikelets** 3–5 mm long, with 2–6 florets; glumes not concealing all florets within, *the upper glume widest just above middle.* **Florets** with lemmas 2.5–4 mm long, *the lemmas with long hairs on marginal veins and keel; callus glabrous*; anthers 0.6–1 mm long. **Leaves** flat or folded, 1–3 mm wide, very soft, prow-shaped at apex; ligules 0.5–3 mm long. **Stems** occasionally rooting at lower nodes and forming large mats. **Rhizomes** absent. **Stolons** occasionally present.
OCCURRENCE: Occasional, widely distributed.

Flat-stemmed Blue Grass • *Poa compressa*

Perennial herb of disturbed areas, wet meadows, and shorelines, 15–60 cm tall. **Inflorescence** a compact, narrow panicle 2–10 cm long, with 1–3 *scabrous branches* at each node. **Spikelets** 3.5–7 mm long, with 3–7 florets; glumes not concealing all florets within, the upper glume widest at or near base. **Florets** with lemmas 2.3–3.5 mm long, the lemmas with short hairs on marginal veins and keel; *callus with sparse cobwebby hairs*; anthers 1.3–1.8 mm long. **Leaves** flat, 1.5–4 mm wide, prow-shaped at apex; ligules 1–3 mm long. **Stems** *strongly flattened above*, decumbent at base. **Rhizomes** *present, elongate, slender.*
OCCURRENCE: Rare, with scattered distribution.
NOTES: See *P. pratensis*.
OTHER NAMES: Canada Bluegrass

Glaucous Blue Grass • *Poa glauca*

Perennial herb of alpine areas, rocky slopes, and shorelines, 10–40 cm tall. **Inflorescence** an *erect panicle 3.5–10 cm long, with 2 or 3 densely scabrous, relatively straight branches at each node, the branches ascending to erect at base.* **Spikelets** 3–7 mm long, with 2–5 florets; glumes not concealing all florets within, the upper glume widest at or near base. **Florets** *with lemmas 2.5–4 mm long, the lemmas with short hairs on marginal veins and keel; callus with long, cobwebby hairs*; anthers 1.2–2.5 mm long. **Leaves** flat or folded, soft, 0.8–2.5 mm wide, prow-shaped at apex; *ligules 1–4 mm long.* **Rhizomes** absent. **Stolons** absent.
OCCURRENCE: Very rare, documented from only one township in Park.
NOTES: See *P. laxa* and *P. nemoralis*. Species is listed as Threatened in Maine.

POACEAE • GRASS FAMILY

Wavy Blue Grass • *Poa laxa*

Perennial herb of alpine areas and rocky slopes, 8–35 cm tall. **Inflorescence** an open, *nodding* panicle *2–8 cm long, with 3–5 smooth or sparsely scabrous, relatively straight branches at each node, the branches ascending to erect at base*. **Spikelets** 4–6 mm long, with 2–5 florets; glumes not concealing all florets within, the upper glume widest at or near base. **Florets** *with lemmas 3–4.6 mm long, the lemmas with short to long hairs on marginal veins and keel; callus with short, sparse, cobwebby hairs; anthers 0.8–1.1 mm long*. **Leaves** flat, soft, 1–2 mm wide, prow-shaped at apex; *ligules 2–4 mm long*. **Rhizomes** absent. **Stolons** absent.
OCCURRENCE: Very rare, documented from only one township in Park.
NOTES: See *P. glauca*. Species is listed as Endangered in Maine.
OTHER NAMES: *Poa fernaldiana*

Wood Blue Grass • *Poa nemoralis*

Perennial herb of forests, 30–80 cm tall. **Inflorescence** an open, nodding panicle 7–16 cm long, with 2–5 scabrous branches at each node. **Spikelets** 3–8 mm long, with 2–5 florets; glumes not concealing all florets within, the upper glume widest at or near base. **Florets** *with lemmas 2.4–4 mm long, the lemmas with sparse short hairs on marginal veins and keel; callus with short, sparse, cobwebby hairs; anthers 0.8–1.9 mm long*. **Leaves** flat, *0.8–3 mm wide*, prow-shaped at apex; *ligules 0.2–0.7 mm long*. **Rhizomes** absent. **Stolons** absent.
OCCURRENCE: Rare, with scattered distribution.
NOTES: See *P. glauca* and *P. palustris*. Species has potential to spread aggressively in some natural habitats.

Fowl Blue Grass • *Poa palustris*

Perennial herb of fields, marshes, and shorelines, 0.5–1.5 m tall. **Inflorescence** *an open, loose, nodding panicle 13–30 cm long, with 2–9 scabrous branches at each node*. **Spikelets** 3–5 mm long, with 2–5 florets; glumes not concealing all florets within, the upper glume widest at or near base. **Florets** *with bronze-tipped lemmas 2–3 mm long, the lemmas with short hairs on marginal veins and keel; callus usually with long, cobwebby hairs; anthers 1.3–1.8 mm long*. **Leaves** flat, *1.5–8 mm wide*, prow-shaped at apex; *ligules 1.5–6 mm long*. **Stems** occasionally branching at base. **Rhizomes** *absent*. **Stolons** *present*.
OCCURRENCE: Uncommon, widely distributed.
NOTES: See *P. nemoralis* and *P. pratensis*.

POACEAE • GRASS FAMILY

Kentucky Blue Grass • *Poa pratensis*

Perennial herb of disturbed sites, shorelines, and alpine areas, 0.8–1 m tall. **Inflorescence** *an open, erect panicle 5–18 cm long*, usually with 3–5 smooth to scabrous branches at each node. **Spikelets** 3.5–6 mm long, with 2–5 florets; glumes not concealing all florets within, the upper glume widest at or near base. **Florets** with lemmas 2.5–4.3 mm long, *the lemmas with long hairs on marginal veins and keel; callus with long, cobwebby hairs*; anthers 1.2–2 mm long. **Leaves** flat or margins rolled inward toward upper surface, *0.4–4 mm wide*, prow-shaped at apex; ligules 1–2 mm long. **Stems** round above. **Rhizomes** *elongate*. **Stolons** absent.
OCCURRENCE: Occasional, widely distributed.
NOTES: See *P. compressa* and *P. palustris*.

Weak Spear Grass • *Poa saltuensis*

Perennial herb of open forests, 20–95 cm tall. **Inflorescence** an open, lax panicle 4–20 cm long, *with 1–3 branches at each node*. **Spikelets** 3–5.6 mm long, with 2–5 florets; glumes not concealing all florets within, the upper glume widest at or near base. **Florets** *with lemmas 2.4–4 mm long, the lemmas glabrous on marginal veins*; anthers 0.4–1.5 mm long. **Leaves** flat, lax, 1–3.6 mm wide; ligules 0.2–3 mm long. **Rhizomes** absent. **Stolons** absent.
OCCURRENCE: Uncommon, widely distributed.
NOTES: See *P. alsodes*.

False Melic Grass • *Schizachne purpurascens*

Perennial herb of moist forests, 50–80 cm tall. **Inflorescence** an open panicle 7–13 cm long. **Spikelets** 11.5–17 mm long, *purplish, with 3–6 florets*; glumes not concealing all florets within. **Florets** with lemmas 8–10 mm long, *the lemmas with an awn 8–15 mm long; callus pubescent*. **Leaves** folded or margins rolled loosely inward toward upper surface, 2–4 mm wide; ligules 0.5–1.5 mm long; *sheaths with united edges nearly to apex*. **Stems** loosely tufted, often decumbent at base.
OCCURRENCE: Uncommon, with scattered distribution.
OTHER NAMES: *Melica purpurascens*

POACEAE • GRASS FAMILY

Pale False Manna Grass • *Torreyochloa pallida*

Perennial, aquatic herb of shallow water of lakes and streams, 0.2–1.4 m tall. **Inflorescence** an open panicle 5–25 cm long and 1.8–16 cm wide with reflexed to erect branches. **Spikelets** 3.6–6.9 mm long, with 2–8 florets; *glumes not concealing all florets within*. **Florets** *with lemmas 2–3.6 mm long, the lemmas with 7–9 prominent veins*. **Leaves** flat, 1.5–17.5 mm wide; sheaths without united edges except at very base; ligules 2–9 mm long. **Stems** *usually decumbent at base, often rooting at nodes*. **Rhizomes** present.
OCCURRENCE: Very rare, documented from only one township in Park.
NOTES: See *Glyceria borealis* and *Glyceria striata*.
OTHER NAMES: Fernald's Mannagrass, *Glyceria fernaldii*.

Narrow False Oat • *Trisetum spicatum*

Perennial herb of alpine areas, forests, rocky ledges, and shorelines, 0.1–1.2 m tall. **Inflorescence** an open to spike-like panicle 20–30 cm long and 1–2.5 cm wide. **Spikelets** 5–7.5 mm long, *with 2 florets; glumes usually concealing all florets within*. **Florets** with lemmas 3–6 mm long, *the lemmas with an awn 3–8 mm long*. **Leaves** flat or margins rolled inward toward upper surface, 1–5 mm wide; ligules 0.5–4 mm long. **Rhizomes** absent.
OCCURRENCE: Rare, with scattered distribution.

Arctic Hair Grass • *Vahlodea atropurpurea*

Perennial herb of alpine areas, rocky ledges, and shorelines, 15–80 cm tall. **Inflorescence** an open to closed panicle 3–20 cm long. **Spikelets** 4–7 mm long, with 2 florets. **Florets** with lemmas 1.8–3 mm long, the lemmas with an awn 2–4 mm long; glumes usually concealing all florets within. **Leaves** flat, 1–5 mm wide; *ligules 0.8–3.5 mm long*.
OCCURRENCE: Very rare, documented from only one township in Park.
NOTES: See *Deschampsia cespitosa*. Species is listed as Endangered in Maine.

GLOSSARY

Achene – a dry, 1-seeded indehiscent fruit.
Acuminate – gradually tapering to a sharp-pointed tip, the sides more or less concave adjacent to the tip.
Acute – tapering to a short, pointed tip, the sides more or less straight to the tip.
Adnate – the fusion of dissimilar parts.
Alternate – structures borne at different levels, or one per node.
Annual – a plant completing its life cycle in one year (germination, flowering, fruiting, and senescence). Look for single flowering stems and the absence of any nonflowering shoots or rhizomes.
Anther – the pollen-bearing portion of a stamen.
Anthesis – the period of flowering when pollen is shed.
Antrorse – angling forward or upward.
Appressed – pressed tightly to another plant organ, usually refers to hairs when pressed close to the stem.
Aril – an appendage (usually fleshy) on a seed; a fleshy seed coat.
Ascending – directed upward or forward.
Auricle – an ear-like lobe usually at the base of a plant organ, usually a leaf blade.
Awn – a terminal, bristle-like appendage.
Axil – the angle formed between the upper side of a plant organ (usually a leaf or bract) and the stem from which it grows.
Axillary – arising in an axil.
Barb – a small, rigid, reflexed, sharp projection, like the barb of a fishhook.
Basal – attached or grouped at the base.
Beak – a terminal projection on a 3-dimensional plant organ.
Berry – a fleshy fruit with 1 or more seeds and without a stony inner coat.
Biennial – completing a life cycle in two years, flowering and fruiting only in its second year.
Bilaterally symmetric – a structure that can be divided through the center into two equal parts along only one plane.
Bipinnate – divided into leaflets (pinnae) and further divided into leafules (pinnules); twice pinnate.
Bipinnate-pinnatifid – divided into leaflets (pinnae) and further divided into lobed leafules (pinnules).
Bisexual – with both stamens and carpels present and functional in a flower; hermaphroditic.
Blade – the broad and flattened portion of a plant organ, such as a leaf.
Bloom – a white, waxy covering on a surface that can be rubbed away.
Bract – a reduced, variously shaped, leaf-like structure subtending a flower or inflorescence.
Bracteole – a small or greatly reduced bract.
Bud – an undeveloped, unexpanded shoot, inflorescence, or solitary flower.

Bulb – an underground swelling at the base of a plant made up of fleshy, overlapping, modified leaves.

Bulbil – a small above ground bulb-like structure, usually formed in a leaf axil, which can fall off and grow into a new plant.

Bulblet – a small bulb.

Calcareous – soils containing a lot of calcium carbonate from underlying rock.

Callus – the firm base of the lemma in the grass family (Poaceae).

Calyx – the sepals of a flower taken collectively.

Capillary – very fine, hair-like.

Capitulescence – the cluster of capitula on a plant.

Capitulum (pl. capitula) – a dense cluster of sessile flowers, subtended by an involucre of bracts; found mostly in the aster family (Asteraceae).

Capsule – a dry fruit, usually several- or many-seeded, splitting into two or more parts to release the seeds.

Carpel – the organ bearing the ovules and seeds, usually composed of an ovary, style, and stigma.

Carpellate – bearing carpels.

Catkin – a dense spike of tiny, pendent, unisexual flowers, found on trees and shrubs.

Cauline – borne on the above ground portion of the stem.

Channeled – with one or more deep grooves.

Cilia – hairs found along the margin of a structure.

Ciliate – with a marginal fringe of hairs.

Clasping – surrounding the stem; usually refers to a sessile leaf with basal lobes projecting around and appearing to clasp the stem.

Compound – branched or composed of more than one like part; regarding a leaf, divided into two or more distinct leaflets.

Concave – bowl-shaped.

Connate – the fusion of similar parts.

Convex – dome-shaped.

Cordate – heart-shaped, with two rounded lobes and a notch between.

Corolla – the petals of a flower taken collectively.

Corymb – an inflorescence in which the outer flower-stalks are much longer than the inner ones, opening from the margins inward, creating a wide, flat- or round-topped inflorescence.

Costa – midvein of a pinna.

Costule – midvein of a pinnule.

Culm – a hollow or pithy above ground stem.

Cyme – an inflorescence in which the terminal flower opens first, followed in succession by lateral and basal flowers.

Deciduous – falling off after the normal function; regarding a woody plant, dropping all of its leaves in autumn, and producing new leaves from buds the next spring.

Decumbent – horizontal at the base but tending to turn upwards at the tip.
Dimorphic – having two forms.
Dioecious – with only staminate flowers or only carpellate flowers on an individual plant.
Disc flower – one of the usually bisexual tubular flowers in a capitulum (flower head) in the aster family (Asteraceae).
Dissected – divided into segments.
Distal – at or toward the tip.
Divergent – spreading.
Divided – separated into distinct parts.
Drupe – a fleshy fruit, resembling a berry, containing a seed enclosed in a hard, stony, inner fruit wall.
Elliptic – widest at the middle and tapering at both ends.
Entire – without teeth or lobes along a margin.
Evergreen – remaining green throughout the winter and functional the following growing season.
Exfoliate – to shed or peel in thin layers; usually refers to bark.
Fascicle – a bundle or compact cluster of leaves, such as the leaves of pines.
Filament – the stalk of a stamen.
Filiform – thread-like.
Floret – an individual flower within a dense cluster; a small flower found in a capitulum in the aster family (Asteraceae).
Follicle – a dry fruit, derived from a single carpel, splitting open along one seam at maturity.
Free – not attached or fused to a different kind of plant organ.
Frond – a fern leaf, including both the blade and the stipe.
Gemma (pl. gemmae) – small vegetative bud.
Girdle – an encircling ridge.
Glabrous – without scales or hairs.
Gland – an enlarged tip to a hair or a depression on a plant organ producing a sticky or oily substance.
Glaucous – covered with a whitish or waxy coating.
Glume – one of the paired bracts at the base of a grass spikelet.
Herbaceous – not woody, dying back to the ground each winter.
Indehiscent – not naturally opening at maturity along lines or pores.
Indusium (pl. indusia) – a thin structure covering and protecting the sorus in ferns.
Inflorescence – an entire flower cluster on a stem or in an axil, including pedicels and bracts.
Involucre – a whorl of bracts at the base of a flowering head, often forming a cup-like structure.
Keel – a central ridge on the underside of a structure; in the pea family (Fabaceae), the two lower petals forming a shape like the underside of a boat.
Lanceolate – narrow overall, but slightly wider below the middle and tapering to the tip.
Leaflet – a first-order division of a compound leaf.

Leafule – a second-order division of a compound leaf.

Legume – a dry, pod-like fruit splitting open along one suture at maturity, characteristic of the pea family (Fabaceae).

Lemma – the lower of the two bracts enclosing a grass floret, often partially enclosing the palea.

Lenticel – a slightly raised pore in the bark of a stem.

Liana – a woody trailing or climbing, vine-like plant.

Ligule – an appendage present on the upper side of a leaf, at the junction of the blade and sheath (as in the grass family, Poaceae); the flattened portion of the ray flower in the aster family (Asteraceae).

Lobed – with large, projecting segments of an organ with sinuses between the segments that do not reach the axis or base.

Megaspore – a large spore; the larger type of spore in the quillwort family (Isoëtaceae).

Monoecious – having separate staminate (pollen-bearing) and carpellate (seed-bearing) flowers on the same plant.

Mucro – a short, sharp, slender point.

Node – the point on a stem where a leaf or leaves are attached.

Nut – a hard, dry, one-seeded, indehiscent fruit.

Nutlet – a small nut.

Oblong – longer than wide and more or less parallel-sided at least in the central portion.

Obtuse – blunt or rounded at apex; with the margins coming together at the apex at more than a 90-degree angle.

Ocrea (pl. ocreae) – the tubular sheath formed by the stipules, at the nodes, characteristic of the buckwheat family (Polygonaceae).

Opposite – arising in pairs, on opposite sides of the stem.

Orbicular – nearly circular in outline.

Oval – widest at the middle and approximately 2-times as long as wide.

Ovary – the carpels of a flower collectively; after pollination and fertilization, the ovary develops into the fruit.

Ovate – more or less egg-shaped in outline.

Ovule – the tiny egg-containing bodies attached inside the ovary; after pollination and fertilization, the ovules develop into the seeds.

Palea – the upper of the two bracts enclosing a grass floret, often partially enclosed by the lemma.

Palmate – with parts radiating from a common point, as with leaflets of some compound leaves.

Panicle – a branched inflorescence, with flowers maturing from the bottom to the top.

Papilla (pl. papillae) – a short, blunt, nipple-shaped bump or projection on a surface.

Papillose – bearing papillae.

Pappus – a crown of hairs or bristles fused to the tip of the achene in the aster family (Asteraceae).

Parasite – a plant deriving all (holoparasite) or part (hemiparisite) of its nourishment from another host plant.
Pedicel – a stalk of a single flower.
Pedicellate – with a pedicel.
Peduncle – a stalk of an inflorescence.
Pendulous – hanging downward.
Perennial – living 3 or more years, generally flowering each year.
Perianth – the sepals and petals of a flower collectively.
Perigynium (pl. perigynia) – a specialized bract of carpellate flowers whose margins are united to form a sac-like structure enclosing the carpel or achene; in the sedge family (Cyperaceae).
Petal – one of the inner whorl of floral blades surrounding stamens or carpels of a flower, usually colored or white.
Petiole – a stalk of a leaf.
Pilose – bearing long, soft hairs.
Pinna (pl. pinnae) – one of the primary subdivisions of a pinnately compound leaf/frond fully divided to the rachis.
Pinnatifid – with the blade lobed but not fully cut to the rachis.
Pinnate – with parts positioned along both sides of an axis, as with leaflets on some leaves.
Pinnate-pinnatifid – divided into deeply lobed leaflets (pinnae).
Pinnule – a distinct sub-division of a leaflet (pinna), fully divided to the costa.
Pinnulet – a subdivision of a pinnule, fully divided to the costule.
Pith – the soft tissue at the center of some stems and roots.
Plumose – feather-like.
Pome – a fleshy berry-like fruit derived from the swollen cup-shaped receptacle of the flower, found in the rose family (Rosaceae).
Prickle – a small, sharp epidermal outgrowth.
Pseudowhorl – a cluster of leaves falsely appearing whorled.
Pubescent – with hairs of any size or texture.
Quadrangular – 4-angled.
Raceme – an elongated inflorescence with individually stalked flowers, opening from the base toward the tip, which can continue to form new buds.
Rachilla (pl. rachillae) – the unbranched stalk or axis of the florets in a grass spikelet.
Rachis – the main axis of a compound leaf or inflorescence.
Radially symmetric – with structures radiating from a common point, like the spokes of a wheel.
Rank – a vertical row.
Ray flower – a flower with one long, flat, strap-shaped petal, found in some species in the aster family (Asteraceae).
Recurved – bent backward or downward in a curve.
Reflexed – bent sharply downward, outward, or backward.

Remote – widely spaced or separated.

Retrorse – angling backward or downward.

Rhizome – an underground, horizontal stem.

Rosette – a dense, radiating cluster of leaves, usually at ground level.

Rugose – wrinkled.

Sagittate – shaped like an arrowhead, with basal lobes pointing downward.

Samara – a dry, indehiscent winged fruit.

Saprophyte – a plant nourishing itself on decomposing humus, usually through a partnership with a fungus.

Scabrous – rough to the touch.

Scale – a small, flat structure that is not leaf-like.

Scape – a leafless flowering stem.

Schizocarp – a fruit splitting into separate 1-seeded segments at maturity.

Secund – with flowers or branches on only one side of an axis.

Sepal – one of the outer whorl of floral blades surrounding the petals of a flower, usually green.

Septate – divided by internal partitions.

Serrate – with sharp, forward-pointing teeth.

Serrulate – with minute, sharp, forward-pointing teeth.

Sessile – without a stalk.

Sheath – a structure surrounding all or part of another, usually a leaf or bract base surrounding a stem.

Silicle – a short, dry, pod-like fruit in the mustard family (Brassicaceae), typically less than 3-times as long as wide.

Silique – an elongate, dry, pod-like fruit in the mustard family (Brassicaceae), typically greater than 3-times as long as wide.

Simple – undivided.

Sinus – the indentation or cut between two lobes of a structure.

Sorus (pl. sori) – a cluster of sporangia, commonly appearing as spots on the underside of a fern blade.

Spadix – a dense, erect, spike-like inflorescence, subtended or enclosed by a spathe, found in the arum family (Araceae).

Spathe – a large bract under or enclosing an inflorescence of the arum family (Araceae).

Spike – an inflorescence of unstalked flowers on an elongate axis, maturing from the bottom toward the tip.

Spikelet – a small spike-like inflorescence or section of an inflorescence; one flower cluster of a grass or sedge.

Sporangium (pl. sporangia) – a small, thin-walled, spore-holding case.

Sporophore – the fertile, spore-bearing portion of the leaf in the Ophioglossaceae.

Sporophyll – a modified leaf bearing a sporangium at its base.

Spur – a hollow projection from the back or base of a petal or sepal, usually containing nectar; a short, slow-growing woody twig projection.
Stamen – the pollen-producing organ of a flower, consisting of the anther and filament.
Staminate – bearing stamens.
Stellate – star-shaped; branching 3 or more times at the base.
Stigma – the pollen-receiving portion of the carpel at the top of the style.
Stipe – a stalk-like structure.
Stipitate – borne on a stipe.
Stipule – a leaf-, scale-, or spine-like appendage at the base of a leaf stalk; found in pairs.
Stolon – a horizontal, above ground stem, rooting at the nodes.
Striate – streaked or marked with fine parallel lines.
Strobilus (pl. strobili) – a specialized region of a stem bearing sporophylls.
Style – the stalk-like structure connecting the stigma to the ovary.
Subtend – to occur below or at the base, often refering to bracts at the base of a flower or inflorescence.
Tepal – one of the units of the perianth when the sepals and petals are not differentiated by size or color.
Terminal – at the tip.
Tripinnate – divided into leaflets (pinnae) and further partioned into divided leafules (pinnulets).
Trophophore – the sterile, photosynthetic portion of the leaf in the Ophioglossaceae.
Truncate – square-ended.
Tubercle – a small swelling or projection.
Umbel – a flat-topped or domed inflorescence with several branches radiating from the same point at the top of the main stem.
Undulate – with a wavy margin.
Unisexual – bearing either carpellate or staminate structures but not both.
Valve – a segment of a fruit, the segments separating at maturity.
Vascular bundle – a cluster of connective tissue conducting nutrients, sugars, and water through a plant.
Velum – a thin, flap of tissue covering the sporangium in *Isoëtes*.
Villous – bearing long, soft, shaggy, unmatted hairs.
Vine – an herbaceous plant trailing along the ground or climbing over other vegetation.
Whorled – with three or more structures originating at the same node on a stalk.

INDEX

A

Abies balsamea 325
Acer pensylvanicum 345
Acer rubrum 346
Acer saccharinum 346
Acer saccharum 347
Acer spicatum 347
Achillea borealis 65
Achillea millefolium 65
ACORACEAE 49
Actaea alba 246
Actaea pachypoda 246
Actaea rubra 246
Adder's-mouth, Green 206
Adiantum pedatum 374
ADOXACEAE 305
Aegopodium podagraria 53
Agalinis, Slender-leaved 215
Agalinis tenuifolia 215
Ageratina altissima 66
Agrimonia gryposepala 256
Agrimonia striata 256
Agrimony, Common 256
Agrimony, Roadside 256
Agropyron trachycaulum 437
Agrostis alba 428
Agrostis canina 426
Agrostis capillaris 426
Agrostis gigantea 426
Agrostis mertensii 427
Agrostis perennans 427
Agrostis scabra 427
Agrostis stolonifera 428
Agrostis tenuis 426
Alder, Black 311
Alder, Green 312
Alder-leaved Buckthorn 255
Alder, Mountain 312
Alder, Speckled 312
Alexanders, Golden 62
Alga-like Pondweed 239
Alisma brevipes 49
ALISMATACEAE 49
Alisma triviale 49
Allegheny Hawkweed 81
Allegheny Monkey-flower 222
ALLIACEAE 51
Allium schoenoprasum 51
Allium sibiricum 51
Alnus alnobetula 312
Alnus crispa 312
Alnus incana 312

Alnus rugosa 312
Alnus viridis 312
Alopecurus pratensis 428
Alpine Arctic-cudweed 87
Alpine-azalea 140
Alpine-bearberry 133
Alpine Bistort 230
Alpine Bitter-cress 111
Alpine Blueberry 147, 149
Alpine Club-rush 419
Alpine Clubsedge 419
Alpine Rattlesnake-root 85
Alpine Speedwell 229
Alpine Sweet Grass 428
Alsike Clover 153
Alsine borealis 125
Alsine graminea 126
Alsine longifolia 126
Alsine media 127
Alternate-flowered Water-milfoil 158
Alternate-leaved Dogwood 317
AMARANTHACEAE 52
Amelanchier arborea 330
Amelanchier bartramiana 330
Amelanchier intermedia 331
Amelanchier laevis 331
Amelanchier nantucketensis 332
Amelanchier spicata 332
Amelanchier stolonifera 332
American Alpine Speedwell 229
American-aster, Calico 102
American-aster, Heart-leaved 101
American-aster, Lance-leaved 101
American-aster, Purple-stemmed 103
American-aster, Rush 100
American-aster, Tradescant's 104
American-aster, Wavy-leaved 104
American Basswood 322
American Beech 321
American Burnweed 73
American Bur-reed 290
American Cow-parsnip 58
American Cow-wheat 218
American Cranberry 148
American Dog Violet 298
American Elm 348
American False Hellebore 186
American Germander 176
American Honeysuckle 119
American Larch 325
American-laurel, Bog 139

American-laurel, Sheep 139
American Linden 322
American Manna Grass 439
American Marsh-pennywort 59
American Mountain-ash 335
American Shinleaf 142
American Speedwell 227
American Spikenard 55
American Spurred-gentian 155
American Trout-lily 181
American Twinflower 118
American Water-horehound 171
American Water-pennywort 59
American Wild Mint 173
American Witch-hazel 322
American Yew 348
ANACARDIACEAE 52, 309
Anacharis nuttallii 161
Anaphalis margaritacea 66
Andromeda caerulea 142
Andromeda glaucophylla 132
Andromeda polifolia 132
Anemone americana 247
Annual Blue Grass 445
Annual Fleabane 74
Antennaria canadensis 67
Antennaria howellii 67
Antennaria neglecta 67
Antennaria neodioica 67
Antennaria parlinii 68
Anthoxanthum monticola 428
Anthoxanthum nitens 429
Anthoxanthum odoratum 429
APIACEAE 53
APOCYNACEAE 62
Apocynum androsaemifolium 62
Appalachian Polypody 373
Appalachian Sedge 382
Apple, Cultivated 333
AQUIFOLIACEAE 310
Aquilegia canadensis 247
Aquilegia vulgaris 248
Arabis drummondii 109
Arabis laevigata 110
ARACEAE 64
Aralia hispida 54
Aralia nudicaulis 55
Aralia racemosa 55
Arborvitae 319
Arceuthobium pusillum 302
Arching Blackberry 271
Arctic Bur-reed 293
Arctic-cudweed, Alpine 87

Arctic-cudweed, Woodland 87
Arctic Hair Grass 448
Arctic Sweet-coltsfoot 89
Arctic Willow 283
Arctium lappa 68
Arctium minus 69
Arctostaphylos alpina 133
Arctostaphylos uva-ursi 132
Arctous alpina 133
Arenaria groenlandica 122
Arenaria lateriflora 122
Arenaria rubra 125
Arethusa bulbosa 201
Arisaema triphyllum 64
Arnica, Lance-leaved 69
Arnica lanceolata 69
Arnica mollis 69
Aronia floribunda 257
Aronia melanocarpa 257
Arrhenatherum elatius 429
Arrowhead, Common 51
Arrowhead, Grass-leaved 50
Arrowhead, Northern 50
Arrowhead Violet 301
Arrowleaf, Broad-leaved 51
Arrow-leaved Tearthumb 234
Arrowwood, Smooth 306
Arrow-wood, Southern 306
Arum, Water 64
Asclepias incarnata 63
Asclepias syriaca 63
Ash, Black 323
Ash, Green 324
Ash, White 323
Aspen, Trembling 338
Aspidium braunii 356
ASPLENIACEAE 349
Asplenium filix-femina 377
Asplenium trichomanes 349
ASTERACEAE 65
Aster acuminatus 86
Aster, Big-leaved 77
Aster, Bog 86
Aster borealis 100
Aster, Calico 102
Aster castaneus 76
Aster cordifolius 101
Aster divaricatus 76
Aster, Goblet 102
Aster, Heart-leaved 101
Aster junciformis 100
Aster lanceolatus 101
Aster lateriflorus 102
Aster macrophyllus 77
Aster, Mountain 86
Aster nemoralis 86

Aster novae-angliae 102
Aster novi-belgii 103
Aster puniceus 103
Aster radula 77
Aster, Rough-leaved 77
Aster saxatilis 104
Aster, Shore 104
Aster tradescantii 104
Aster umbellatus 73
Aster undulatus 104
Aster vimineus 104
Athyrium angustum 377
Athyrium thelypteroides 378
Atlantic Manna Grass 440
Auricled Twayblade 207
Autumn Bentgrass 427
Avens, Floodplain 261
Avens, Large-leaved 261
Avens, Purple 262
Avens, Water 262
Avens, White 260
Avens, Yellow 260
Awl-fruited Sedge 406
Azalea lapponica 146
Azalea procumbens 140

B

Baby's-breath, Low 121
Balmony 224
Balsam Fir 325
Balsam Groundsel 88
BALSAMINACEAE 107
Balsam Poplar 337
Balsam Willow 344
Baneberry, Red 246
Baneberry, White 246
Barbarea stricta 108
Barbarea vulgaris 109
Barber-pole Bulrush 418
Basswood, American 322
Bayberry, Small 188
Bayonet Rush 422
Beaked Hazelnut 316
Beaked Sedge 404
Beak-rush, Brown 416
Beak-rush, White 416
Beaksedge, Brown 416
Beaksedge, White 416
Bearberry, Common 132
Bearberry, Mountain 133
Bearberry, Red 132
Bearberry Willow 284
Bear Sedge 382
Bebb's Sedge 384
Bedstraw, Boreal 275
Bedstraw, Fragrant 278

Bedstraw, Marsh 276
Bedstraw, Northern 274
Bedstraw, Rough 274
Bedstraw, Scratch 273
Bedstraw, Small 277
Bedstraw, Smooth 276
Bedstraw, Three-petaled 278
Bedstraw, Whorled 276
Bedstraw, Wood 277
Beech, American 321
Beech-drops 216
Beech-drops, False 138
Beewort 49
Beggar-ticks, Devil's 71
Beggar-ticks, Nodding 70
Beggar-ticks, Purple-stemmed 70
Bellflower, Creeping 115
Bellflower, Roving 115
Bellflower, Scotch 115
Bell-heather, Moss 138
Bellwort, Sessile-leaved 128
Bentgrass, Autumn 427
Bentgrass, Creeping 428
Bentgrass, Dog 426
Bentgrass, Northern 427
Bentgrass, Redtop 426
Bentgrass, Rough 427
BERBERIDACEAE 107
Berchtold's Pondweed 238
Betula alleghaniensis 313
BETULACEAE 312
Betula cordifolia 313
Betula glandulosa 314
Betula lutea 313
Betula minor 314
Betula papyrifera 315
Betula populifolia 315
Betula pubescens 314
Bidens cernua 70
Bidens connata 70
Bidens frondosa 71
Bigelow's Sedge 384
Big-leaved Aster 77
Big-leaved Pondweed 238
Big-toothed Poplar 337
Billings' Sedge 384
Bindweed, Black 231
Bindweed, Climbing 231
Bindweed, False 129
Bindweed, Fringed 230
Bindweed, Hedge 129
Birch, Canoe 315
Birch, Dwarf 314
Birch, Fire 315
Birch, Glandular 314
Birch, Gray 315

Birch, Paper 315
Birch, White 315
Birch, Yellow 313
Bird Cherry 334
Bird's-eye Pearlwort 123
Bird's-foot-trefoil, Garden 151
Bird's Nest 58
Bird Vetch 154
Bishop's-cap, Naked 287
Bishop's Goutweed 53
Bistort, Alpine 230
Bistorta vivipara 230
Bitter-cress, Alpine 111
Bitter-cress, Pennsylvania 112
Bitter-cress, Small-flowered 112
Bitter Dock 236
Black Alder 311
Black Ash 323
Blackberry, Arching 271
Blackberry, Bristly 268
Blackberry, Canada 267
Blackberry, Common 266
Blackberry, Pennsylvania 270
Blackberry, Showy 268
Blackberry, Smooth 267
Blackberry, Vermont 271
Black Bindweed 231
Black-bindweed, Fringed 230
Black Cherry 334
Black Chokeberry 257
Black Crowberry 135
Black Currant 155
Black Elderberry 305
Black-eyed Coneflower 92
Black-eyed Susan 92
Black-girdled Wool-grass 417
Black-girdled Woolsedge 417
Black Highbush Blueberry 320
Black Huckleberry 137
Black Raspberry 269
Black Spruce 326
Black Willow 341
Bladder Campion 124
Bladder-pod Lobelia 117
Bladderwort, Creeping 177
Bladderwort, Flat-leaved 178
Bladderwort, Floating 179
Bladderwort, Greater 180
Bladderwort, Horned 176
Bladderwort, Humped 177
Bladderwort, Inflated 179
Bladderwort, Lesser 178
Bladderwort, Mixed 177
Bladderwort, Northern 178
Bladderwort, Purple 179
Bladderwort, Resupinate 180

Bland Sweet-cicely 59
BLECHNACEAE 350
Blood-root 222
Bloodwort 65
Blue-bead Lily 181
Blueberry, Alpine 147, 149
Blueberry, Bog 149
Blueberry, Dwarf 147
Blueberry, Highbush 319
Blueberry, Lowbush 146
Blueberry, Northern 147
Blueberry, Sourtop 148
Blueberry, Swamp 319
Blueberry, Velvet-leaved 148
Blue Cohosh 107
Blue-eyed-grass, Needle-tipped 169
Blue-eyed-grass, Strict 169
Blue Grass 445
Bluegrass, Canada 445
Blue Ground-cedar 365
Blue Iris 168
Bluejoint, Canada 431
Blue Lettuce 82
Blue Marsh Violet 297
Blue Ridge Sedge 396
Blue Skullcap 174
Bluet, Little 279
Blue Vervain 296
Blunt Broom Sedge 408
Blunt-leaved Bog-orchid 212
Blunt-leaved Grove-sandwort 122
Blunt-leaved Pondweed 242
Blunt-leaved Sandwort 122
Blunt-lobed Hepatica 247
Blunt Spikesedge 413
Boechera drummondii 109
Boechera laevigata 110
Boechera stricta 109
Boehmeria cylindrica 294
Bog American-laurel 139
Bog Aster 86
Bogbean 187
Bog Blueberry 149
Bog-candle 210
Bog-clubmoss, Northern 365
Bog Cranberry 149
Bog Goldenrod 99
Bog Laurel 139
Bog Muhly 442
Bog Nodding-aster 86
Bog-orchid, Blunt-leaved 212
Bog-orchid, Club-spur 209
Bog-orchid, Green 211
Bog-orchid, Hooker's 211
Bog-orchid, Northern 210

Bog-orchid, Purple-fringed 210
Bog-orchid, Round-leaved 212
Bog-orchid, White-fringed 209
Bog Rose 265
Bog-rosemary 132
Bog Spruce 326
Bog Willow 342
Bog Willow-herb 198
Bog Yellow-eyed-grass 303
Boneset, Estuary 76
Boneset Thoroughwort 76
Boott's Rattlesnake-root 85
BORAGINACEAE 108
Boreal Bedstraw 275
Boreal Bog Sedge 397
Boreal Stitchwort 125
Borodinia laevigata 110
Botrychium angustisegmentum 369
Botrychium matricariifolium 369
Botrychium multifidum 370
Botrychium tenebrosum 370
Botrychium virginianum 371
Boulder Fern 350
Bouncing Bet 123
Brachyelytrum aristosum 430
Brachyelytrum septentrionale 430
Bracken Fern 351
Brasenia schreberi 192
BRASSICACEAE 108
Braun's Holly Fern 356
Bristle-stalk Sedge 395
Bristly Blackberry 268
Bristly Black Currant 157
Bristly Crowfoot 253
Bristly Locust 320
Bristly Sarsaparilla 54
Bristly Swamp Currant 157
Brittle-stemmed Hemp-nettle 170
Broad-leaved Arrowleaf 51
Broad-leaved Cat-tail 293
Broad-leaved Enchanter's-nightshade 195
Broad-leaved Helleborine 205
Broad-leaved Twayblade 207
Broad Loose-flowered Sedge 394
Brome, Cheat 431
Brome, Fringed 430
Brome-like Sedge 385
Brome, Smooth 430
Bromus ciliatus 430
Bromus inermis 430
Bromus tectorum 431
Brook Lobelia 117
Broom-rape, One-flowered 218
Brown Beak-rush 416

Brown Beaksedge 416
Brown Cudweed 79
Brown-fruited Rush 422
Brownish Sedge 385
Buck-bean 187
Buckbrush 258
Buckthorn, Alder-leaved 255
Budding Pondweed 240
Bugleweed 172
Bugs-on-a-plate 287
Bulblet-bearing Water-hemlock 56
Bulbostylis capillaris 381
Bullhead Pond-lily 193
Bull Thistle 72
Bulrush, Barber-pole 418
Bulrush, Hard-stemmed 416
Bulrush, Mosquito 418
Bulrush, Red-tinged 418
Bulrush, Soft-stemmed 417
Bulrush, Water 417
Bunchberry 129
Bunchberry, Canada 129
Burdock, Common 69
Burdock, Great 68
Burdock, Lesser 69
Bur-marigold, Nodding 70
Burnweed, American 73
Bur-reed, American 290
Bur-reed, Arctic 293
Bur-reed, Floating 292
Bur-reed, Great 292
Bur-reed, Lesser 290
Bur-reed, Narrow-leaved 291
Bur-reed, Simple-stemmed 291
Bursa bursa-pastoris 110
Bush-honeysuckle 118
Butter-and-eggs Toadflax 226
Buttercup, Tall 250
Butter-weed 74

C

Calamagrostis canadensis 431
Calamagrostis stricta 431
Calico American-aster 102
Calico Aster 102
Calla palustris 64
Calla, Wild 64
Callitriche heterophylla 223
Callitriche palustris 223
Calopogon pulchellus 201
Calopogon tuberosus 201
Calypso bulbosa 202
Calystegia sepium 129
CAMPANULACEAE 115
Campanula rapunculoides 115

Campion, Bladder 124
Campion, Moss 124
Canada Blackberry 267
Canada Bluegrass 445
Canada Bluejoint 431
Canada Bunchberry 129
Canada Clearweed 295
Canada Dwarf-dogwood 129
Canada Fleabane 74
Canada Germander 176
Canada Goldenrod 94
Canada Gooseberry 156
Canada Hawkweed 79
Canada Lettuce 83
Canada Lily 182
Canada Mayflower 280
Canada Reed Grass 431
Canada Rockcress 109
Canada Rush 421
Canada Thistle 71
Canada Wood Nettle 294
Canada Yew 348
Canadian Single-spike Sedge 405
Cancerroot, One-flowered 218
Candleberry 188
Canescent Whitlow-mustard 113
Canoe Birch 315
Capnoides sempervirens 220
CAPRIFOLIACEAE 118
Capsella bursa-pastoris 110
Caraway 56
Cardamine bellidifolia 111
Cardamine diphylla 111
Cardamine parviflora 112
Cardamine pensylvanica 112
Cardinal-flower 116
Carex albicans 381
Carex angustior 390
Carex appalachica 382
Carex aquatilis 382
Carex arcta 382
Carex arctata 383
Carex artitecta 381
Carex atlantica 383
Carex atratiformis 383
Carex bebbii 384
Carex bigelowii 384
Carex billingsii 384
Carex bromoides 385
Carex brunnescens 385
Carex canescens 385
Carex chordorrhiza 386
Carex communis 386
Carex conoidea 386
Carex convoluta 403
Carex crawfordii 387

Carex crinita 387
Carex cryptolepis 387
Carex debilis 388
Carex deflexa 388
Carex deweyana 388
Carex diandra 389
Carex digitalis 389
Carex disperma 389
Carex echinata 390
Carex emmonsii 381
Carex exilis 390
Carex flava 390
Carex foenea 391
Carex folliculata 391
Carex gracillima 391
Carex gynandra 392
Carex haydenii 392
Carex hirtifolia 392
Carex hystericina 393
Carex interior 393
Carex intumescens 393
Carex josselynii 390
Carex lacustris 394
Carex lasiocarpa 394
Carex laxiflora 394
Carex lenticularis 395
Carex leporina 395
Carex leptalea 395
Carex leptonervia 396
Carex limosa 396
Carex longirostris 406
Carex lucorum 396
Carex lupulina 397
Carex lurida 397
Carex magellanica 397
Carex michauxiana 398
Carex muehlenbergii 398
Carex nigra 398
Carex normalis 399
Carex novae-angliae 399
Carex oligosperma 399
Carex ormostachya 400
Carex oronensis 400
Carex pallescens 400
Carex paucifolia 401
Carex paupercula 397
Carex pedunculata 401
Carex pensylvanica 401
Carex projecta 402
Carex pseudocyperus 402
Carex radiata 402
Carex rariflora 403
Carex retrorsa 403
Carex rosea 403
Carex rostrata 404
Carex saxatilis 404

Carex scabrata 404
Carex scirpoidea 405
Carex scoparia 405
Carex siccata 405
Carex sprengelii 406
Carex stipata 406
Carex stricta 406
Carex tenera 407
Carex tenuiflora 407
Carex tincta 407
Carex torta 408
Carex tribuloides 408
Carex trisperma 408
Carex umbellata 409
Carex utriculata 409
Carex vesicaria 409
Carex viridula 410
Carex vulpinoidea 410
Carex wiegandii 410
Carolina Spring-beauty 237
Carrot, Wild 58
Carum carvi 56
CARYOPHYLLACEAE 121
Cassandra 133
Cassandra calyculata 133
Cassiope hypnoides 138
Castilleja septentrionalis 215
Catberry 310
Cat Spruce 326
Cat-tail, Broad-leaved 293
Caulophyllum thalictroides 107
Celandine, Greater 221
Cerastium fontanum 121
Cerastium vulgatum 121
Chain Fern 350
Chamaedaphne calyculata 133
Chamaenerion angustifolium 194
Chamaepericlymenum canadense 129
Chamerion angustifolium 194
Chamomile, Rayless 84
Cheat Brome 431
Checkerberry 137
Checkered Rattlesnake-plantain 206
Chelidonium majus 221
Chelone glabra 224
Chenopodium album 52
Chenopodium lanceolatum 52
Cherry, Bird 334
Cherry, Black 334
Cherry, Choke 335
Cherry, Fire 334
Cherry, Pin 334
Cherry, Rum 334
Chickweed, Mouse-ear 121

Chimaphila umbellata 134
Chinese Hemlock-parsley 57
Chiogenes hispidula 136
Chiogenes serpyllifolia 136
Chives, Wild 51
Chokeberry, Black 257
Chokeberry, Purple 257
Choke Cherry 335
Christmas Fern 356
Chrysanthemum leucanthemum 83
Chrysosplenium americanum 286
Cicuta bulbifera 56
Cicuta maculata 57
Cinna arundinacea 432
Cinna latifolia 432
Cinnamon Fern 372
Cinquefoil, Norwegian 263
Cinquefoil, Old-field 264
Cinquefoil, Rough 263
Cinquefoil, Rough-fruited 263
Cinquefoil, Shrubby 258
Cinquefoil, Silver-leaved 262
Cinquefoil, Sulphur 263
Circaea alpina 194
Circaea canadensis 195
Cirsium arvense 71
Cirsium muticum 72
Cirsium vulgare 72
CISTACEAE 127
Cladium mariscoides 411
Clammy Hedge-hyssop 225
Clasping-leaved Pondweed 243
Clasping-leaved Twistedstalk 183
Claytonia caroliniana 237
Clearweed, Canada 295
Cleavers, Spring 273
Clematis occidentalis 248
Clematis virginiana 249
Cliff Fern 379
Climbing Bindweed 231
Clintonia borealis 181
Clinton's Wood Fern 352
Clover, Alsike 153
Clover, Hop 152
Clover, Rabbit-foot 151
Clover, Red 153
Clover, Stone 151
Clover, White 154
Clubmoss, Common 366
Clubmoss, One-cone 366
Club-rush, Alpine 419
Clubsedge, Alpine 419
Clubsedge, Tufted 419
Club-spur Bog-orchid 209
Coeloglossum bracteatum 202

Coeloglossum viride 202
Cohosh, Blue 107
COLCHICACEAE 128
Coltsfoot 106
Columbine, European 248
Columbine, Red 247
Columbine, Wild 247
COMANDRACEAE 128
Comandra lividum 128
Comandra, Northern 128
Comarum palustre 258
Common Agrimony 256
Common Arrowhead 51
Common Bearberry 132
Common Blackberry 266
Common Burdock 69
Common Clubmoss 366
Common Dandelion 105
Common Duckweed 65
Common Eastern Wild-rye 437
Common Elderberry 305
Common Evening-primrose 199
Common Eyebright 216
Common Fox 410
Common Golden Alexanders 62
Common Goldenrod 94
Common Grapefern 371
Common Grass-leaved-goldenrod 78
Common Hawkweed 80
Common Interrupted-clubmoss 367
Common Juniper 318
Common Lilac 324
Common Lowbush Blueberry 146
Common Mare's-tail 225
Common Milkweed 63
Common Mullein 289
Common Plantain 226
Common Poison-ivy 52
Common Reed 444
Common Selfheal 173
Common Snowberry 120
Common Soapwort 123
Common Soft Rush 421
Common Speedwell 228
Common Spikesedge 413
Common Stitchwort 127
Common St. John's-wort 166
Common Thistle 72
Common Timothy 444
Common Wall Hawkweed 80
Common Winterberry 311
Common Winter-cress 109
Common Wood Rush 424

Common Wool-grass 418
Common Woolsedge 418
Common Wrinkle-leaved Gold-
    enrod 98
Common Yarrow 65
Common Yellow-cress 114
Common Yellow Wood Sorrel
    220
Comptonia peregrina 188
Coneflower, Black-eyed 92
Conioselinum chinense 57
CONVOLVULACEAE 129
Convolvulus sepium 129
Conyza canadensis 74
Coptis groenlandica 249
Coptis trifolia 249
Corallorhiza maculata 203
Corallorhiza trifida 203
Coral-root, Early 203
Coral-root, Spotted 203
CORNACEAE 129, 317
Cornel, Dwarf 129
Cornus alternifolia 317
Cornus canadensis 129
Cornus rugosa 317
Cornus sericea 318
Cornus stolonifera 318
Corydalis, Pink 220
Corydalis sempervirens 220
Corylus cornuta 316
Cotton-grass, Five-nerved 415
Cotton-grass, Tall 414
Cotton-grass, Tawny 415
Cottonsedge, Few-nerved 415
Cottonsedge, Slender 414
Cottonsedge, Tall 414
Cottonsedge, Tawny 415
Cottonsedge, Tussock 415
Cowbane, Spotted 57
Cowberry 150
Cow-parsnip, American 58
Cow Vetch 154
Cow-wheat, American 218
Crabgrass, Smooth 436
Crackerberry 129
Cranberry, American 148
Cranberry, Bog 149
Cranberry, Large 148
Cranberry, Mountain 150
Cranberry, Rock 150
Cranberry, Small 149
Crataegus keepii 333
Crawford's Sedge 387
Creeping Bellflower 115
Creeping Bentgrass 428
Creeping Bladderwort 177

Creeping Crowfoot 252
Creeping Snowberry 136
Creeping Sow-thistle 100
Creeping Spearwort 252
Creeping Spicy-wintergreen 136
Creeping Thistle 71
Creeping Wild-rye 436
Creeping Yellow-loosestrife 190
Crested Wood Fern 353
Crowberry, Black 135
Crowberry, Red 135
Crowfoot, Bristly 253
Crowfoot, Creeping 252
Crowfoot, Hooked 253
Crowfoot, Kidney-leaved 250
Crowfoot, Small-flowered 250
Crowfoot, Spot-leaved 254
Crowfoot, Swamp 251
Crowfoot, Tall 250
Crowfoot, Water 252
Crunchberry 129
Cucumber Root 182
Cudweed, Brown 79
Cudweed, Low 79
Cultivated Apple 333
CUPRESSACEAE 318
Curlewberry 135
Curly Dock 235
Currant, Black 155
Currant, Red 157
Currant, Skunk 156
Currant, Swamp 157
Currant, Swamp Red 158
Cushion Pink 124
Cushion-plant 130
Cut-leaved Water-horehound 171
Cutler's Goldenrod 96
CYPERACEAE 381
Cyperus dentatus 411
Cyperus diandrus 411
Cypress-like Sedge 402
Cypripedium acaule 204
Cypripedium parviflorum 204
Cystopteris fragilis 377
Cystopteris tenuis 378

D

Dactylis glomerata 432
Daisy Fleabane 74
Daisy-leaved Moonwort 369
Daisy, Ox-eye 83
Daisy, White 83
Dalibarda repens 267
Dandelion, Common 105
Dandelion, Red-seeded 105
Danthonia compressa 433

Danthonia spicata 433
Dasiphora floribunda 258
Dasiphora fruticosa 258
Daucus carota 58
Day-lily, Orange 161
Deep Water Quillwort 360
Deer Grass 419
Deer's Hair 419
Deer-tongue Rosette-panicgrass
    434
Delicate Quill Sedge 407
Dendrolycopodium dendroideum
    362
Dendrolycopodium hickeyi 362
Dendrolycopodium obscurum
    363
DENNSTAEDTIACEAE 350
Dennstaedtia punctilobula 350
Dentaria diphylla 111
Dentaria incisa 111
Deparia acrostichoides 378
Deschampsia cespitosa 433
Deschampsia flexuosa 434
Devil's Beggar-ticks 71
Dewdrop 267
DIAPENSIACEAE 130
Diapensia lapponica 130
Dicentra cucullaria 221
Dichanthelium acuminatum ssp.
    columbianum 435
Dichanthelium acuminatum ssp.
    fasciculatum 435
Dichanthelium acuminatum ssp.
    implicatum 435
Dichanthelium acuminatum ssp.
    lindheimeri 436
Dichanthelium boreale 434
Dichanthelium clandestinum 434
Dichanthelium columbianum 435
Dichanthelium implicatum 435
Dichanthelium lanuginosum 435
Dichanthelium lanuginosum var.
    fasciculatum 435
Dichanthelium lindheimeri 436
Diervilla diervilla 118
Diervilla lonicera 118
Digitaria ischaemum 436
Diphasiastrum complanatum 363
Diphasiastrum digitatum 364
Diphasiastrum sitchense 364
Diphasiastrum tristachyum 365
Dirca palustris 290
Dock, Bitter 236
Dock, Curly 235
Dock-leaved Smartweed 233
Dockmackie 306

Dock, Sheep 234
Dock, Water 235
Dock, Yard 236
Dock, Yellow 235
Doellingeria umbellata 73
Dogbane, Spreading 62
Dog Bentgrass 426
Dogwood, Alternate-leaved 317
Dogwood, Pagoda 317
Dogwood, Red-osier 318
Dogwood, Round-leaved 317
Doll's Eyes 246
Downy Goldenrod 97
Downy Serviceberry 330
Downy Shadbush 330
Downy Willow-herb 199
Draba breweri 113
Draba cana 113
Draba, Lance-leaved 113
Draba lanceolata 113
Dragon's-mouth 201
Drooping Woodland Sedge 383
DROSERACEAE 130
Drosera intermedia 130
Drosera rotundifolia 131
Drummond's Rockcress 109
Dry Land Bitter-cress 112
Dry Land Sedge 405
Drymocallis arguta 259
DRYOPTERIDACEAE 351
Dryopteris campyloptera 351
Dryopteris carthusiana 352
Dryopteris clintoniana 352
Dryopteris cristata 353
Dryopteris filix-mas 353
Dryopteris fragrans 354
Dryopteris intermedia 355
Dryopteris marginalis 355
Dryopteris noveboracensis 374
Dryopteris phegopteris 375
Dryopteris spinulosa 351
Dryopteris thelypteris 376
Duck-potato 51
Duckweed, Common 65
Dulichium arundinaceum 412
Dusty Goldenrod 97
Dutchman's-breeches 221
Dwarf Birch 314
Dwarf Blueberry 147
Dwarf Cornel 129
Dwarf-dogwood, Canada 129
Dwarf Enchanter's-nightshade 194
Dwarf Ginseng 60
Dwarf Mistletoe 302
Dwarf Raspberry 270

Dwarf Rattlesnake-plantain 205
Dwarf Shadbush 332
Dwarf St. John's-wort 166

E

Early Coral-root 203
Early Goldenrod 95
Early Meadow-rue 254
Early Small-flowered Saxifrage 287
Eastern Black Currant 155
Eastern Hay-scented Fern 350
Eastern Hemlock 329
Eastern Leatherwood 290
Eastern Purple Bladderwort 179
Eastern Rough Sedge 404
Eastern Spicy-wintergreen 137
Eastern Star Sedge 402
Eastern White Pine 329
Eastern Wild-rye 437
Eastern Willow-herb 196
ELATINACEAE 131
Elatine minima 131
Elderberry, Black 305
Elderberry, Common 305
Elderberry, Red 305
Elder, Red-berried 305
Elder, Stinking 305
Eleocharis acicularis 412
Eleocharis erythropoda 412
Eleocharis obtusa 413
Eleocharis ovata 413
Eleocharis palustris 413
Eleocharis robbinsii 414
Elliptic-leaved Shinleaf 144
Elliptic St. John's-wort 164
Elm, American 348
Elodea nuttallii 161
Elymus repens 436
Elymus trachycaulus 437
Elymus virginicus 437
Empetrum atropurpureum 135
Empetrum eamesii 135
Empetrum nigrum 135
Enchanter's-nightshade, Broad-leaved 195
Enchanter's-nightshade, Dwarf 194
Enchanter's-nightshade, Small 194
Epifagus virginiana 216
Epigaea repens 136
Epilobium alpinum 195
Epilobium anagallidifolium 195
Epilobium angustifolium 194
Epilobium ciliatum 196

Epilobium coloratum 196
Epilobium glandulosum 196
Epilobium hornemannii 197
Epilobium lactiflorum 197
Epilobium leptophyllum 198
Epilobium lineare 198
Epilobium palustre 198
Epilobium strictum 199
Epipactis helleborine 205
EQUISETACEAE 357
Equisetum arvense 357
Equisetum fluviatile 357
Equisetum sylvaticum 358
Equisetum variegatum 358
Erechtites hieraciifolius 73
ERICACEAE 132, 319
Erigeron annuus 74
Erigeron canadensis 74
Erigeron philadelphicus 75
Erigeron strigosus 75
ERIOCAULACEAE 150
Eriocaulon aquaticum 150
Eriocaulon septangulare 150
Eriophorum angustifolium 414
Eriophorum gracile 414
Eriophorum spissum 415
Eriophorum tenellum 415
Eriophorum vaginatum 415
Eriophorum virginicum 415
Erythronium americanum 181
Estuary Boneset 76
Eupatoriadelphus maculatus 78
Eupatorium maculatum 78
Eupatorium perfoliatum 76
Eupatorium purpureum 78
Euphrasia americana 216
Euphrasia nemorosa 216
Euphrasia oakesii 217
Euphrasia officinalis 217
Euphrasia stricta 217
Euphrasia williamsii 217
European Columbine 248
European Hawkweed 80
European Mountain-ash 336
Eurybia divaricata 76
Eurybia macrophylla 77
Eurybia radula 77
Euthamia graminifolia 78
Eutrochium maculatum 78
Evening-primrose, Common 199
Evening-primrose, Little 200
Evening-primrose, Northern 200
Evening-primrose, Small-flowered 200
Evening-primrose, Yellow 199
Evergreen Wood Fern 355

Eyebright, Common 216
Eyebright, Oakes' 217
Eyebright, Strict 217

F

FABACEAE 151, 320
FAGACEAE 321
Fagus grandifolia 321
Fairy-slipper 202
Fairy Spuds 60
Fall-dandelion 93
Fallopia cilinodis 230
Fallopia convolvulus 231
Fallopia scandens 231
False Beech-drops 138
False Bindweed 129
False Hellebore 186
False Melic Grass 447
False Miterwort 288
False Nettle 294
False Solomon's-seal 281
False Spikenard 281
False Spleenwort 378
False Toadflax 128
Farwell's Water-milfoil 159
Feathery False Solomon's-seal 281
Fernald's Mannagrass 448
Fern, Boulder 350
Fern, Bracken 351
Fern, Chain 350
Fern, Christmas 356
Fern, Cinnamon 372
Fern, Cliff 379
Fern, Fiddlehead 368
Fern, Fragile 377
Fern, Hay-scented 350
Fern, Interrupted 371
Fern, Lady 377
Fern, Maidenhair 374
Fern, Marsh 376
Fern, Massachusetts 375
Fern, Oak 379
Fern, Ostrich 368
Fern, Rattlesnake 371
Fern, Royal 372
Fern, Sensitive 368
Fescue, Proliferous 438
Fescue, Red 438
Festuca capillata 437
Festuca filiformis 437
Festuca prolifera 438
Festuca rubra 438
Few-flowered Sedge 401
Few-nerved Cottonsedge 415
Few-seeded Sedge 399
Fibrous-rooted Sedge 386

Fiddlehead Fern 368
Field Hawkweed 90
Field Horsetail 357
Field Meadow-foxtail 428
Field Pussytoes 67
Field Sow-thistle 100
Fine-leaved Sheep Fescue 437
Fir, Balsam 325
Fire Birch 315
Fire Cherry 334
Fireweed, Narrow-leaved 194
Firmoss, Mountain 359
Firmoss, Shining 359
Five-finger 264
Five-nerved Cotton-grass 415
Flaccid Manna Grass 439
Flannel-plant 289
Flat-branched Tree-clubmoss 363
Flat-leaved Bladderwort 178
Flatsedge, Toothed 411
Flatsedge, Umbrella 411
Flat-stemmed Blue Grass 445
Flat-stem Pondweed 245
Flattened Oatgrass 433
Flat-topped Goldenrod 78
Flat-topped White-aster 73
Flat-topped White Aster 73
Fleabane, Annual 74
Fleabane, Canada 74
Fleabane, Daisy 74
Fleabane, Philadelphia 75
Fleabane, Rough 75
Floating Bladderwort 179
Floating Bur-reed 292
Floating-heart, Little 187
Floating Pondweed 241
Floodplain Avens 261
Flowering Fern 372
Fly-away Grass 427
Fly Honeysuckle 119
Foam-flower 288
Forest Licorice Bedstraw 275
Forget-me-not, Woodland 108
Fowl Blue Grass 446
Fowl Manna Grass 440
Fragaria virginiana 259
Fragile Fern 377
Fragrant Bedstraw 278
Fragrant Water-lily 193
Fragrant Wood Fern 354
Fraser's St. John's-wort 164
Fraxinus americana 323
Fraxinus nigra 323
Fraxinus pennsylvanica 324
Free-flowered Waterweed 161
Fringed Bindweed 230

Fringed Black-bindweed 230
Fringed Brome 430
Fringed Sedge 387
Fringed Willow-herb 196
Fringed Yellow-loosestrife 190
Frog Orchid 202

G

Galeopsis bifida 170
Galeopsis tetrahit 170
Gale, Sweet 189
Galium aparine 273
Galium asprellum 274
Galium boreale 274
Galium circaezans 275
Galium kamtschaticum 275
Galium mollugo 276
Galium palustre 276
Galium sylvaticum 277
Galium tinctorium 277
Galium trifidum 278
Galium triflorum 278
Gall-of-the-earth 85
Garden Bird's-foot-trefoil 151
Garden Red Currant 157
Garden Yellow-rocket 109
Gaultheria hispidula 136
Gaultheria procumbens 137
Gaylussacia baccata 137
GENTIANACEAE 155
Geocaulon lividum 128
Gerardia tenuifolia 215
Germander, American 176
Germander, Canada 176
Geum aleppicum 260
Geum canadense 260
Geum laciniatum 261
Geum macrophyllum 261
Geum rivale 262
Ghost-flower 141
Gill-over-the-ground 171
Ginseng, Dwarf 60
Glandular Birch 314
Glandular Willow-herb 196
Glandular Wood Fern 355
Glaucous Blue Grass 445
Glaucous King-devil 91
Glechoma hederacea 171
Glyceria borealis 438
Glyceria canadensis 439
Glyceria fernaldii 448
Glyceria grandis 439
Glyceria laxa 439
Glyceria melicaria 440
Glyceria obtusa 440
Glyceria striata 440

Gnaphalium supinum 87
Gnaphalium sylvaticum 87
Gnaphalium uliginosum 79
Goatsbeard, Showy 106
Goatsbeard, Yellow 106
Goblet Aster 102
Golden Alexanders 62
Golden Hedge-hyssop 224
Goldenrod, Bog 99
Goldenrod, Canada 94
Goldenrod, Common 94
Goldenrod, Cutler's 96
Goldenrod, Downy 97
Goldenrod, Dusty 97
Goldenrod, Early 95
Goldenrod, Flat-topped 78
Goldenrod, Gray 97
Goldenrod, Hairy 95
Goldenrod, Lance-leaved 78
Goldenrod, Large-leaved 96
Goldenrod, Rand's 98
Goldenrod, Rough-stemmed 98
Goldenrod, Squarrose 99
Goldenrod, Swamp 99
Goldenrod, White 93
Goldenrod, Wrinkle-leaved 98
Goldenrod, Zig-zag 94
Golden Saxifrage 286
Goldthread, Three-leaved 249
Goodenough's Sedge 398
Goodyera repens 205
Goodyera tesselata 206
Gooseberry, Canada 156
Gooseberry, Hairy-stemmed 156
Gooseberry, Smooth 156
Goose-foot Maple 345
Goosefoot, White 52
Goutweed, Bishop's 53
Graceful Sedge 391
Grapefern, Common 371
Grapefern, Leathery 370
Grass, Blue 445
Grass, Fly-away 427
Grass-leaved Arrowhead 50
Grass-leaved-goldenrod, Common 78
Grass-leaved Stitchwort 126
Grass, Millet 441
Grass, Orchard 432
Grass-pink, Tuberous 201
Grassy Pondweed 240
Gratiola aurea 224
Gratiola lutea 224
Gratiola neglecta 225
Gray Birch 315
Gray Goldenrod 97

Great Burdock 68
Great Bur-reed 292
Greater Bladder Sedge 393
Greater Bladderwort 180
Greater Canada St. John's-wort 165
Greater Celandine 221
Greater Purple Bladderwort 179
Greater Purple-fringed Bog-orchid 210
Greater Straw Sedge 399
Greater Water Dock 235
Greater Water-starwort 223
Greater Yellow Water Crowfoot 251
Great-spurred Violet 301
Great Willow-herb 194
Green Adder's-mouth 206
Green Alder 312
Green Ash 324
Green Bog-orchid 211
Green-flowered Pyrola 143
Green-flowered Shinleaf 143
Green Osier 317
Green Woodland Orchis 209
GROSSULARIACEAE 155
Ground-cedar, Blue 365
Ground-cedar, Northern 363
Ground-cedar, Sitka 364
Ground-cedar, Southern 364
Ground-hemlock 348
Ground Juniper 318
Groundsel, Balsam 88
Grove Blue Grass 444
Grove-sandwort, Blunt-leaved 122
Gymnocarpium dryopteris 379
Gypsophila muralis 121

H

Habenaria blephariglottis 209
Habenaria bracteata 202
Habenaria clavellata 209
Habenaria dilatata 210
Habenaria fimbriata 210
Habenaria grandiflora 210
Habenaria hookeri 211
Habenaria huronensis 211
Habenaria obtusata 212
Habenaria orbiculata 212
Habenaria psycodes 213
Hackmatack 325
Hairgrass 427
Hair-sedge, Tufted 381
Hairy Goldenrod 95
Hairy Hedge-nettle 175

Hairy Solomon's-seal 282
Hairy-stemmed Gooseberry 156
Hairy Wood Rush 424
Halenia deflexa 155
Halenia heterantha 155
HALORAGACEAE 158
HAMAMELIDACEAE 322
Hamamelis virginiana 322
Hard-stemmed Bulrush 416
Harebell 115
Harestail 415
Harrimanella hypnoides 138
Hawkweed, Allegheny 81
Hawkweed, Canada 79
Hawkweed, Common 80
Hawkweed, European 80
Hawkweed, Field 90
Hawkweed, Kalm's 79
Hawkweed, Panicled 81
Hawkweed, Rough 82
Hawkweed, Savoy 81
Hawkweed, Spotted 80
Hawkweed, Sticky 82
Hawthorn, Keep's 333
Hayden's Sedge 392
Hay-scented Fern 350
Hazelnut, Beaked 316
Heal-all 173
Heart-leaved American-aster 101
Heart-leaved Aster 101
Heart-leaved Paper Birch 313
Heart-leaved Twayblade 208
Heart-leaved Willow 340
Hedge Bindweed 129
Hedge False Bindweed 129
Hedge-hyssop, Clammy 225
Hedge-hyssop, Golden 224
Hedge-mustard, Tumbling 114
Hedge-nettle, Hairy 175
Hedge-nettle, Marsh 175
Hellebore, False 186
Helleborine, Broad-leaved 205
HEMEROCALLIDACEAE 161
Hemerocallis fulva 161
Hemlock, Eastern 329
Hemlock-parsley, Chinese 57
Hemp, Indian 62
Hemp-nettle, Brittle-stemmed 170
Hemp-nettle, Split-lipped 170
Hepatica americana 247
Hepatica, Blunt-lobed 247
Heracleum lanatum 58
Heracleum maximum 58
Herb Gerard 53
Hickey's Tree-clubmoss 362

Hieracium aurantiacum 89
Hieracium caespitosum 90
Hieracium canadense 79
Hieracium flagellare 90
Hieracium florentinum 91
Hieracium kalmii 79
Hieracium lachenalii 80
Hieracium maculatum 80
Hieracium paniculatum 81
Hieracium pilosella 91
Hieracium piloselloides 91
Hieracium praealtum 92
Hieracium pratense 90
Hieracium sabaudum 81
Hieracium scabrum 82
Hieracium vulgatum 80
Hierochloe alpina 428
Hierochloe monticola 428
Hierochloe odorata 429
Highbush Blueberry 319
Highbush-cranberry 309
Highland-rush 425
Hippuris vulgaris 225
Hoary Sedge 385
Hobblebush 307
Hogbrake 351
Holly, Mountain 310
Honeysuckle, American 119
Honeysuckle, Fly 119
Honeysuckle, Mountain 120
Honeysuckle, Swamp 119
Hooded Ladies'-tresses 214
Hooded Skullcap 174
Hooked Crowfoot 253
Hooker's Bog-orchid 211
Hook-spurred Violet 296
Hop Clover 152
Hop-hornbeam 316
Hop Sedge 397
Horned Bladderwort 176
Hornemann's Willow-herb 197
Horsetail, Field 357
Horsetail, River 357
Horsetail, Water 357
Horsetail, Wood 358
Horsetail, Woodland 358
Houstonia caerulea 279
Howell's Pussytoes 67
Huckleberry, Black 137
Humped Bladderwort 177
Huperzia appalachiana 359
Huperzia appressa 359
HUPERZIACEAE 359
Huperzia lucidula 359
HYDROCHARITACAE 161
Hydrocotyle americana 59

HYPERICACEAE 163
Hypericum boreale 163
Hypericum canadense 163
Hypericum ellipticum 164
Hypericum fraseri 164
Hypericum gentianoides 165
Hypericum majus 165
Hypericum mutilum 166
Hypericum perforatum 166
Hypericum prolificum 167
Hypericum punctatum 167
Hypericum spathulatum 167
Hypericum subpetiolatum 167
Hypericum virginicum 168
Hypopitys monotropa 138

I

Ilex mucronata 310
Ilex verticillata 311
Impatiens capensis 107
Indian Cucumber Root 182
Indian Hemp 62
Indian Paintbrush 89
Indian-pipe, One-flowered 141
Indian-tobacco 117
Indian Turnip 64
Inflated Bladderwort 179
Inland Sedge 393
Intermediate Pinweed 127
Intermediate Shadbush 331
Interrupted-clubmoss, Common 367
Interrupted-clubmoss, Northern 367
Interrupted Fern 371
IRIDACEAE 168
Iris, Blue 168
Iris versicolor 168
Ironwood 316
ISOËTACEAE 360
Isoëtes echinospora 360
Isoëtes lacustris 360
Isoëtes macrospora 360
Isoëtes muricata 360
Isoëtes prototypus 361
Isoëtes tuckermanii 361

J

Jack-in-the-pulpit 64
Jack Pine 327
Joe-Pye Weed 78
Joint-leaved Rush 420
JUNCACEAE 420
Juncus articulatus 420
Juncus brachycephalus 420

Juncus brevicaudatus 420
Juncus bufonius 421
Juncus canadensis 421
Juncus effusus 421
Juncus filiformis 422
Juncus militaris 422
Juncus pelocarpus 422
Juncus pylaei 423
Juncus stygius 423
Juncus tenuis 423
Juncus trifidus 425
Juniper, Common 318
Juniper, Ground 318
Juniperus communis 318

K

Kalmia angustifolia 139
Kalmia polifolia 139
Kalmia procumbens 140
Kalm's Hawkweed 79
Keep's Hawthorn 333
Kentucky Blue Grass 447
Kidney-leaved Crowfoot 250
Kidney-leaved Violet 300
King-devil, Glaucous 91
King-devil, Mouse-ear 91
King-devil, Orange 89
King-devil, Tall 92
King-devil, Whip 90
King-devil, Yellow 90
King-of-the-meadow 255
Kinnikinnick 132

L

Labrador-tea 145
Labrador Willow 283
Lactuca biennis 82
Lactuca canadensis 83
Ladies'-tresses, Hooded 214
Ladies'-tresses, Nodding 214
Lady Fern 377
Lady's-slipper, Pink 204
Lady's-slipper, Yellow 204
Lady's-thumb Smartweed 233
Lake Huron Green Bog-orchid 211
Lake Quillwort 360
Lake Shore Sedge 395
Lakeside Sedge 394
Lamb's Quarters 52
LAMIACEAE 170
Lance-leaved American-aster 101
Lance-leaved Arnica 69
Lance-leaved Draba 113
Lance-leaved Goldenrod 78

Lance-leaved Twistedstalk 183
Lance-leaved Violet 298
Lapland Rosebay 146
Laportea canadensis 294
Larch, American 325
Large Cranberry 148
Large-leaved Avens 261
Large-leaved Goldenrod 96
Large-leaved White Violet 297
Large-leaved Wood-aster 77
Large-pod Pinweed 127
Large Purple-fringed Orchis 210
Large Sweet Grass 429
Large Water-plantain 49
Larix laricina 325
Laurel, Bog 139
Laurel, Pale 139
Laurel, Sheep 139
Leafy Saxifrage 286
Leatherleaf 133
Leatherwood, Eastern 290
Leathery Grapefern 370
Lechea intermedia 127
Ledum groenlandicum 145
Leersia oryzoides 441
Lemna minor 65
LENTIBULARIACEAE 176
Leontodon autumnalis 93
Leontodon erythrospermum 105
Leontodon taraxacum 105
Lepidium virginicum 113
Lesser Bladder Sedge 409
Lesser Bladderwort 178
Lesser Burdock 69
Lesser Bur-reed 290
Lesser Daisy Fleabane 75
Lesser Hop Clover 152
Lesser Purple-fringed Bog-orchid 213
Lesser Rattlesnake-plantain 205
Lesser Stitchwort 126
Lesser St. John's-wort 163
Lesser Tussock Sedge 389
Lettuce, Blue 82
Lettuce, Canada 83
Lettuce, Tall 83
Lettuce, Wild 83
Leucanthemum vulgare 83
Leverwood 316
Lilac, Common 324
LILIACEAE 181
Lilium canadense 182
Lily, Blue-bead 181
Lily, Canada 182
Lily-of-the-valley, Wild 280
Linaria vulgaris 226

Linden, American 322
Lindheimer's Rosette-panicgrass 436
Lingonberry 150
Linnaea americana 118
Linnaea borealis 118
Listera auriculata 207
Listera convallarioides 207
Listera cordata 208
Little Bluet 279
Little Club-spur Bog-orchid 209
Little Evening-primrose 200
Little Floating-heart 187
Little Green Sedge 410
Little Merrybells 128
Little Shinleaf 144
Little Yellow-rattle 219
Lobelia, Bladder-pod 117
Lobelia, Brook 117
Lobelia cardinalis 116
Lobelia dortmanna 116
Lobelia inflata 117
Lobelia kalmii 117
Lobelia, Red 116
Lobelia, Water 116
Locust, Bristly 320
Locust, Mossy 320
Loiseleuria procumbens 140
Long-beaked Sedge 406
Long-beaked Water Crowfoot 252
Long-beaked Willow 338
Long Beech Fern 375
Long-bracted Green Orchid 202
Long-leaved Pondweed 241
Long-leaved Stitchwort 126
Long-stalked Sedge 401
Lonicera canadensis 119
Lonicera oblongifolia 119
Lonicera villosa 120
Loose-flowered Alpine Sedge 403
Loosestrife, Purple 184
Lotus corniculatus 151
Low Baby's-breath 121
Lowbush Blueberry 146
Low Cudweed 79
Low Sweet Blueberry 146
Luzula acuminata 424
Luzula confusa 424
Luzula hyperborea 424
Luzula multiflora 424
Luzula parviflora 425
Luzula saltuensis 424
Luzula spicata 425
Lychnis saponaria 123
LYCOPODIACEAE 362

Lycopodiella inundata 365
Lycopodium annotinum 367
Lycopodium canadense 367
Lycopodium clavatum 366
Lycopodium complanatum 363
Lycopodium dendroideum 362
Lycopodium digitatum 364
Lycopodium flabelliforme 364
Lycopodium hickeyi 362
Lycopodium inundatum 365
Lycopodium lagopus 366
Lycopodium lucidulum 359
Lycopodium obscurum 363
Lycopodium sitchense 364
Lycopodium tristachyum 365
Lycopus americanus 171
Lycopus uniflorus 172
Lycopus virginicus 172
Lysimachia borealis 189
Lysimachia ciliata 190
Lysimachia nummularia 190
Lysimachia terrestris 191
Lysimachia thyrsiflora 191
LYTHRACEAE 184
Lythrum salicaria 184

M

Mackay's Fragile Fern 378
Mad Dog Skullcap 174
Maianthemum canadense 280
Maianthemum racemosum 281
Maianthemum stellatum 281
Maianthemum trifolium 282
Maidenhair Fern 374
Maidenhair Spleenwort 349
Maiden's-tears 124
Malaxis unifolia 206
Male Wood Fern 353
Mallow, Musk 184
Malus pumila 333
Malus sylvestris 333
MALVACEAE 184, 322
Malva moschata 184
Mandarin, Rose 183
Mannagrass, Fernald's 448
Maple, Goose-foot 345
Maple-leaved Viburnum 306
Maple, Mountain 347
Maple, Red 346
Maple, Rock 347
Maple, Silver 345
Maple, Striped 345
Maple, Sugar 347
Maple, Swamp 346
Mare's-tail, Common 225
Marginal Wood Fern 355

Marsh Bedstraw 276
Marsh Blue Violet 297
Marsh-cinquefoil 258
Marsh Fern 376
Marsh-five-finger 258
Marsh Hedge-nettle 175
Marsh-pennywort, American 59
Marsh Skullcap 174
Marsh Speedwell 228
Marsh Willow-herb 198
Maryland Sanicle 61
Massachusetts Fern 375
Matricaria discoidea 84
Matricaria matricarioides 84
Matteuccia struthiopteris 368
Mayflower 136
Mayflower, Canada 280
Mayweed 84
Meadow-beauty, Virginia 186
Meadow-foxtail, Field 428
Meadow Goat's Beard 106
Meadow-rue, Early 254
Meadow-rue, Tall 255
Meadowsweet, Rosy 273
Meadowsweet, White 272
Meadow Willow 343
Meager Sedge 390
Medeola virginiana 182
Melampyrum lineare 218
MELANTHIACEAE 185
MELASTOMATACEAE 186
Melica purpurascens 447
Mentha arvensis 173
Mentha canadensis 173
MENYANTHACEAE 187
Menyanthes trifoliata 187
Merrybells, Little 128
Michaux's Sedge 398
Micranthes foliolosa 286
Micranthes virginiensis 287
Milfoil 65
Milium effusum 441
Milkweed, Common 63
Milkweed, Silky 63
Milkweed, Swamp 63
Millet Grass 441
Mimulus ringens 222
Mint, Wild 173
Minuartia groenlandica 122
Mistletoe, Dwarf 302
Mitchella repens 279
Mitella nuda 287
Miterwort, False 288
Mitterwort, Naked 287
Mixed Bladderwort 177
Moccasin Flower 204

Moehringia lateriflora 122
Moneses uniflora 140
Monkey-flower, Allegheny 222
Mononeuria groenlandica 122
Monotropa hypopithys 138
Monotropa uniflora 141
Moonwort, Daisy-leaved 369
Moonwort, Swamp 370
Moor Rush 423
Mooseberry 307
Moosewood 345
Morella caroliniensis 188
Morning-glory, Wild 129
Mosquito Bulrush 418
Moss Bell-heather 138
Moss Campion 124
Moss-plant 138
Mossy Locust 320
Mountain Alder 312
Mountain-ash, American 335
Mountain-ash, European 336
Mountain-ash, Showy 336
Mountain Aster 86
Mountain Bearberry 133
Mountain Cranberry 150
Mountain Firmoss 359
Mountain Fly Honeysuckle 120
Mountain-heath, Purple 142
Mountain Holly 310
Mountain Honeysuckle 120
Mountain Maple 347
Mountain Paper Birch 313
Mountain Pyrola 144
Mountain Sandplant 122
Mountain Sandwort 122
Mountain Serviceberry 330
Mountain Shadbush 330
Mountain Timothy 443
Mountain Wood Fern 351
Mouse-ear Chickweed 121
Mouse-ear King-devil 91
Mud Sedge 396
Muhlenbergia glomerata 441
Muhlenbergia sylvatica 442
Muhlenbergia uniflora 442
Muhlenberg's Sedge 398
Muhly, Bog 442
Muhly, Spike 441
Muhly, Woodland 442
Mullein, Common 289
Musk Mallow 184
Musquash-root 57
Myosotis sylvatica 108
MYRICACEAE 188
Myrica gale 189
Myrica pensylvanica 188

Myrica peregrina 188
Myriophyllum alterniflorum 158
Myriophyllum farwellii 159
Myriophyllum sibiricum 159
Myriophyllum tenellum 160
Myriophyllum verticillatum 160
MYRSINACEAE 189

N

Nabalus altissimus 84
Nabalus boottii 85
Nabalus trifoliolatus 85
Naiad, Slender 162
Najas caespitosa 162
Najas flexilis 162
Naked Bishop's-cap 287
Naked-bulbil Small-flowered Saxifrage 286
Naked Mitterwort 287
Nannyberry 308
Nantucket Shadbush 332
Narrow False Oat 448
Narrow-leaved Bur-reed 291
Narrow-leaved Fireweed 194
Narrow-leaved Speedwell 228
Narrow-leaved Sundew 130
Narrow-leaved Willow-herb 198
Narrow Triangle Moonwort 369
Necklace Sedge 402
Necklace Spike Sedge 400
Needle Spikesedge 412
Needle-tipped Blue-eyed-grass 169
Neglected Reed Grass 431
Nemopanthus mucronatus 310
Neottia auriculata 207
Neottia convallarioides 207
Neottia cordata 208
Nepeta hederacea 171
Nerveless Woodland Sedge 396
Nettle, False 294
Nettle, Stinging 295
New England American-aster 102
New England Groundsel 88
New England Hawkweed 81
New England Sedge 399
New York American-aster 103
New York Aster 103
New York Fern 374
Noble Prince's-pine 134
Nodding-aster, Bog 86
Nodding-aster, Sharp-toothed 86
Nodding Beggar-ticks 70
Nodding Bur-marigold 70
Nodding Ladies'-tresses 214
Nodding Sedge 392
Nodding Smartweed 233

Northeastern Manna Grass 440
Northeastern Sedge 387
Northern Arrowhead 50
Northern Bedstraw 274
Northern Bentgrass 427
Northern Bladderwort 178
Northern Blueberry 147
Northern Blue Flag 168
Northern Bog Aster 100
Northern Bog-clubmoss 365
Northern Bog-orchid 210
Northern Bush-honeysuckle 118
Northern Cluster Sedge 382
Northern Comandra 128
Northern Evening-primrose 200
Northern Ground-cedar 363
Northern Interrupted-clubmoss 367
Northern Lady Fern 377
Northern Long-awned Wood Grass 430
Northern Long Sedge 391
Northern Maidenhair Fern 374
Northern Manna Grass 438
Northern Oak Fern 379
Northern Painted-cup 215
Northern Red Oak 321
Northern Rosette-panicgrass 434
Northern Sedge 388
Northern Snail-seed Pondweed 245
Northern St. John's-wort 163
Northern Sweet-coltsfoot 89
Northern Water-horehound 172
Northern Water-milfoil 159
Northern Waternymph 162
Northern Water-plantain 49
Northern White-cedar 319
Northern Willow 283
Northern Wood Rush 424
Northern Wood Sorrel 219
Northern Yellow-eyed-grass 304
North Wind Bog-orchid 208
Norwegian Cinquefoil 263
Nuphar microphylla 192
Nuphar variegata 193
NYMPHAEACEAE 192
Nymphaea odorata 193
Nymphoides cordata 187

O

Oakes' Eyebright 217
Oakes' Pondweed 242
Oak Fern 379
Oak, Red 321
Oatgrass, Flattened 433

Oatgrass, Poverty 433
Oats, Wild 128
Oclemena acuminata 86
Oclemena nemoralis 86
Oenothera biennis 199
Oenothera cruciata 200
Oenothera parviflora 200
Oenothera perennis 200
Old-field Cinquefoil 264
OLEACEAE 323
Omalotheca supina 87
Omalotheca sylvatica 87
ONAGRACEAE 194
One-cone Clubmoss 366
One-flowered Broom-rape 218
One-flowered Cancerroot 218
One-flowered Indian-pipe 141
One-flowered Pyrola 140
One-flowered-shinleaf 140
One-sided Pyrola 141
One-sided-shinleaf 141
ONOCLEACEAE 368
Onoclea sensibilis 368
Open-field Sedge 386
OPHIOGLOSSACEAE 369
Orange Day-lily 161
Orange-grass St. John's-wort 165
Orange King-devil 89
Orange Paintbrush 89
Orange Touch-me-not 107
Orchard Grass 432
ORCHIDACEAE 201
Orchid, Frog 202
Oreojuncus trifidus 425
OROBANCHACEAE 215
Orobanche uniflora 218
Orono Sedge 400
Orthilia secunda 141
Oryzopsis asperifolia 442
Osier, Green 317
Osmorhiza claytonii 59
OSMUNDACEAE 371
Osmunda cinnamomea 372
Osmunda claytoniana 371
Osmunda regalis 372
Osmundastrum cinnamomeum 372
Ostrich Fern 368
Ostrya virginiana 316
Oval Sedge 395
Ovoid Spikesedge 413
OXALIDACEAE 219
Oxalis europaea 220
Oxalis montana 219
Oxalis stricta 220
Ox-eye Daisy 83

P

Packera paupercula 88
Packera schweinitziana 88
Pagoda Dogwood 317
Painted-cup, Northern 215
Painted Trillium 185
Painted Wakerobin 185
Pale False Manna Grass 448
Pale Laurel 139
Pale Sedge 400
Pale St. John's-wort 164
Palmate Hop Clover 152
Panax trifolius 60
Panicgrass, Tuckerman's 443
Panicled Hawkweed 81
Panicum acuminatum var. implicatum 435
Panicum boreale 434
Panicum columbianum 435
Panicum lanuginosum var. fasciculatum 435
Panicum lanuginosum var. implicatum 435
Panicum lanuginosum var. lindheimeri 436
Panicum lanuginosum var. septentrionale 436
Panicum tuckermanii 443
PAPAVERACEAE 220
Paper Birch 315
Parasol Sedge 409
Parathelypteris noveboracensis 374
Parathelypteris simulata 375
Parlin's Pussytoes 68
Parsnip, Wild 60
Parthenocissus quinquefolia 303
Partridge-berry 279
Pastinaca sativa 60
Path Rush 423
Pearlwort, Bird's-eye 123
Pearly Everlasting 66
Pedicellate Wool-grass 419
Pennsylvania Bitter-cress 112
Pennsylvania Blackberry 270
Pennsylvania Sedge 401
Pentaphylloides floribunda 258
Peppergrass, Wild 113
Pepperweed, Poor-man's 113
Persicaria amphibia 232
Persicaria hydropiper 232
Persicaria lapathifolia 233
Persicaria maculosa 233
Persicaria persicaria 233
Persicaria sagittata 234

Persicaria vivipara 230
Persicaria vulgaris 233
Petasites frigidus 89
Petasites palmatus 89
Phalaris arundinacea 443
Phegopteris connectilis 375
Philadelphia Fleabane 75
Phleum alpinum 443
Phleum pratense 444
Photinia floribunda 257
Photinia melanocarpa 257
Phragmites australis 444
Phragmites communis 444
PHRYMACEAE 222
Phyllodoce caerulea 142
Picea glauca 326
Picea mariana 326
Picea rubens 327
Pickerelweed 237
Pick-pocket 110
Pilea pumila 295
Pilewort 73
Pilosella aurantiaca 89
Pilosella caespitosa 90
Pilosella flagellaris 90
Pilosella officinarum 91
Pilosella piloselloides 91
Pilosella praealta 92
Pimbina 309
Pimpernel Willow-herb 195
PINACEAE 325
Pin Cherry 334
Pineapple-weed 84
Pine, Jack 327
Pine, Pitch 328
Pine, Red 328
Pine-sap, Yellow 138
Pine, Scrub 327
Pineweed 165
Pine, White 329
Pink-bells 118
Pink Corydalis 220
Pink Lady's-slipper 204
Pink Shinleaf 143
Pinus banksiana 327
Pinus resinosa 328
Pinus rigida 328
Pinus strobus 329
Pinweed, Intermediate 127
Pinweed, Large-pod 127
Pinweed, Round-fruited 127
Pipewort, Seven-angled 150
Pipsissewa 134
Pitcherplant, Purple 285
Pitch Pine 328
PLANTAGINACEAE 223

Plantago major 226
Plantago rugelii 227
Plantain, Common 226
Plantain, Rugel's 227
Platanthera aquilonis 208
Platanthera blephariglottis 209
Platanthera clavellata 209
Platanthera dilatata 210
Platanthera grandiflora 210
Platanthera hookeri 211
Platanthera huronensis 211
Platanthera obtusata 212
Platanthera orbiculata 212
Platanthera psycodes 213
Poa alsodes 444
Poa annua 445
POACEAE 426
Poa compressa 445
Poa fernaldiana 446
Poa glauca 445
Poa laxa 446
Poa nemoralis 446
Poa palustris 446
Poa pratensis 447
Poa saltuensis 447
Pod-grass 289
Pogonia ophioglossoides 213
Pogonia, Rose 213
Pointed Broom Sedge 405
Poison-ivy 52
Poison-ivy, Common 52
POLYGONACEAE 230
Polygonatum pubescens 282
Polygonum amphibium 232
Polygonum cilinode 230
Polygonum convolvulus 231
Polygonum hydropiper 232
Polygonum lapathifolium 233
Polygonum persicaria 233
Polygonum sagittatum 234
Polygonum scandens 231
Polygonum viviparum 230
POLYPODIACEAE 373
Polypodium appalachianum 373
Polypodium virginianum 373
Polypody, Appalachian 373
Polypody, Rock 373
Polystichum acrostichoides 356
Polystichum braunii 356
Pond-lily, Bullhead 193
Pond-lily, Small-leaved 192
Pondweed, Alga-like 239
Pondweed, Berchtold's 238
Pondweed, Big-leaved 238
Pondweed, Blunt-leaved 242
Pondweed, Budding 240

Pondweed, Clasping-leaved 243
Pondweed, Flat-stem 245
Pondweed, Floating 241
Pondweed, Grassy 240
Pondweed, Long-leaved 241
Pondweed, Oakes' 242
Pondweed, Ribbon-leaved 239
Pondweed, Richardson's 244
Pondweed, Robbins' 244
Pondweed, Spiral 245
Pondweed, Tuckerman's 239
Pondweed, White-stemmed 243
PONTEDERIACEAE 237
Pontederia cordata 237
Poor-man's Pepperweed 113
Poplar, Balsam 337
Poplar, Big-toothed 337
Poplar, Quaking 338
Populus balsamifera 337
Populus grandidentata 337
Populus tremuloides 338
Porcupine Sedge 393
PORTULACACEAE 237
POTAMOGETONACEAE 238
Potamogeton amplifolius 238
Potamogeton berchtoldii 238
Potamogeton confervoides 239
Potamogeton epihydrus 239
Potamogeton gemmiparus 240
Potamogeton gramineus 240
Potamogeton natans 241
Potamogeton nodosus 241
Potamogeton oakesianus 242
Potamogeton obtusifolius 242
Potamogeton perfoliatus 243
Potamogeton praelongus 243
Potamogeton richardsonii 244
Potamogeton robbinsii 244
Potamogeton spirillus 245
Potamogeton zosteriformis 245
Potentilla argentea 262
Potentilla arguta 259
Potentilla fruticosa 258
Potentilla norvegica 263
Potentilla palustris 258
Potentilla recta 263
Potentilla simplex 264
Potentilla tridentata 272
Poverty Oatgrass 433
Prairie Willow 340
Prenanthes altissima 84
Prenanthes bootii 85
Prenanthes nana 85
Prenanthes trifoliolata 85
Prickly Bog Sedge 383
Prickly Tree-Clubmoss 362

Prince's-pine, Noble 134
Proliferous Fescue 438
Prototype Quillwort 361
Prunella vulgaris 173
Prunus pensylvanica 334
Prunus serotina 334
Prunus virginiana 335
Psammophiliella muralis 121
PTERIDACEAE 374
Pteridium aquilinum 351
Pteridium latiusculum 351
Pubescent Sedge 392
Purple Avens 262
Purple Bladderwort 179
Purple Chokeberry 257
Purple-fringed Bog-orchid 210
Purple Loosestrife 184
Purple Mountain-heath 142
Purple Pitcherplant 285
Purple-stemmed American-aster 103
Purple-stemmed Beggar-ticks 70
Purple-veined Willow-herb 196
Purple Virgin's-bower 248
Purple Wen-dock 192
Pussytoes, Field 67
Pussytoes, Howell's 67
Pussytoes, Parlin's 68
Pussytoes, Small 67
Pussy Willow 339
Pylae's Soft Rush 423
Pyrola americana 142
Pyrola asarifolia 143
Pyrola chlorantha 143
Pyrola elliptica 144
Pyrola, Green-flowered 143
Pyrola minor 144
Pyrola, Mountain 144
Pyrola, One-flowered 140
Pyrola, One-sided 141
Pyrola rotundifolia 142
Pyrola, Round-leaved 142
Pyrola secunda 141
Pyrola virens 143
Pyrus americana 335
Pyrus aucuparia 336
Pyrus decora 336
Pyrus floribunda 257
Pyrus malus 333
Pyrus melanocarpa 257

Q

Quaker Ladies 279
Quaking Poplar 338
Queen Anne's Lace 58
Quercus rubra 321

Quillwort, Lake 360
Quillwort, Prototype 361
Quillwort, Spiny-spored 360
Quillwort, Tuckerman's 361

R

Rabbit-foot Clover 151
Ragwort, Swamp 88
Rand's Goldenrod 98
RANUNCULACEAE 246
Ranunculus abortivus 250
Ranunculus acris 250
Ranunculus aquatilis 252
Ranunculus caricetorum 251
Ranunculus flabellaris 251
Ranunculus flammula 252
Ranunculus longirostris 252
Ranunculus pensylvanicus 253
Ranunculus recurvatus 253
Ranunculus repens 254
Ranunculus reptans 252
Raspberry, Black 269
Raspberry, Dwarf 270
Raspberry, Red 269
Rattlesnake Fern 371
Rattlesnake Manna Grass 439
Rattlesnake-plantain, Checkered 206
Rattlesnake-plantain, Dwarf 205
Rattlesnake-plantain, Lesser 205
Rattlesnake-root, Alpine 85
Rattlesnake-root, Boott's 85
Rattlesnake-root, Tall 84
Rattlesnake-root, Three-leaved 85
Rayless Chamomile 84
Red Baneberry 246
Red Bearberry 132
Red-berried Elder 305
Red Clover 153
Red Columbine 247
Red Crowberry 135
Red Currant 157
Red Elderberry 305
Red Fescue 438
Red-footed Spikesedge 412
Red Lobelia 116
Red Maple 346
Red Oak 321
Red-osier Dogwood 318
Red Pine 328
Red Raspberry 269
Red Sand-spurry 125
Red-seeded Dandelion 105
Red Sorrel 234
Red Spruce 327
Red Sumac 309

Red-tinged Bulrush 418
Redtop Bentgrass 426
Red Trillium 185
Red Wakerobin 185
Reed Canary Grass 443
Reed, Common 444
Resupinate Bladderwort 180
Retrorse Sedge 403
RHAMNACEAE 255
Rhamnus alnifolia 255
Rhexia virginica 186
Rhinanthus crista-galli 219
Rhinanthus minor 219
Rhode Island Bentgrass 426
Rhododendron canadense 145
Rhododendron groenlandicum 145
Rhododendron lapponicum 146
Rhodora 145
Rhodora canadensis 145
Rhus glabra 309
Rhus hirta 310
Rhus radicans 52
Rhus typhina 310
Rhynchospora alba 416
Rhynchospora fusca 416
Ribbon-leaved Pondweed 239
Ribes americanum 155
Ribes glandulosum 156
Ribes hirtellum 156
Ribes lacustre 157
Ribes rubrum 157
Ribes triste 158
Rice Cut Grass 441
Richardson's Pondweed 244
River Horsetail 357
Roadside Agrimony 256
Roadside Sand-spurrey 125
Robbins' Pondweed 244
Robbins' Spikesedge 414
Robinia hispida 320
Rock Cranberry 150
Rockcress, Canada 109
Rockcress, Drummond's 109
Rockcress, Smooth 110
Rock-harlequin 220
Rock Maple 347
Rock Polypody 373
Rope-root Sedge 386
Rorippa palustris 114
Rosa blanda 264
ROSACEAE 256, 260, 330
Rosa nitida 265
Rosa palustris 265
Rosa virginiana 266
Rose-acacia 320

Rose, Bog 265
Rose Mandarin 183
Rose Pogonia 213
Rose, Shining 265
Rose, Smooth 264
Rose, Swamp 265
Rosette-panicgrass, Deer-tongue 434
Rosette-panicgrass, Lindheimer's 436
Rosette-panicgrass, Northern 434
Rosette-panicgrass, Sand 435
Rosette-panicgrass, Tangled 435
Rosette-panicgrass, Woolly 435
Rose Twisted-stalk 183
Rose, Virginia 266
Rosybells 183
Rosy Meadowsweet 273
Rosy Sedge 403
Rough Bedstraw 274
Rough Bentgrass 427
Rough Cinquefoil 263
Rough Fleabane 75
Rough-fruited Cinquefoil 263
Rough Hawkweed 82
Rough-leaved Aster 77
Rough-stemmed Goldenrod 98
Rough Wood-aster 77
Round-fruited Pinweed 127
Round-fruited Short-scaled Sedge 388
Round-leaved Bog-orchid 212
Round-leaved Dogwood 317
Round-leaved Pyrola 142
Round-leaved Sundew 131
Round-leaved Violet 300
Roving Bellflower 115
Rowan 336
Royal Fern 372
RUBIACEAE 273
Rubus allegheniensis 266
Rubus canadensis 267
Rubus dalibarda 267
Rubus elegantulus 268
Rubus hispidus 268
Rubus idaeus 269
Rubus occidentalis 269
Rubus pensilvanicus 270
Rubus pubescens 270
Rubus recurvicaulis 271
Rubus vermontanus 271
Rudbeckia hirta 92
Rudbeckia serotina 92
Rugel's Plantain 227
Rum Cherry 334
Rumex acetosella 234

Rumex britannica 235
Rumex crispus 235
Rumex domesticus 236
Rumex longifolius 236
Rumex obtusifolius 236
Rumex orbiculatus 235
Running Shadbush 332
RUSCACEAE 280
Rush American-aster 100
Rush, Bayonet 422
Rush, Brown-fruited 422
Rush, Canada 421
Rush, Joint-leaved 420
Rush, Moor 423
Rush, Path 423
Rush, Short-tailed 420
Rush, Small-headed 420
Rush, Thread 422
Rush, Toad 421
Russet Sedge 404
Rusty Cliff Fern 380

S

Sage Willow 339
Sagina procumbens 123
Sagittaria cuneata 50
Sagittaria graminea 50
Sagittaria latifolia 51
SALICACEAE 283, 337
Salix arctophila 283
Salix argyrocarpa 283
Salix bebbiana 338
Salix candida 339
Salix discolor 339
Salix eriocephala 340
Salix herbacea 284
Salix humilis 340
Salix lucida 341
Salix nigra 341
Salix pedicellaris 342
Salix pellita 342
Salix petiolaris 343
Salix planifolia 343
Salix pyrifolia 344
Salix sericea 344
Salix uva-ursi 284
Sallow Sedge 397
Sambucus canadensis 305
Sambucus nigra 305
Sambucus pubens 305
Sambucus racemosa 305
Sandplant, Mountain 122
Sand Rosette-panicgrass 435
Sand-spurrey, Roadside 125
Sand-spurry, Red 125
Sandwort, Blunt-leaved 122

Sandwort, Mountain 122
Sandwort, Smooth 122
Sanguinaria canadensis 222
Sanicle, Maryland 61
Sanicula canadensis 61
Sanicula marilandica 61
SAPINDACEAE 345
Saponaria officinalis 123
SARRACENIACEAE 285
Sarracenia purpurea 285
Sarsaparilla, Bristly 54
Sarsaparilla, Wild 55
Satiny Willow 342
Savoy Hawkweed 81
Saw-sedge, Smooth 411
Saxifraga aizoon 288
SAXIFRAGACEAE 286
Saxifraga foliolosa 286
Saxifraga paniculata 288
Saxifraga stellaris 286
Saxifraga virginiensis 287
Saxifrage, Golden 286
Saxifrage, Leafy 286
Saxifrage, Small-flowered 286, 287
Scabrous Black Sedge 383
SCHEUCHZERIACEAE 289
Scheuchzeria palustris 289
Schizachne purpurascens 447
Schoenoplectus acutus 416
Schoenoplectus subterminalis 417
Schoenoplectus tabernaemontani 417
Scirpus acutus 416
Scirpus atrocinctus 417
Scirpus cespitosus 419
Scirpus cyperinus 418
Scirpus hattorianus 418
Scirpus hudsonianus 419
Scirpus microcarpus 418
Scirpus pedicellatus 419
Scirpus rubrotinctus 418
Scirpus subterminalis 417
Scirpus validus 417
Scorzoneroides autumnalis 93
Scotch Bellflower 115
Scouring-rush, Variegated 358
Scratch Bedstraw 273
SCROPHULARIACEAE 289
Scrub Pine 327
Scutellaria epilobiifolia 174
Scutellaria galericulata 174
Scutellaria lateriflora 174
Sedge, Appalachian 382
Sedge, Awl-fruited 406
Sedge, Beaked 404

Sedge, Bear 382
Sedge, Bebb's 384
Sedge, Bigelow's 384
Sedge, Billings' 384
Sedge, Bristle-stalk 395
Sedge, Brome-like 385
Sedge, Brownish 385
Sedge, Crawford's 387
Sedge, Cypress-like 402
Sedge, Few-flowered 401
Sedge, Few-seeded 399
Sedge, Fibrous-rooted 386
Sedge, Fringed 387
Sedge, Goodenough's 398
Sedge, Graceful 391
Sedge, Hayden's 392
Sedge, Hoary 385
Sedge, Hop 397
Sedge, Inland 393
Sedge, Lakeside 394
Sedge, Long-beaked 406
Sedge, Long-stalked 401
Sedge, Meager 390
Sedge, Michaux's 398
Sedge, Mud 396
Sedge, Muhlenberg's 398
Sedge, Necklace 402
Sedge, Nodding 392
Sedge, Northeastern 387
Sedge, Northern 388
Sedge, Open-field 386
Sedge, Orono 400
Sedge, Oval 395
Sedge, Pale 400
Sedge, Parasol 409
Sedge, Pennsylvania 401
Sedge, Porcupine 393
Sedge, Pubescent 392
Sedge, Retrorse 403
Sedge, Rope-root 386
Sedge, Rosy 403
Sedge, Russet 404
Sedge, Sallow 397
Sedge, Sparse-flowered 407
Sedge, Star 390
Sedge, Straw 391
Sedge, Swollen-beaked 409
Sedge, Three-seeded 408
Sedge, Three-way 412
Sedge, Tinged 407
Sedge, Tussock 406
Sedge, Twisted 408
Sedge, Water 382
Sedge, White-edged 388
Sedge, White-tinged 381
Sedge, Wiegand's 410

Sedge, Woolly-fruited 394
Sedge, Yellow-green 390
Selfheal, Common 173
Senecio balsamitae 88
Senecio hieraciifolius 73
Senecio pauperculus 88
Senecio robbinsii 88
Senecio schweinitzianus 88
Sensitive Fern 368
Serviceberry, Downy 330
Serviceberry, Mountain 330
Sessile-leaved Bellwort 128
Seven-angled Pipewort 150
Shadbush, Downy 330
Shadbush, Dwarf 332
Shadbush, Intermediate 331
Shadbush, Mountain 330
Shadbush, Nantucket 332
Shadbush, Running 332
Shadbush, Smooth 331
Shadbush, Thicket 332
Sharp-toothed Nodding-aster 86
Sheep American-laurel 139
Sheep Dock 234
Sheep Laurel 139
Sheep Sorrel 234
Shepherd's-purse 110
Shining Firmoss 359
Shining Rose 265
Shining Willow 341
Shinleaf, American 142
Shinleaf, Elliptic-leaved 144
Shinleaf, Green-flowered 143
Shinleaf, Little 144
Shinleaf, Pink 143
Shore Aster 104
Short-tailed Rush 420
Showy Blackberry 268
Showy Goatsbeard 106
Showy Mountain-ash 336
Shrubby Cinquefoil 258
Shrubby St. John's-wort 167
Sibbaldia tridentata 272
Sibbaldiopsis tridentata 272
Silene acaulis 124
Silene cucubalus 124
Silene inflata 124
Silene vulgaris 124
Silkweed 63
Silky Milkweed 63
Silky Willow 344
Silver Glade Fern 378
Silver-leaved Cinquefoil 262
Silver Maple 346
Silverrod 93
Silvery False Spleenwort 378

Simpler's-joy 296
Simple-stemmed Bur-reed 291
Single Delight 140
Sisymbrium altissimum 114
Sisyrinchium intermedium 169
Sisyrinchium montanum 169
Sisyrinchium mucronatum 169
Sitka Ground-cedar 364
Sium suave 61
Skullcap, Blue 174
Skullcap, Hooded 174
Skullcap, Marsh 174
Skunk Currant 156
Slender Cottonsedge 414
Slender-leaved Agalinis 215
Slender Naiad 162
Slender Stinging Nettle 295
Slender Water-milfoil 160
Slender Wheatgrass 437
Slender Wild-rye 437
Slender Woodland Sedge 389
Slender Wood-reed 432
Slipper, Venus' 202
Small Bayberry 188
Small Bedstraw 277
Small Cranberry 149
Small Enchanter's-nightshade 194
Small-flowered Bitter-cress 112
Small-flowered Crowfoot 250
Small-flowered Evening-primrose 200
Small-flowered Saxifrage 286, 287
Small-flowered Wood Rush 425
Small-headed Rush 420
Small-leaved Pond-lily 192
Small Purple-fringed Orchis 213
Small Pussytoes 67
Small-spiked False Nettle 294
Small Sundrops 200
Small Waterwort 131
Small White Violet 299
Smartweed, Dock-leaved 233
Smartweed, Lady's-thumb 233
Smartweed, Nodding 233
Smartweed, Water 232
Smartweed, Water-pepper 232
Smilacina racemosa 281
Smilacina stellata 281
Smilacina trifolia 282
Smooth Arrowwood 306
Smooth Bedstraw 276
Smooth Blackberry 267
Smooth Black Sedge 398
Smooth Brome 430
Smooth Cliff Fern 379

Smooth Crabgrass 436
Smooth Gooseberry 156
Smooth Rockcress 110
Smooth Rose 264
Smooth Sandwort 122
Smooth Saw-sedge 411
Smooth Shadbush 331
Smooth Sumac 309
Smooth White Violet 299
Snakeberry 246
Snakemouth 213
Snakeroot, Whitespot 66
Snapdragon, Wild 226
Snapweed 107
Snow-bed Willow 284
Snowberry, Common 120
Snowberry, Creeping 136
Soapwort, Common 123
Soft-leaved Sedge 389
Soft-stemmed Bulrush 417
Solidago bicolor 93
Solidago canadensis 94
Solidago cutleri 96
Solidago flexicaulis 94
Solidago graminifolia 78
Solidago hispida 95
Solidago juncea 95
Solidago latifolia 94
Solidago leiocarpa 96
Solidago macrophylla 96
Solidago nemoralis 97
Solidago puberula 97
Solidago randii 98
Solidago rugosa 98
Solidago simplex 98
Solidago squarrosa 99
Solidago uliginosa 99
Solidago virgaurea 96
Solomon's-seal, False 281
Solomon's-seal, Hairy 282
Sonchus arvensis 100
Sorbus americana 335
Sorbus aucuparia 336
Sorbus decora 336
Sorrel, Red 234
Sorrel, Sheep 234
Sorrel, Wood 219, 220
Sourtop Blueberry 148
Southern Arrow-wood 306
Southern Ground-cedar 364
Sow-thistle, Creeping 100
Sow-thistle, Field 100
Sparganium americanum 290
Sparganium angustifolium 291
Sparganium chlorocarpum 291
Sparganium emersum 291

Sparganium erectum 291
Sparganium eurycarpum 292
Sparganium fluctuans 292
Sparganium minimum 293
Sparganium natans 293
Sparse-flowered Sedge 407
Spatterdock 193
Spatulate-leaved Sundew 130
Spear Thistle 72
Spearwort, Creeping 252
Speckled Alder 312
Speedwell, Alpine 229
Speedwell, American 227
Speedwell, Common 228
Speedwell, Marsh 228
Speedwell, Narrow-leaved 228
Speedwell, Thyme-leaved 229
Spergularia rubra 125
Spicy-wintergreen, Creeping 136
Spicy-wintergreen, Eastern 137
Spiked Wood Rush 425
Spike Muhly 441
Spikenard, American 55
Spikenard, False 281
Spikesedge, Blunt 413
Spikesedge, Common 413
Spikesedge, Needle 412
Spikesedge, Ovoid 413
Spikesedge, Red-footed 412
Spikesedge, Robbins' 414
Spinulose Wood Fern 352
Spinulum annotinum 367
Spinulum canadense 367
Spiny-spored Quillwort 360
Spiny Swamp Currant 157
Spiraea alba 272
Spiraea latifolia 272
Spiraea septentrionalis 272
Spiraea tomentosa 273
Spiral Pondweed 245
Spiranthes cernua 214
Spiranthes romanzoffiana 214
Spleenwort, False 378
Spleenwort, Maidenhair 349
Split-lipped Hemp-nettle 170
Spot-leaved Crowfoot 254
Spotted Coral-root 203
Spotted Cowbane 57
Spotted Hawkweed 80
Spotted Joe-Pye Weed 78
Spotted St. John's-wort 167
Spotted Touch-me-not 107
Spotted Water-hemlock 57
Spreading Dogbane 62
Spreading Wood Fern 351
Spring-beauty, Carolina 237

Spring Cleavers 273
Spruce, Black 326
Spruce, Bog 326
Spruce, Cat 326
Spruce, Red 327
Spruce, White 326
Spurred-gentian, American 155
Squarrose Goldenrod 99
Squashberry 307
Stachys borealis 175
Stachys palustris 175
Stachys pilosa 175
Staghorn Sumac 310
Stalked Woolsedge 419
Starflower 189
Star-like False Solomon's-seal 281
Star Sedge 390
Steeple-bush 273
Stellaria borealis 125
Stellaria graminea 126
Stellaria longifolia 126
Stellaria media 127
Stick-tight 71
Sticky Hawkweed 82
Stiff Three-petaled Bedstraw 277
Stinging Nettle 295
Stinking Benjamin 185
Stinking Elder 305
Stitchwort, Boreal 125
Stitchwort, Common 127
Stitchwort, Grass-leaved 126
Stitchwort, Lesser 126
Stitchwort, Long-leaved 126
St. John's-wort 163
Stone Clover 151
Strawberry 259
Strawberry, Wild 259
Straw Sedge 391
Streptopus amplexifolius 183
Streptopus lanceolatus 183
Streptopus roseus 183
Strict Blue-eyed-grass 169
Strict Eyebright 217
Striped Maple 345
Sugar Maple 347
Sulphur Cinquefoil 263
Sumac, Red 309
Sumac, Smooth 309
Sumac, Staghorn 310
Sumac, Velvet 310
Sundew, Narrow-leaved 130
Sundew, Round-leaved 131
Sundew, Spatulate-leaved 130
Sundrops, Small 200
Swamp Blueberry 319
Swamp Candles 191

Swamp Crowfoot 251
Swamp Currant 157
Swamp Goldenrod 99
Swamp Honeysuckle 119
Swamp Maple 346
Swamp Milkweed 63
Swamp Moonwort 370
Swamp Ragwort 88
Swamp Red Currant 158
Swamp Red Raspberry 270
Swamp Rose 265
Swamp Thistle 72
Swamp Willow-herb 198
Swamp Yellow-loosestrife 191
Sweet-cicely, Bland 59
Sweet-coltsfoot, Arctic 89
Sweet-fern 188
Sweet Gale 189
Sweet Hurts 147
Sweet-scabious 74
Sweet Vernal Grass 429
Sweet White Violet 297
Sweet Wood-reed 432
Swertia deflexa 155
Swida alternifolia 317
Swida rugosa 317
Swida sericea 318
Swida stolonifera 318
Swollen-beaked Sedge 409
Symphoricarpos albus 120
Symphyotrichum boreale 100
Symphyotrichum cordifolium 101
Symphyotrichum lanceolatum 101
Symphyotrichum lateriflorum 102
Symphyotrichum novae-angliae 102
Symphyotrichum novi-belgii 103
Symphyotrichum puniceum 103
Symphyotrichum tradescantii 104
Symphyotrichum undulatum 104
Syringa vulgaris 324

T

Tall Blue Lettuce 82
Tall Buttercup 250
Tall Cotton-grass 414
Tall Cottonsedge 414
Tall Crowfoot 250
Tall King-devil 92
Tall Lettuce 83
Tall Meadow-rue 255
Tall Oat Grass 429
Tall Rattlesnake-root 84
Tall White-aster 73
Tall White Lettuce 84

Tall Wood-beauty 259
Tamarack 325
Tangled Rosette-panicgrass 435
Tangle-legs 307
Tape-grass 162
Taraxacum erythrospermum 105
Taraxacum laevigatum 105
Taraxacum officinale 105
Tawny Cotton-grass 415
Tawny Cottonsedge 415
TAXACEAE 348
Taxus canadensis 348
Teaberry 137
Tea-leaved Willow 343
Tearthumb, Arrow-leaved 234
Terrel Grass 437
Tetragonanthus deflexus 155
Teucrium canadense 176
Thalictrum dioicum 254
Thalictrum polygamum 255
Thalictrum pubescens 255
THELYPTERIDACEAE 374, 376
Thelypteris noveboracensis 374
Thelypteris palustris 376
Thelypteris phegopteris 375
Thelypteris simulata 375
Thicket Shadbush 332
Thistle, Bull 72
Thistle, Canada 71
Thistle, Common 72
Thistle, Creeping 71
Thistle, Spear 72
Thistle, Swamp 72
Thoroughwort, Boneset 76
Thread Rush 422
Three-leaved False Solomon's-seal 282
Three-leaved Goldthread 249
Three-leaved Rattlesnake-root 85
Three-petaled Bedstraw 278
Three-seeded Sedge 408
Three-toothed-cinquefoil 272
Three-way Sedge 412
Thuja occidentalis 319
THYMELAEACEAE 290
Thyme-leaved Speedwell 229
Tiarella cordifolia 288
Ticklegrass 427
Tilia americana 322
Timothy, Common 444
Timothy, Mountain 443
Tinged Sedge 407
Tissa rubra 125
Toadflax, Butter-and-eggs 226
Toadflax, False 128

Toad Rush 421
Toothed Flatsedge 411
Toothed Wood Fern 352
toothwort, Two-leaved 111
Torreyochloa pallida 448
Touch-me-not, Orange 107
Touch-me-not, Spotted 107
Toxicodendron radicans 52
Toxicodendron rydbergii 53
Tradescant's American-aster 104
Tragopogon pratensis 106
Trailing-arbutus 136
Tree-clubmoss, Flat-branched 363
Tree-clubmoss, Hickey's 362
Tree-Clubmoss, Prickly 362
Trembling Aspen 338
Triadenum fraseri 164
Triadenum virginicum 168
Trichophorum alpinum 419
Trichophorum cespitosum 419
Trientalis borealis 189
Trifolium agrarium 152
Trifolium arvense 151
Trifolium aureum 152
Trifolium dubium 152
Trifolium hybridum 153
Trifolium pratense 153
Trifolium repens 154
Trillium erectum 185
Trillium erythrocarpum 185
Trillium, Painted 185
Trillium purpureum 185
Trillium, Red 185
Trillium undulatum 185
Trisetum spicatum 448
Trout-lily, American 181
Tsuga canadensis 329
Tuberous Grass-pink 201
Tuckerman's Panicgrass 443
Tuckerman's Pondweed 239
Tuckerman's Quillwort 361
Tufted Clubsedge 419
Tufted Hair Grass 433
Tufted Hair-sedge 381
Tufted Vetch 154
Tufted Yellow-loosestrife 191
Tumbling Hedge-mustard 114
Turnip, Indian 64
Turritis drummondii 109
Turritis laevigata 110
Turritis stricta 109
Turtlehead, White 224
Tussilago farfara 106
Tussock Cottonsedge 415
Tussock Sedge 406
Twayblade, Auricled 207

Twayblade, Broad-leaved 207
Twayblade, Heart-leaved 208
Twinflower, American 118
Twisted Sedge 408
Twistedstalk, Clasping-leaved 183
Twistedstalk, Lance-leaved 183
Twisted-stalk, Rose 183
Two-leaved toothwort 111
TYPHACEAE 290
Typha latifolia 293

U

ULMACEAE 348
Ulmus americana 348
Umbrella Flatsedge 411
Upright Yellow-rocket 108
Urtica canadensis 294
URTICACEAE 294, 295
Urtica cylindrica 294
Urtica gracilis 295
Urtica pumila 295
Utricularia cornuta 176
Utricularia geminiscapa 177
Utricularia gibba 177
Utricularia inflata 179
Utricularia intermedia 178
Utricularia macrorhiza 180
Utricularia minor 178
Utricularia purpurea 179
Utricularia radiata 179
Utricularia resupinata 180
Utricularia vulgaris 180
Uvularia amplexifolia 183
Uvularia sessilifolia 128

V

Vaccinium angustifolium 146
Vaccinium atrococcum 320
Vaccinium boreale 147
Vaccinium cespitosum 147
Vaccinium corymbosum 319
Vaccinium fuscatum 320
Vaccinium macrocarpon 148
Vaccinium myrtilloides 148
Vaccinium oxycoccos 149
Vaccinium uliginosum 149
Vaccinium vitis-idaea 150
Vagabond 381
Vahlodea atropurpurea 448
Vallisneria americana 162
Vallisneria spiralis 162
Vanilla Sweet Grass 429
Variegated Scouring-rush 358
Velvet-leaved Blueberry 148
Velvet Sumac 310

Venus' Slipper 202
Veratrum viride 186
Verbascum thapsus 289
VERBENACEAE 296
Verbena hastata 296
Vermont Blackberry 271
Vernal Water-starwort 223
Veronica alpina 229
Veronica americana 227
Veronica officinalis 228
Veronica scutellata 228
Veronica serpyllifolia 229
Veronica wormskjoldii 229
Vervain, Blue 296
Vetch, Bird 154
Vetch, Cow 154
Vetch, Tufted 154
Viburnum acerifolium 306
Viburnum alnifolium 307
Viburnum cassinoides 308
Viburnum dentatum 306
Viburnum edule 307
Viburnum lantanoides 307
Viburnum lentago 308
Viburnum, Maple-leaved 306
Viburnum nudum 308
Viburnum opulus 309
Viburnum recognitum 306
Viburnum trilobum 309
Vicia cracca 154
Vinegar-tree 310
Viola adunca 296
Viola blanda 297
VIOLACEAE 296
Viola cucullata 297
Viola labradorica 298
Viola lanceolata 298
Viola macloskeyi 299
Viola pallens 299
Viola pubescens 299
Viola renifolia 300
Viola rotundifolia 300
Viola sagittata 301
Viola selkirkii 301
Viola septentrionalis 302
Viola sororia 302
Violet, Arrowhead 301
Violet, Great-spurred 301
Violet, Hook-spurred 296
Violet, Kidney-leaved 300
Violet, Lance-leaved 298
Violet, Round-leaved 300
Virginia Chain Fern 350
Virginia-creeper 303
Virginia Lyme Grass 437
Virginia Meadow-beauty 186

Virginia Rose 266
Virginia St. John's-wort 168
Virginia Virgin's-bower 249
Virginia Water-horehound 172
Virgin's-bower, Purple 248
Virgin's-bower, Virginia 249
VISCACEAE 302
VITACEAE 303

W

Wakerobin, Painted 185
Wakerobin, Red 185
Wapato 51
Water Arum 64
Water Avens 262
Water Bulrush 417
Water Carpet 286
Water-celery 162
Water Crowfoot 252
Water Dock 235
Water-gladiole 116
Water-hemlock, Bulblet-bearing 56
Water-hemlock, Spotted 57
Water-horehound, American 171
Water-horehound, Cut-leaved 171
Water-horehound, Northern 172
Water-horehound, Virginia 172
Water Horsetail 357
Water-lily, Fragrant 193
Water-lily, White 193
Water Lobelia 116
Water-milfoil, Alternate-flowered 158
Water-milfoil, Farwell's 159
Water-milfoil, Northern 159
Water-milfoil, Slender 160
Water-milfoil, Whorled 160
Waternymph, Northern 162
Waternymph, Wavy 162
Water-parsnip 61
Water-pennywort, American 59
Water-pepper Smartweed 232
Water-plantain, Large 49
Water-plantain, Northern 49
Water Sedge 382
Water-shield 192
Water Smartweed 232
Water-starwort, Greater 223
Water-starwort, Vernal 223
Waterweed, Free-flowered 161
Waterwort, Small 131
Wavy Blue Grass 446
Wavy Hair Grass 434
Wavy-leaved American-aster 104
Wavy Waternymph 162

Weak Spear Grass 447
Wen-dock, Purple 192
Western Poison-ivy 53
Wheatgrass, Slender 437
Whip King-devil 90
White Ash 323
White-aster, Flat-topped 73
White-aster, Tall 73
White Avens 260
White Baneberry 246
White Beak-rush 416
White Beaksedge 416
White Birch 315
White-buttons 150
White-cedar, Northern 319
White Clover 154
White Daisy 83
White-edged Sedge 388
White-flowered Willow-herb 197
White-fringed Bog-orchid 209
White Goldenrod 93
White Goosefoot 52
White-grained Rice Grass 442
White Meadowsweet 272
White Mountain Saxifrage 288
White Northern Bog-orchid 210
White Pine 329
Whitespot Snakeroot 66
White Spruce 326
White-stemmed Pondweed 243
White-tinged Sedge 381
White Turtlehead 224
White Water-lily 193
White Wood-aster 76
Whitlow-mustard, Canescent 113
Whorled Bedstraw 276
Whorled Water-milfoil 160
Whorled Wood Aster 86
Wickup 194
Wiegand's Sedge 410
Wild Calla 64
Wild Carrot 58
Wild Chives 51
Wild Columbine 247
Wild Lettuce 83
Wild Lily-of-the-valley 280
Wild Mint 173
Wild Morning-glory 129
Wild Oats 128
Wild Parsnip 60
Wild Peppergrass 113
Wild-raisin 308
Wild-rye, Creeping 436
Wild-rye, Eastern 437

Wild-rye, Slender 437
Wild Sarsaparilla 55
Wild Snapdragon 226
Wild Strawberry 259
Willow, Arctic 283
Willow, Balsam 344
Willow, Bearberry 284
Willow, Black 341
Willow, Bog 342
Willow, Heart-leaved 340
Willow-herb, Bog 198
Willow-herb, Downy 199
Willow-herb, Eastern 196
Willow-herb, Fringed 196
Willow-herb, Glandular 196
Willow-herb, Great 194
Willow-herb, Hornemann's 197
Willow-herb, Marsh 198
Willow-herb, Narrow-leaved 198
Willow-herb, Pimpernel 195
Willow-herb, Purple-veined 196
Willow-herb, Swamp 198
Willow-herb, White-flowered 197
Willow, Labrador 283
Willow, Long-beaked 338
Willow, Meadow 343
Willow, Northern 283
Willow, Prairie 340
Willow, Pussy 339
Willow, Sage 339
Willow, Satiny 342
Willow, Shining 341
Willow, Silky 344
Willow, Snow-bed 284
Willow, Tea-leaved 343
Winterberry, Common 311
Winter-cress, Common 109
Witch-hazel, American 322
Witch-hobble 307
Withe-rod 308
Wood-aster, Large-leaved 77
Wood-aster, Rough 77
Wood-aster, White 76
Wood-beauty, Tall 259
Wood Bedstraw 277
Wood Blue Grass 446
Wood Fern 351
Wood Horsetail 358
Woodland Arctic-cudweed 87
Woodland Forget-me-not 108
Woodland Horsetail 358
Woodland Muhly 442
Wood-reed, Slender 432
Wood-reed, Sweet 432

WOODSIACEAE 377
Woodsia glabella 379
Woodsia ilvensis 380
Wood Sorrel 219, 220
Woodwardia virginica 350
Wool-grass, Black-girdled 417
Wool-grass, Common 418
Wool-grass, Pedicellate 419
Woolly Blue Violet 302
Woolly-fruited Sedge 394
Woolly Rosette-panicgrass 435
Woolsedge, Black-girdled 417
Woolsedge, Common 418
Woolsedge, Stalked 419
Wrinkle-leaved Goldenrod 98

X

XYRIDACEAE 303
Xyris difformis 303
Xyris montana 304

Y

Yard Dock 236
Yarrow, Common 65
Yellow Avens 260
Yellow Birch 313
Yellow Blue-bead Lily 181
Yellow-cress, Common 114
Yellow Dock 235
Yellow Evening-primrose 199
Yellow-eyed-grass, Bog 303
Yellow-eyed-grass, Northern 304
Yellow Forest Violet 299
Yellow Goatsbeard 106
Yellow-green Sedge 390
Yellow King-devil 90
Yellow King Devil 90
Yellow Lady's-slipper 204
Yellow-loosestrife, Creeping 190
Yellow-loosestrife, Fringed 190
Yellow-loosestrife, Swamp 191
Yellow-loosestrife, Tufted 191
Yellow Pine-sap 138
Yellow-rattle, Little 219
Yellow-rocket, Garden 109
Yellow-rocket, Upright 108
Yew, American 348
Yew, Canada 348

Z

Zig-zag Goldenrod 94
Zizia aurea 62

ABOUT THE AUTHORS

Glen Mittelhauser is the botanist, ecologist, and Executive Director for Maine Natural History Observatory, based in Gouldsboro, Maine. He has also been a contributing author for several botanical guides and is currently the managing editor for the *Northeastern Naturalist* and *Southeastern Naturalist* journals.

Jensen Bissell has worked at Baxter State Park for twenty-nine years and currently serves as Park Director.

Don Cameron works as a botanist and ecologist at the Maine Natural Areas Program where he conducts botanical inventories and manages Maine's Official List of Endangered and Threatened Plants.

Dr. Alison C. Dibble is a Conservation Biologist with focus on plants and pollinators. She is Assistant Research Scientist in the School of Biology and Ecology, University of Maine, Orono. She also operates a consulting firm, Stewards LLC, in Brooklin, Maine, through which she develops pollinator habitat plans for farms throughout Maine, and ecological inventories for public lands, land trusts, and private landowners.

Arthur Haines is a research botanist for the New England Wild Flower Society. He has authored several floristic works for the northeast, including *Flora of Maine* and *Flora Novae Angliae*.

Jean Hoekwater is a graduate of College of the Atlantic and has served Baxter State Park as Park Naturalist for twenty-eight years.

Marilee Lovit is an independent researcher who lives in Addison, Maine.

Aaron Megquier is a botanist and conservation biologist from Belmont, Maine. He is the Executive Director of Friends of Baxter State Park.

FIELD NOTES

FIELD NOTES

FIELD NOTES

FIELD NOTES

FIELD NOTES

FIELD NOTES

Wildflowers and low shrubs .... Page 49

Trees and tall shrubs .... Page 305

Ferns and other spore-producing plants .... Page 349

Sedges, rushes, and grasses .... Page 381